PHYSICAL PROCESSES IN THE COASTAL ZONE

PHYSICAL PROCESSES IN THE COASTAL ZONE

COMPUTER MODELLING AND REMOTE SENSING

Proceedings of the Forty Ninth Scottish
Universities Summer School in Physics,
Dundee, August 1997.

Edited by

A P Cracknell — University of Dundee
E S Rowan — University of Dundee

Series Editor

P Osborne — University of Edinburgh

Copublished by
Scottish Universities Summer School in Physics &
Institute of Physics Publishing, Bristol and Philadelphia

British Library Cataloguing-in-Publication Data:

A catalogue record for this book is available
from the British Library

ISBN 0-7503-0563-0

Library of Congress Cataloging-in-Publication Data are available.

Copublished by

SUSSP Publications
The Department of Physics, Edinburgh University,
The King's Buildings, Mayfield Road, Edinburgh EH9 3JZ, Scotland.

and

Institute of Physics Publishing, wholly owned by
The Institute of Physics, London.
Institute of Physics Publishing, Dirac House, Temple Back, Bristol BS1 6BE, UK.
US Office: Institute of Physics Publishing, The Public Ledger Building,
Suite 1035, 150 South Independence Mall West, Philadelphia, PA 19106, USA.

Printed in Great Britain by J W Arrowsmith Ltd, Bristol.

SUSSP Proceedings

1	1960	Dispersion Relations
2	1961	Fluctuation, Relaxation and Resonance in Magnetic Systems
3	1962	Polarons and Excitons
4	1963	Strong Interactions and High Energy Physics
5	1964	Nuclear Structure and Electromagnetic Interactions
6	1965	Phonons in Perfect and Imperfect Lattices
7	1966	Particle Interactions at High Energy
8	1967	Methods in Solid State and Superfluid Theory
9	1968	Physics of Hot Plasmas
10	1969	Quantum Optics
11	1970	Hadronic Interactions of Photons and Electrons
12	1971	Atoms and Molecules in Astrophysics
13	1972	Properties of Amorphous Semiconductors
14	1973	Phenomenology of Particles at High Energy
15	1974	The Helium Liquids
16	1975	Non-linear Optics
17	1976	Fundamentals of Quark Models
18	1977	Nuclear Structure Physics
19	1978	Metal Non-metal Transitions in Disordered Solids
20	1979	Laser-Plasma Interactions: 1
21	1980	Gauge Theories and Experiments at High Energy
22	1981	Magnetism in Solids
23	1982	Laser-Plasma Interactions: 2
24	1982	Lasers: Physics, Systems and Techniques
25	1983	Quantitative Electron Microscopy
26	1983	Statistical and Particle Physics
27	1984	Fundamental Forces
28	1985	Superstrings and Supergravity
29	1985	Laser-Plasma Interactions: 3
30	1985	Synchrotron Radiation
31	1986	Localisation and Interaction
32	1987	Computational Physics
33	1987	Astrophysical Plasma Spectroscopy
34	1988	Optical Computing

/continued

SUSSP Proceedings (continued)

35	1988	Laser-Plasma Interactions: 4
36	1989	Physics of the Early Universe
37	1990	Pattern Recognition and Image Processing
38	1991	Physics of Nanostructures
39	1991	High Temperature Superconductivity
40	1992	Quantitative Microbeam Analysis
41	1992	Spatial Complexity in Optical Systems
42	1993	High Energy Phenomenology
43	1994	Determination of Geophysical Parameters from Space
44	1994	Simple Quantum Systems
45	1994	Laser-Plasma Interactions 5: Inertial Confinement Fusion
46	1995	General Relativity
47	1995	Laser Sources and Applications
48	1996	Generation and Application of High power Microwaves
49	1997	Physical Processes in the Coastal Zone
50	1998	Semiconductor Quantum Optoelectronics
51	1998	Muon Science
52	1998	Advances in Lasers and Applications

1	V. Klemas	17	R. Gouda	33	I. Sargent
2	W.G. Rees	18	J.A. Gomez-Sanchez	34	D. Akimov
3	S.K.Newcombe	19	O. Puentedura	35	J. Laanearu
4	S. Victorov	20	K. Bradtke	36	V. E. Brando
5	A.P. Cracknell	21	V. Subramanian	37	F. Nerozzi
6	N. Lo Presti	22	K. Maeno	38	P.S. Crawford
7	L.M. Morales	23	R. Feron	39	T. Malthus
8	A. Black	24	E. Ricchetti	40	D. Sirjacobs
9	M. Jimenez Michavila	25	M. Kleih	41	J. Monk
10	A.O. Tooke	26	D. Burska	42	M.Zoller-Shimoni
11	A. Reinart	27	A. Brady	43	A.Schwiebus
12	A. Di Iorio	28	M. Benvenuti	44	P. Albert
13	M.L. Nirala	29	D.H.Z. Mhango	45	F. Durand
14	M.C. Pizzoferrato	30	G. Abbate	46	I. Lee-Bapty
15	I.M.J. van Enckevort	31	P. Cunningham		
16	S. Naruekhatpichai	32	A. Jacob		

Executive Committee

Professor Arthur Cracknell	University of Dundee	*Director and Co-Editor*
Dr. Robin Vaughan	University of Dundee	*Director*
Ms. Sheila Newcombe	University of Dundee	*Secretary*
Dr. Andrew Folkard	University of Strathclyde	*Treasurer*
Dr. Tony Tooke	University of Dundee	*Social Secretary*
Ms. Elaine Rowan	University of Dundee	*Co-Editor*

International Advisory Committee

Professor Arthur Cracknell	University of Dundee, Scotland
Professor Peter Davies	University of Dundee, Scotland
Professor Vic Klemas	University of Delaware, USA
Professor John McManus	University of St. Andrews, Scotland
Professor Arthur Roberts	Simon Fraser University, Canada
Dr. Dave Sloggett	Anite Systems Ltd., UK
Professor Costas Varotsos	University of Athens, Greece
Dr. Robin Vaughan	University of Dundee, Scotland
Dr. Serge Victorov	VNIIKAM, St. Petersburg, Russia

Lecturers

Professor Arthur Cracknell	University of Dundee
Dr. Andrew Folkard	University of Strathclyde
Professor Arthur Roberts	Simon Fraser University
Mr. Keiran Millard	HR Wallingford
Dr. Gareth Rees	University of Cambridge
Professor Vic Klemas	University of Delaware
Dr. Raymond Feron	Rijkswaterstaat
Professor Jin-Eong Ong	Universiti Sains Malaysia
Dr. Dave Sloggett	Anite Systems Ltd.
Dr. Serge Victorov	VNIIKAM, St. Petersburg
Dr. George Chronopoulos	University of Athens

Seminar Speakers

Dr. Paul Crawford	University of Dundee
Dr. Ping Dong	University of Dundee
Mr. Yan Gu	University of Dundee
Dr. Tim Malthus	University of Edinburgh
Professor John McManus	University of St. Andrews
Mr. Mohan Nirala	University of Dundee
Dr. Steve Parkes	University of Dundee
Dr. Silke Wewetzer	University of Dundee

Directors' Preface

This volume is based on lectures and seminars which were presented at the 49th Scottish Universities Summer School in Physics for which we had chosen the title 'Physical Processes in the Coastal Zone: Computer Modelling and Remote Sensing'. The study of the coastal zone is becoming increasingly important as more and more people realise the importance of environmental issues and of the particular problems associated with environmental damage and environmental pollution in the coastal zone. Moreover we are likely to see increasing amounts of new environmental legislation at regional, national and European level which is likely to be introduced over the next few years and much of this is going to be related to the coastal zone. These factors are going to lead to extensive requirements for information and monitoring systems for the coastal zone. In the development and understanding of such systems there is going to be a need for a good understanding of the important physical processes in the coastal zone and there is going to be a need for access to new and comprehensive sources of data, such as are only available by using remote sensing techniques. The present text is therefore concerned with the integration of the understanding of the basic physics of the various processes that occur in the coastal zone and the gathering of extensive and frequent information about the coastal zone by using data from remote sensing systems flown on aircraft and spacecraft.

We take as our definition of the coastal zone the definition used in the LOICZ (Land–Ocean Interactions in the Coastal Zone) programme of the IGBP (International Geosphere–Biosphere Programme). On the seaward side, this involves going to the edge of the continental shelf and, on the landward side, to the contour that is 200 m above sea level. With this definition, we see that the coastal zone is of far greater importance than would be suggested by its area in relation to that of the whole surface of the Earth. Moreover one can also quite reasonably argue that anthropogenically-driven change is more intense in coastal regions than elsewhere (see the first article).

Our own particular interest is in remote sensing and there are going to be great opportunities in the near future for remote sensing in coastal zones—particularly from InSAR (interferometric SAR) and the new high resolution (1 metre resolution) optical band systems that are soon to be flown in space. But remote sensing is not everything and will not do everything. In addition to remotely-sensed data we need *in situ* data for validation of the remotely-sensed data, physical models for the study of coastal processes and prediction of future evolution and a GIS (Geographical Information System), or more generally a RSIM (Remote Sensing Information Model), into which to incorporate data.

There are, in a sense, two extremes of coastal zone studies. On the one hand there is that of the global climate modeller whose aim is to be able to obtain as good a representation as possible for the entire coastal regions of the Earth so as to provide the most realistic representation of those regions in a general circulation model being used in climate studies. On the other hand there is the ecologist, environmentalist, policy maker, media reporter or decision maker, who is usually interested in monitoring or predicting change in a very small and local area of the coastal zone. The approaches of these two types of person are bound to be very different. The global climate modeller is bound—for the foreseeable future at least—to consider the coastline at a spatial resolution, and with a data format, that is quite unsuitable and inappropriate for the local ecologist, environmentalist, etc. In more specialised terms, different people are likely to require quite different spatial scales for the GISs or RSIMs that they use and we shall need to consider these user requirements separately.

The choice of this topic also fitted in very well with recent research developments in our own research programme because at the time of our original planning for this school we had just set up the Dundee Centre for Coastal Zones Research (DCCZR) in collaboration with Anite Systems Ltd (formerly Cray Systems Ltd), particularly under the inspiration of Dr. D R Sloggett of Anite Systems Ltd. Why did we actually choose to concentrate on this activity and especially why the coastal zone? The reasons were several.

When we set out in remote sensing in the late 1970s and early 1980s we looked at a number of potential applications of remote sensing and we demonstrated their feasibility for environmental monitoring; these applications included gas flare monitoring and agricultural straw burning, pollution monitoring, etc. But at that stage there was no interest whatsoever among potential users of remotely-sensed data.

For the next 10–15 years we did not bother about end users or real applications for remotely-sensed data. We just worked on those scientific applications that happened to be of interest to us and our students.

It was in late 1994 and early 1995 that Dave Sloggett approached us and tried to convince us that we should look seriously again at real applications and end users for remotely-sensed data. Dave argued that the last real frontier for applications of remote sensing lies in the coastal zone. This is largely because spatial resolution and therefore mixed pixels, as well as cloud cover and cost, have been against the use of satellite remote sensing in the coastal zone in the past; but this is changing in terms of the technical situation.

The other factor is that in Dundee we have our own direct access to data at a variety of scales: (a) satellite data from NOAA AVHRR (from 1978 and ongoing), CZCS (previously) and SeaWiFS, (b) airborne data from our own low-cost VIFIS hyperspectral imaging system and (c) *in situ* data on water quality parameters and water circulation from the in-shore craft of the Tay Estuary Research Centre. In Dundee we had done a fair amount of research work using these sources of remotely-sensed data and *in situ* data in the coastal zone. Thus we came to our involvement with Anite Systems and to the formation of the Dundee Centre for Coastal Zones Research.

The programme of lectures and seminars that were presented at the Summer School, and which are included in this book, cover a variety of topics: fractals and coastline definition, hydrodynamical modelling, remote sensing (both from space and using airborne systems), beach nourishment, pollution (including the question of the clean-up of industrial pollution in the post-Soviet era in eastern Europe), wetlands, trace gases and atmospheric pollution.

We are very grateful to the various lecturers and seminar speakers who gave of their valuable time to come to Dundee and present their lecture material. We are also grateful to the participants who attended the summer school, as without them the exercise would have been pointless, to our colleagues on the organising committee and to the technical and secretarial staff (especially Mrs Pat Cunningham) for their assistance in making the summer school a success.

The summer school was supported by DG-XII of the European Commission under its Conference Support and Accompanying Measures Programme, contract ERBFM-MACT960175, and by the Scottish Universities.

Arthur P Cracknell and Robin A Vaughan

Dundee, March 1998

Editors' Note

We would like to thank the lecturers for producing their contributions so quickly after the end of the summer school. We are also grateful to the publications officer of the Scottish Universities Summer Schools in Physics, Dr. Peter Osborne, for considerable help in the later stages of the editorial work.

There are a number of colour figures in the text and these have been referred to as plates, rather than figures, and the plates will be found collected together in a single signature at one point in the book. Some of the sets of lectures presented at the summer school included far more images and diagrams than it has been possible to reproduce here; however, we have tried to include the more important ones in this set of Proceedings.

The text of the first article relies very heavily on one of the reports of the LOICZ Programme and we are very grateful to the LOICZ Programme for granting us permission to make such extensive use of their material.

Arthur P Cracknell and Elaine S Rowan

Dundee, March 1998

Contents

/continued

Overview of physical processes and remote sensing in the coastal zone

Arthur P Cracknell

University of Dundee

1. The importance of the coastal zone

It is not a completely trivial exercise to identify what is meant by, or what is defined to be, the *coastal zone*. However, as a working basis we take a diagram, Figure 1, from the LOICZ (Land-Ocean Interactions in the Coastal Zone) Implementation Plan (Pernetta and Milliman 1995) of the IGBP (International Geosphere Biosphere Programme). On the ocean side it involves going out to the edge of the continental shelf. Obviously the actual distance (measured in km) will vary, but at least the boundary is fairly clearly defined. The question of the definition of the landward boundary of the coastal zone is much more difficult. In the report to which we have already referred (Pernetta and Milliman 1995), this boundary is (fairly arbitrarily) taken to be at the contour that is 200m above sea level.

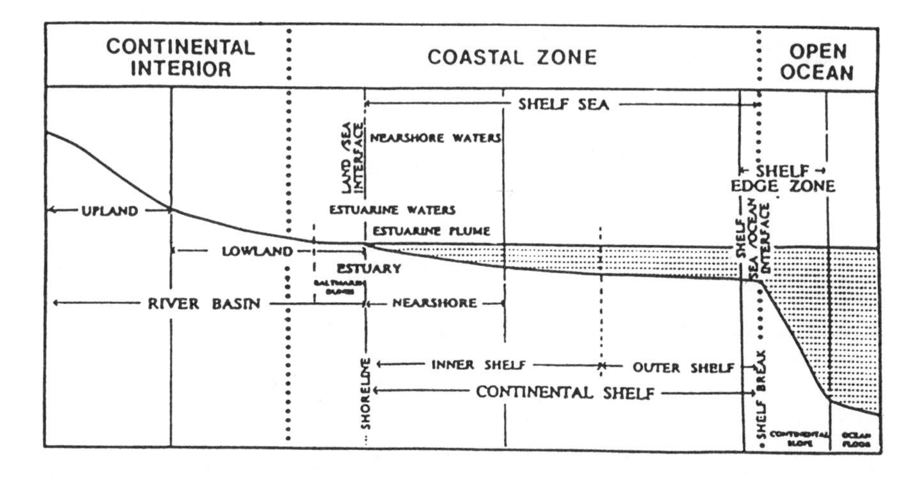

Figure 1. Diagrammatic representation of the coastal zone, continental shelf and other features of the land-ocean boundary. The relative sizes and proportions of these components show considerable geographic variability (Pernetta and Milliman 1995).

With the definition of the coastal domain (Figure 1) as 200m above to 200m below sea level we note that it :

- occupies 18% of the surface of the globe,
- is the area where around a quarter of global primary productivity occurs,
- ie where around 60% of the human population lives,
- is where two thirds of cities with populations of over 1.6 million people are located,
- supplies approximately 90% of world fish catch.

The coastal ocean accounts for:

- 8% of the ocean surface,
- <0.5% of the ocean volume,
- around 14% of global ocean production,
- up to 50% of global oceanic denitrification,
- 80% of the global organic matter burial,
- 90% of the global sedimentary mineralisation,
- 75–90% of the global sink of suspended river load and its associated elements/pollutants,
- in excess of 50% of present day global carbonate deposition.

Thus, with this definition of the coastal zone, we see that it is of far greater importance than would be suggested by its area in relation to that of the whole surface of the Earth. Moreover one can also quite reasonably argue that anthropogenically-driven change is more intense in coastal regions than elsewhere and various reasons for this are also listed by Pernetta and Milliman (1995):

- That is where the majority of people presently reside, resulting in comparatively high existing population densities.
- Rates of growth in coastal populations are usually greater than rates of growth in inland areas due to migration of permanent residents.
- Most international tourism is coastal.
- 90% of land-based pollution including sewage, nutrients and toxic materials remain in the coastal ocean.
- Much of the fertile agricultural land in tropical countries is coastal, hence land-use and cover change are most intense in coastal margins.
- Earlier this century, the anthropogenically driven flux of sediment from land to ocean exceeded natural rates of flux.
- On continental scales the rate of sediment flux is now considerably reduced, due to dam construction and large-scale irrigation schemes (a reversal of previous trends).
- The anthropogenic flux of dissolved nutrients from land to coastal ocean is now equal to, and in some areas greatly in excess of, the natural flux.

Our own particular interest is in remote sensing and there are going to be great opportunities in the near future for remote sensing in coastal zones, particularly from InSAR, (interferometric SAR) and the new high resolution (1m resolution) optical band systems. But remote sensing is not everything and will not do everything. In addition to remotely-sensed data we need

- *in situ* data for validation of the remotely-sensed data,
- physical models for the study of coastal processes and prediction of future evolution,
- a GIS (Geographical Information System), or more generally a RSIM (Remote Sensing Information Model) into which to incorporate the remotely-sensed data.

There are, in a sense, two extremes of coastal zone studies. On the one hand there is that of the global climate modeller whose aim is to be able to obtain as good a representation as possible for the entire coastal regions of the Earth so as to provide the most realistic representation of those regions in a general circulation model being used in climate studies. On the other hand there is the ecologist, environmentalist, policy maker, media reporter, or decision maker who is usually interested in monitoring or predicting change in a very small and local area of the coastal zone. The approaches of these two types of person are bound to be very different. The global climate modeller is bound –for the foreseeable future at least–to consider the coastline at a spatial resolution, and with a data format, that is quite unsuitable and inappropriate for the local ecologist, environmentalist, etc. In more specialised terms, different people are likely to require quite different spatial scales for the GISs or RSIMs that they use and we shall need to consider these user requirements separately.

In this chaper we shall follow quite closely the IGBP (International Geosphere-Biosphere Programme) report on LOICZ (Land-Ocean Interactions in the Coastal Zone) (Pernetta and Milliman 1995). This provides a very good framework for studying the main processes that occur in the coastal zones.

2. Land-ocean interactions in the coastal zone

2.1 The IGBP and LOICZ

The IGBP was established by ICSU (the International Council of Scientific Unions) at its General Assembly meeting in Berne in 1986. The overall aim of the IGBP is "To describe and understand the interactive physical, chemical and biological processes that regulate the total Earth system, the unique environment that it provides for life, the changes that are occurring in this system, and the manner in which they are influenced by human actions".

The IGBP originally listed 6 key research questions (IGBP 1990) and each is being addressed by a Core Project. Since then two other Core Projects and three over-arching and integrative projects have been added.

1. How is the chemistry of the global atmosphere regulated, and what is the role of the biological processes in producing and consuming trace gases?
   Core project: International Global Atmospheric Chemistry Project (IGAC).
2. How will global changes affect terrestrial ecosystems?
   Core project: Global Change and the Terrestrial Ecosystem (GCTE).

3. How does vegetation interact with the physical processes of the hydrological cycle? Core project: Biospheric Aspects of the Hydrological Cycle (BAHC).
4. How will changes in land-use, sea-level rise and climate alter coastal ecosystems, and what are their wider consequences?
Core project: Land-Ocean Interactions in the Coastal Zone (LOICZ).
5. How do ocean biogeochemical processes influence and respond to climate change? Core project: Joint Global Ocean Flux Study (JGOFS).
6. What significant climatic and environmental changes occurred in the past, and what were their causes?
Core project: Past Global Changes (PAGES).
7. Core project: Land Use / Cover Change (LUCC).
8. Core project: The Global Euphotic Zone Study (GOEZS).
9. Core project: Global Analysis, Interpretation and Modelling (GIAM).
10. The IGBP Data Information System (IGBP-DIS).
11. The Global Change System for Analysis, Research and Training (START).

It is clear that there are overlaps but the Core Project most relevant to us is the Land-Ocean Interactions in the Coastal Zone (LOICZ) project. This Core Project is described in some detail in the LOICZ Implementation Plan (Pernetta and Milliman 1995).

LOICZ is one of seven Core Projects which comprise the International Geosphere-Biosphere Programme: A Study of Global Change. The LOICZ project, which commenced in 1993, was scheduled to run for 10 years and focuses on the area of the Earth's surface where land, ocean and atmosphere meet. The overall aim of the project "is to determine at regional and global scales: the nature of that dynamic interaction (of land, ocean and atmosphere); how changes in various compartments of the Earth system are affecting coastal zones and altering their roles in global cycles; to assess how future changes in these areas will affect their use by people; and to provide a sound scientific basis for future integrated management of coastal areas on a sustainable basis." Although this volume has no special connection with LOICZ, much of what is covered fits into the framework of ideas involved in LOICZ and so it should be useful to consider it.

2.2 Background

In comparison with the relatively uniform environment of the sunlit zone of the open ocean, or the rapidly mixed environment of the atmosphere, the spatial and temporal heterogeneity of the world's coastal zones is considerable. There are, as a consequence, considerable methodological problems associated with developing global perspectives of the role of this component in the functioning of the total Earth system.

Planet Earth is not a steady state system. The external energy sources are variable, arising from variations in the energy output from the Sun, variations in the Earth's orbit and variations in its attitude (i.e. in its orientation relative to the axis of that orbit) and irregular impacts by extra-terrestrial objects such as meteorites or comets. The internal energy, augmented by the decay of radioactive elements, provides the energy for volcanic activity and plate tectonic movements in addition to the more steady radial flux of geothermal energy all over the Earth's surface. While energy is arriving at the Earth's surface from external and internal sources, energy is also leaving the Earth's surface by reflection, transmission and emission to the atmosphere and thence into space.

As we are generally aware, the Earth's climate has varied con-siderably over geological time. At present we are experiencing a relatively warm and stable period, called the Holocene, which is about 2°C warmer (in terms of global average temperature), than the Last Glacial. The Holocene started about 11,000 to 11,700 years ago when the last cold period, the younger Dryas, ended in only a few years. It is not known very precisely what happened, but it is thought that shifting ocean currents in the north Atlantic played a key role. Consequently the northern hemisphere heated rapidly, large ice sheets melted, there was a rise of global sea level by about 100m and there was a large change in the trace gas composition of the Earth's atmosphere.

What is different now from previous situations in geological time is that our species has developed the capacity to modify the conditions at the Earth's surface to such a large extent as to influence the Earth's climate. We, as a species, have developed a large technologically based society which uses energy at a vast rate and which moves material within the Earth system at rates comparable to, or surpassing, the natural rates of biogeochemical cycling. Predicting the next dramatic shift in the Earth's climate, which could happen very suddenly arising from the combined effect of natural causes and of human activities, is not an easy thing to do. Apart from global changes in the climate, which would have dramatic consequences for large parts of the population, there are other changes which are of less significance in global terms but which can still be very serious in regional or local terms; these include a rise in mean sea level, the loss of biodiversity, the expansion of the deserts, the loss of agricultural soil by erosion, the accumulation of nitrate in fresh water supplies, a general shortage of clean drinking water, etc.

One of the very important things to realise when thinking about climate change is to appreciate that there are natural processes and that there are human activities. The actual change will be a consequence of both of these and to separate - at the observational level - one from the other is very difficult.

The goals of LOICZ, as taken from pages 18-19 of the report which has been mentioned several times already (Pernetta and Milliman 1995) are:

1. To determine at global and regional scales:
    a) the fluxes of material between land, sea and atmosphere through the coastal zone,
    b) the capacity of coastal systems to transform and store particulate and dissolved matter and
    c) the effects of changes in external forcing conditions on the structure and functioning of coastal ecosystems.
2. To determine how changes in land use, climate, sea level and human activities alter the fluxes and retention of particulate matter in the coastal zone, and affect coastal morphodynamics.
3. To determine how changes in coastal systems, including responses to varying terrestrial and oceanic inputs of organic matter and nutrients, will affect the global carbon cycle and the trace gas composition of the atmosphere.

4. To assess how responses of coastal systems to global change will affect the habitation and usage by humans of coastal environments, and to develop further the scientific and socio-economic bases for the integrated management of the coastal environment.

2.3 The LOICZ foci

A Scientific Planning Committee for LOICZ was established in 1990 and this Committee produced the Science Plan (Holligan and de Boois 1993) which identified a number of research activities. Subsequently these research activities were collected into a number of groups under the headings of Foci. The structure of the LOICZ Implementation Plan (Pernetta and Milliman 1995) is built around the scheme shown in the following table.

Focus 1: Effects of Changes in External Forcing or Boundary Conditions on Coastal Fluxes

Activity 1.1	*Catchment basin dynamics and delivery*
Activity 1.2	*Atmospheric inputs to the coastal zone*
Activity 1.3	*Exchanges of energy and matter at the shelf edge*
Activity 1.4	*Development of coupled models for coastal systems*

Focus 2: Coastal Biogeomorphology and Global Change

Activity 2.1	*The role of ecosystems in determining coastal morphodynamics under varying environmental conditions*
Activity 2.2	*Coastal biogeomorphological responses to anthropogenic activities*
Activity 2.3	*Reconstruction and prediction of coastal zone evolution as a consequence of global change*

Focus 3: Carbon Fluxes and Trace Gas Emissions

Activity 3.1	*Cycling of organic matter within coastal systems*
Activity 3.2	*Estimation of net fluxes of N_2O and CH_4 in the coastal zone*
Activity 3.3	*Estimation of global coastal emissions of dimethylsulphide*

Focus 4: Economic and Social Impacts of Global Change in Coastal Systems

Activity 4.1	*Evolution of coastal systems under different scenarios of global change.*
Activity 4.2	*Effects of changes to coastal systems on social and economic activities*
Activity 4.3	*Development of improved strategies for the management of coastal resources*

Note that Activity 1.4 above encompasses activities 1.4 and 1.6 of the LOICZ Science Plan, and Activity 2.3 encompasses activities 1.5 and 2.3 of the LOICZ Science Plan.

2.4 The LOICZ Framework Activities

There is a chapter in the LOICZ Implementation Plan entitled "LOICZ Framework Activities". Although it makes interesting reading, its formulation does seem rather ambitious. The tone of the discussion in this section is not intended to be scornful. It is simply intended to express a certain amount of healthy scepticism as to whether the effort and resources needed to carry out this ambitious plan will actually be available. Consider one example. The LOICZ Framework Activity 2 is the "Development of a Coastal Typology". It is defined as having an overall objective "to categorise the world's coastal zone on the basis of natural features, into a realistic number of geographical units, which will serve as a framework for

- overall co-ordination and planning of LOICZ research activities,
- organisation of data bases,
- selection of regions for extensive studies (remote sensing, long term monitoring),
- selection of appropriate sites for new local and regional coastal zone field and modelling studies,
- scaling local and regional models to regional and global scales,
- analysis, compilation and reporting of LOICZ results in the form of regional and global syntheses,
- interfacing with the regional research nodes.

If one looks in a simple textbook on the coastal zone (e.g. Viles and Spencer 1995) it can be found that the coastline is classified into 5 types

- sandy coastline: beaches and dunes,
- rocky coasts: cliffs and platforms,
- coastal wetlands,
- coral reefs,
- cold coasts: permafrost, glaciers, sea ice and fjords.

This classification is considered far too coarse for LOICZ which says that "of the order of 100-200 divisions of the world's coastal zone are anticipated" although, in fairness, it does go on to say that many of them "will have similar properties, such that they can be assumed to be identical in function from the Earth system perspective." This is consistent with the rationale of the development of a system for formulating the LOICZ global coastal typologies: "The global scope of LOICZ and the constraints of human and financial resources, necessitate the development of an objective typology of coastal units to serve as a sampling framework and to determine the appropriate weighting for preparing global syntheses, scenarios and models on the basis of limited spatial and temporal research data".

The specific objectives of this Activity (Activity 2) were defined to be

Short-term

- to develop a framework global coastal zone typology based upon existing scientific information,

- to use the typology to guide the development of the LOICZ Core Project.

Long-term

- to refine and develop the typology according to the evolving needs of the Project and the individual Foci,
- to apply the typology in preparing regional and global syntheses, and in developing scenarios and models.

The outputs were defined to be

- a functional typology of the world's coastal zone,
- a framework for the selection of sites for LOICZ research activities,
- a framework for the compilation of regional and global syntheses of data, scenarios and models of the role of the coastal sub-system in Earth system functioning.

This activity is one example of the six Framework Activities within LOICZ; the complete set of Framework Activities is as follows:

Framework Activity 1	Scientific Networking
Framework Activity 2	Development of Coastal Typology
Framework Activity 3	Data System Plan
Framework Activity 4	Measurement Standards, Protocols and Methods
Framework Activity 5	Modelling in LOICZ
Framework Activity 6	Determination of the Rates, Causes and Impacts of Sea Level Change.

Before leaving these Framework Activities, we devote some attention to the Framework Activity 5 - Modelling in LOICZ - because there are some very useful points in that. At the outset it is emphasised that studying the coastal zone cannot be confined to observational studies alone, at whatever scale. Numerical modelling is important because:

- the process of developing models helps to improve the scientific understanding of how coastal ecosystems function and respond to changing forcing functions,
- the models themselves, if properly constructed and validated, can be used to predict the effects of future change.

Three general categories of numerical models are identified:

- **Budget models** - Mass balance calculations of specific variables (e.g. carbon, nitrogen, phosphorus, sediment, etc.) within defined spatial and temporal scales. They include all major sources, sinks, and transformation processes and are descriptive, with no internal dynamics.
- **Process models** - Describe specific environmental processes (e.g. flushing, sedimentation, photosynthesis, respiration, etc.) and how they are influenced by major forcing functions. They are usually constructed from empirical studies.

- **System models** - Based upon process models. They integrate what are considered to be the most important variables and processes for the system in question within defined spatial and temporal scales. Particular attention is usually given to the interaction among state variables. They can be either descriptive (e.g. network analysis) or dynamic (e.g. simulation models).

Note that numerical modelling can be done on different spatial and temporal scales:

- **Local** - Scale of the order of 1-100km. These would address specific habitats such as saltmarshes, mangrove forests, deltas, coral reefs, estuaries, bays and fishing banks.
- **Regional** - of the order of 100-10,000km. These incorporate a variety of near-shore and off-shore habitats out to the 200m isobath. This scale will coincide with the units defined in the coastal typology exercise.
- **Global** - These incorporate several regional models representing the entire coastal zone of the Earth.

The third and first of these, respectively, correspond to the two extremes at which the coastal zones can be studied and which were mentioned towards the end of Section 1.

3 LOICZ foci in detail

Each of the four LOICZ Foci in the table of Section 2.3 is given a separate chapter in the LOICZ Implementation Plan and we summarise each of these four chapters in turn. There is, fortunately, a common general structure to each of these chapters. In each case we summarise the Activities of the Focus and then give a breakdown of the Activities into tasks. For each task we note the *rationale, specific objectives* and *outputs.*

Focus 1: The effects of changes in external forcing or boundary conditions on coastal fluxes

Focus 1	Activity 1.1	Catchment basin dynamics and delivery
•	Task 1.1.1	Development of regional scale river delivery models in theory and practice
•	Task 1.1.2	Determination of temporal variations in discharge characteristics for selected and representative rivers
•	Task 1.1.3	Chemical composition: gradients from terrestrial sources to marine burial
•	Task 1.1.4	Groundwater discharge and coastal erosion
Focus 1	Activity 1.2	Atmospheric inputs to the coastal zone
Focus 1	Activity 1.3	Exchanges of energy and matter at the shelf edge
•	Task 1.3.1	Synthesis of results of existing or on-going regional or national research projects on energy and matter exchanges at the shelf edge
•	Task 1.3.2	Dissolved and suspended matter exchanges at the shelf edge
•	Task 1.3.3	Inter-annual variability in energy and matter exchanges at the shelf edge
•	Task 1.3.4	The role of long-term fluctuations in shelf-edge exchange rates in global cycles
Focus 1	Activity 1.4	Development of coupled models for coastal systems
•	Task 1.4.1	The development of new local and regional ecosystems models

Activity 1.1 Catchment basin dynamics and delivery

The overall objectives for the river input to the coastal zone activity of LOICZ are to determine at regional and global scales:

- the quantities and chemical attributes of the river water delivered to the ocean,
- the quantities and chemical attributes of the bulk sediments, and particulate and dissolved chemical species (C, N and P) delivered to the ocean,
- the temporal variation in quantities and chemical attributes of riverine inputs to the ocean,
- the relationship between these fluxes and changes in the upstream environment,
- the physical and chemical controls operating at continental-scales on these fluxes.

Task 1.1.1 Development of regional scale river delivery models in theory and practice

Rationale

The systematic data required to make global estimates of land to ocean flux via rivers are limited, and there are large uncertainties concerning the magnitude of the material transport. The classic approach based on a river-by-river inventory to produce a global estimate of flux is inadequate, and an approach based on the identification of the central properties of rivers, as a function of human use, climate, and geography, is required.

Short-term specific objectives

- to develop the theoretical basis for regional-scale river delivery models to the coastal ocean,
- to develop a global typology of river systems based on the central properties of rivers, as a function of human use, climate and geography,
- to develop quantitative models for specific river catchment types,
- to assemble an electronic, global database of existing river discharge data.

Long-term specific objectives

- to assemble a GIS global digital database of river networks delivering to the sea,
- to apply the quantitative models to the digital database to determine present riverine delivery to the sea,
- to develop scenarios of future delivery under conditions of changed climate and human land-use and freshwater use.

Outputs

- a global database of existing river discharge data,
- a global typology of river systems based on the central properties of rivers, as a function of human use, climate and geography,

- algorithms for describing river discharge at specific points in the drainage system,
- quantitative coupled models of river chemistry, river networks and river discharge,
- a global estimate of the present export of water and river borne materials to the coastal zone encompassing those regions which are poorly described by actual monitoring networks.

Task 1.1.2 Determination of temporal variations in discharge characteristics for selected and representative rivers

Rationale

One of the major weaknesses of present global syntheses of river flux to the oceans is lack of adequate seasonal and inter-annual records for many rivers and an inadequate data and information base on which to develop estimates of temporal variations in rivers of different type.

Short-term specific objectives

- to provide input data on temporal variations in river fluxes to the development of a global typology of river systems,
- to define the seasonal and inter-annual variation in the input of rivers representative of the major drainage basin types.

Long-term specific objectives

- to provide regional and global estimates of seasonal and inter-annual variation in river discharge to the oceans,
- to encourage the development of appropriate river observation and measurement systems for those classes of rivers where present data on variables such as discharge (Q), particulate organic carbon (POC), dissolved organic carbon (DOC), free dissolved CO_2 (pCO_2), and total suspended sediments (TSS) are inadequate.

Outputs

- data sets describing intra-annual and inter-annual variation in river discharge for identified classes of river system,
- input data on Q, POC, DOC, pCO_2 and TSS for the definition of the global river typology,
- protocols for future programmes of measurement, including sampling and analytical methods,
- regional and global estimates of temporal variation in river discharge to the coastal ocean,
- input data for developing scenarios of future regional and global river fluxes to the coastal ocean.

Task 1.1.3 Chemical Composition: Gradients from Terrestrial Sources to Marine Burial

Rationale

Organic matter is subjected to within-channel transport and reactive processes before it is finally discharged into coastal waters. To construct mass balances for coastal systems at regional and global scales, not only are data concerning the quantity, and timing of delivery of river borne materials essential, but data are required, concerning specific chemical attributes of the organic matter in transport, in order to understand the physical and chemical controls operating at regional and global scales, and to determine the origin of the river-borne material, and the dynamics of downstream routing.

Short-term specific objectives

- to define the specific chemical attributes relating to the sources and condition of organic matter in transport and buried in temporary marine depocentres,
- to define the required chemical information relative to sources, partitioning and diagenesis, and on how refractory or labile different fractions will be in particular environments.

Long-term specific objectives

- to determine how catchment type influences the regional mass balance budgets for POC, TSS, C, N, and Si in the contiguous coastal zone,
- to define the chemical composition of riverine inputs to coastal waters on regional and global scales.

Outputs

- definition of chemical indicators of source and history of river borne organic matter,
- representative datasets for development of regional and global mass balance budgets.
- regional and global datasets on the chemical composition of riverine inputs to coastal waters.

Task 1.1.4 Groundwater discharge and coastal erosion

Rationale

Global estimates of land-ocean exchange remain incomplete without a consideration of sub-surface water exchange, and coastal erosion. Groundwater discharge directly into the seas may be important in the hydrochemical, thermal and biogenic regime of the coastal zone. Neither of these sources of input to the coastal ocean is adequately understood at present.

Short-term specific objectives

- to develop databases of sub-surface exchange and coastal erosion,
- to assess the regional and global significance of ionic groundwater discharge in the coastal zone,
- to assess the likely changes in groundwater discharge under the impact of climate and anthropogenic factors causing change,

- to make regional and global estimates of the impact of sea water intrusion on the groundwater regime in the coastal zone.

Long-term specific objectives

- to develop and apply coastal erosion models based on climatic zonation, geology, geomorphology, palaeography and hydrodynamics for key regions and various climatic zones,
- to define (quantitatively and qualitatively) the contribution of coastal erosion to the total terrigenous inputs,
- to develop models of coastal vulnerability to erosion under differing scenarios of human land-use, climate change and sea level rise.

Outputs

- regional and global estimates and models of groundwater discharge, dissolved solids, eroded sediments transported with the ground water and from cliffs into the sea,
- a global database and synthesis of erosion and groundwater inputs and exchange between land and ocean.

Activity 1.2 Atmospheric inputs to the coastal zone

The overall objective of this activity is to quantify, at regional and global scales, atmospheric deposition in the coastal zone of naturally and anthropogenically derived materials and to evaluate the impact of such deposition on the chemical and biological fluxes and processes in coastal systems. More specifically, this activity aims to determine:

- the proportion of atmospheric deposition that represents a net input to the coastal zone and that which represents a recycling of materials between coastal seas and land due to the emission, transport, and deposition of sea-spray aerosols,
- the bioactive fraction of atmospheric deposition,
- the contribution of atmospheric deposition on watersheds to the chemical composition of waters draining into coastal waters,
- the role of local, regional, or global transport processes in controlling atmospheric deposition in coastal areas,
- the spatial and temporal variation in atmospheric deposition to coastal areas,
- the regional and global importance of atmospheric deposition as a natural and/or anthropogenic forcing function in the coastal zone.

It was noted that, ideally, the use of atmospheric transport and deposition models, combined with inventory data concerning the land-based sources, should allow the prediction of atmospheric concentrations and flux fields. But such an approach has not been successfully applied to date, primarily owing to a lack of accurate source emission data for most elements and/or substances of interest. A major sampling programme is thus needed to document temporal and spatial variability in atmospheric inputs. Such a programme must encompass short-term pollution episodes and sporadic but intense transport and deposition events of mineral dust from soil erosion determined by local meteorology, including air movements and precipitation events; This sampling programme must also regard longer term, seasonal and inter-annual variability which depend on the temporal patterns of local climatology. A coherent sampling programme should include:

- measurements of the total and bioactive concentration and flux of major nutrients and micronutrients (N, P and essential trace metals Fe, Mn, Al),
- identification of suitable tracers of natural and anthropogenic sources,
- monitoring of relevant meteorological parameters, including wind fields and precipitation.

No particular tasks were identified within this Activity.

Activity 1.3 Exchanges of energy and matter at the shelf edge

The overall objective of this activity is to quantify at regional and global scales the exchange of energy and matter across the continental shelf edges; to estimate the seasonal and inter-annual variations in these fluxes; and to provide the basis for predicting future fluxes due to long-term changes in external forcing functions such as climate or land-use. This longer-term objective can only be met through incremental advances in scientific understanding. More specifically this activity aims to:

- clarify scientific understanding of the basic processes that regulate energy and matter exchanges at the continental shelf edge,
- elucidate the seasonal and inter-annual variability in the fluxes of energy and matter across the continental shelf edge,
- quantify energy and matter fluxes across the shelf edge in representative shelf seas.

It was accepted that there is a lack of scientific knowledge concerning the fundamental processes that control and regulate exchanges at the continental shelf edge and that initially it is necessary to develop further the body of scientific knowledge and information relating to processes that determine the magnitude of the fluxes involved. These processes include mixing due to tidal forcing, frontal eddies, upwelling and the density of wind-driven currents. It was proposed that in parallel with basic studies of these processes there should be representative case studies undertaken as well and suggested case studies were

- wide shelf seas with a strong western boundary current such as the East China Sea,
- wide shelf seas without a strong boundary current such as the North Sea or Indonesian Seas,
- narrow shelf seas with a strong western boundary current such as the South Atlantic Bight,
- narrow shelf seas without a strong boundary current such as the western coast of Canada,
- strong coastal upwelling regions such as the Peruvian Coast,
- coastal boundary current regions such as the western coast of Norway,
- shelves with strong seasonal surface currents such as the Grand Banks off eastern Canada.

Task 1.3.1 Synthesis of results of existing or on-going regional or national research projects on energy and matter exchanges at the shelf edge.

It is at about this point in the Report (Pernetta and Milliman 1995) that the rationale, objectives and outputs become more vague. Nevertheless we mention them.

Rationale

No regional or global review of the processes regulating and controlling exchanges at the shelf edge has been produced to date; hence the relative importance of different processes, their frequency of occurrence, temporal and spatial variability cannot be evaluated.

Specific objectives

- to update present knowledge on the processes regulating and controlling energy and matter exchanges at the shelf edge,
- to prepare and publish a review of processes and extent of shelf edge exchanges based on the results of existing or on-going regional or national research projects related to Activity 1.3.

It was also recognised that resources would not be available for LOICZ to initiate and conduct independent shelf-edge studies and that successful achievement of LOICZ objectives in this field will only be possible through the agreement and close cooperation with JGOFS and a variety of national and regional studies.

Output

- published review volume on global state of knowledge concerning shelf edge exchange processes.

Task 1.3.2 Dissolved and suspended matter exchanges at the shelf edge

Rationale

The present data concerning shelf edge exchanges of energy and matter are inadequate for the purposes of preparing regional and global estimates of fluxes between the coastal seas and open oceans and a concerted and co-ordinated programme of measurements is required to develop such a database.

Specific objectives

- to clarify our understanding of the processes that regulate the exchanges of dissolved and suspended matter at the shelf edge,
- to estimate the fluxes of dissolved and suspended matter across the shelf edge; programmes for the implementation of this task were proposed,
- in situ observations of water temperature, salinity, turbidity, concentration of dissolved matter (macro and trace nutrients, DOC), suspended matter (plant pigments, POC, PON and POP etc.) at least four, and preferably six, times a year in the shelf sea and the adjacent open ocean,
- obtaining synchronous satellite images (NOAA, SeaWiFS, OCTS, etc.) for the time of the field observations,

- deployment of moorings with time-series sediment traps, current meters and nephelometers at some stations around the shelf edge for the period of at least one year,
- to establish a box model which can estimate material budgets in the shelf sea and fluxes across the shelf edge.

Outputs

- quantitative assessment of the role of stationary processes such as tidal current and barotropic or baroclinic instabilities and event-like processes such as wind or density-driven currents in the exchange of dissolved and suspended matter at the shelf edge,
- quantitative estimate of the dissolved and suspended matter fluxes across the shelf edge and their seasonal variability,
- input data concerning the exchanges across the shelf edge for developing the coastal ecosystem models envisaged under Task 1.4.

Task 1.3.3 Inter-annual variability in energy and matter exchanges at the shelf edge.

Rationale

At present virtually no data exist concerning inter-annual variations in shelf edge exchanges of energy and matter. Such data are vital to developing global estimates of the role of coastal and shelf seas in the global carbon cycle and are a pre-requisite for developing future scenarios of shelf edge exchange under changed climatic conditions and anthropogenic, land-based changes to the land-ocean flux of energy and matter.

Specific objective

- to investigate and quantify for representative shelf edge types, the inter-annual variation in fluxes of energy and matter at the shelf edge.

The implementation of this task was envisaged as involving temporal extensions of the studies proposed under Task 1.3.2.

Output

- quantitative estimate of inter-annual variations in energy and matter exchanges at the shelf edge.

Task 1.3.4 Role of long-term fluctuations in shelf-edge exchange rates, in global cycles.

Rationale

Exchanges at the shelf edge between coastal and open ocean are subject to open ocean forcing of long periodicity such as El Niño/Southern Oscillation (ENSO) events or land use changes such as dam construction. The impact of such events on energy and material flux across the continental shelf edge is at present unknown.

Specific objectives

- to quantify the role of long-term basin scale phenomena such as El Niño on variations in the exchange of energy and matter at the shelf edge,

- to quantify the role of anthropogenic forcing through large scale changes in land use in changes to the exchanges of energy and matter at the shelf edge,
- to quantify the role of such phenomena in the global cycling of matter and their impact on the coastal environment.

Outputs

- improved understanding of the role of changes in basin scale dynamics on the flux of materials across the shelf edge,
- improved understanding of the role of changes in land and water use on the flux of materials across the shelf edge,
- input data for developing scenarios of future global cycling of matter in relation to changes in exchanges of energy and matter at the shelf edge under conditions of global change.

Activity 1.4 Development of coupled models for coastal systems

It is acknowledged that previous modelling work in the coastal zone has tended to focus on particular processes such as run-off, hydrodynamics, nutrient cycling, primary production etc. However, it is argued that the stage has now been reached where it is possible to imagine that the development of coupled land-estuarine-atmosphere ocean numerical models should be possible. The overall objectives of this activity are thus:

- to develop budget models for carbon, nitrogen, phosphorus and silicate in selected coastal units using the procedures recommended under Task F.5.1 (see Figure 7 of Pernetta and Milliman 1995),
- to develop carbon flow system models for a more limited number of coastal units, where the necessary resources are available, using the modelling procedures recommended under Task F.5.1 (see Figure 7 of Pernetta and Milliman 1995).

It is acknowledged that models developed under this activity will be of local and regional scale.

Task 1.4.1 The development of new local and regional ecosystem models

Rationale

Given the high spatial and temporal variability of the world's coastal zone, new models will have to be developed in order to describe adequately the functioning of the full range of coastal ecosystems. Some models can be built using existing information while others will require the collection of new data which will be done under various Activities in Foci 1-3.

Specific objectives

- to develop budget models for carbon, nitrogen, phosphorus and silicate in selected coastal units using the recommended techniques,
- to develop carbon flow system models for a more limited number of coastal units using the recommended techniques.

Outputs

- budget models for carbon, nitrogen, phosphorus and silicate for a large number of representative coastal regions around the world (1996 onward),
- system models of carbon flow for representative coastal regions around the world (1998 onward).

Focus 2: Coastal biogeomorphology and global change

This Focus is even more complicated than Focus 1. It contains not only Tasks but Sub-tasks as well and concentrates particularly on coral reefs and mangrove swamps. We shall not work through Focus 2 in the same kind of detail that we did in the case of Focus 1.

Focus 2	Activity 2.1	The role of ecosystems in determining coastal morphodynamics under varying environmental conditions
•	Task 2.1.1	Sediment budgets in coastal ecosystems
	---- Subtask 2.1.1.1	Coral reef system sediment budgets
	---- Subtask 2.1.1.2	Determination of sediment budgets on mangrove shorelines
•	Task 2.1.2	Coastal ecosystem characteristics, processes and interactions
	---- Subtask 2.1.2.1	Coral reef ecosystem processes and external interactions
	---- Subtask 2.1.2.2	Mangrove ecosystem interactions and feedbacks
•	Task 2.1.3	Episodic events
	---- Subtask 2.1.3.1	Episodic disturbances in coral reef communities
	---- SubTask 2.1.3.2	Extreme events and their impact on mangrove shorelines
Focus 2	Activity 2.2	Coastal biogeomorphological responses to anthropogenic activities
•	Task 2.2.1	Direct impacts of human activities on the coastal zone
	---- Subtask 2.2.1.1	Direct impacts of human activities on coral reefs
	---- Subtask 2.2.1.2	Direct impacts of human activities on mangrove shorelines
•	Task 2.2.2	Indirect impacts of human activities on the coastal zone through changes in land and freshwater use
	---- Subtask 2.2.2.1	Indirect effects of land use and hydrologic changes on coral reef communities
	---- Subtask 2.2.2.2	Indirect impacts of human activities on mangrove shorelines
Focus 2	Activity 2.3	Reconstruction and prediction of coastal zone evolution as a consequence of global change
•	Task 2.3.1	Palaeographical, palaeoenvironmental and historical reconstructions of the coastal zone
	---- Subtask 2.3.1.1	Reef palaeogeography and palaeoenvironments
	---- Subtask 2.3.1.2	Palaeogeographic and palaeoenvironmental reconstructions of mangrove shorelines
•	Task 2.3.2	Systems modelling and future scenarios

Focus 3: Carbon fluxes and trace gas emissions

Despite its title we find that only the carbon fluxes part of this Focus had been developed in any detail in the LOICZ Implementation Plan. The question of trace gas emissions (N_2O, NH_4 and DMS (dimethylsulphide) was left for later development. Thus it was only for Activity 3.1 that a number of tasks were identified as follows.

Focus 3	Activity 3.1	Cycling of organic matter within coastal systems
•	Task 3.1.1	Synoptic distribution of key carbon variables (data collection)
•	Task 3.1.2	Carbon dynamics of coastal systems investigated by case studies
•	Task 3.1.3	Sensitivity studies
•	Task 3.1.4	Evaluation and validation of budget and system models
Focus 3	Activity 3.2	Estimation of net fluxes of N_2O and CH_4 in the coastal zone
Focus 3	Activity 3.3	Estimation of global coastal emissions of dimethylsulphide

One of the currently unanswered questions of considerable importance to an understanding of global change is the role of the coastal seas in both the natural and the disturbed carbon cycle. Coastal seas serve as final receivers of natural and anthropogenic organic matter and nutrients derived from the land and brought in by river and groundwater discharge and through direct inputs of sewage and other wastes. It is estimated that roughly 0.4Gt of organic carbon are delivered annually by rivers to the coastal sea, an unknown part of which could be respired, constituting a natural source of CO_2 to the atmosphere. At the same time anthropogenically mobilised nutrients are delivered to the coastal seas. Consequently coastal seas may simultaneously sequester some carbon from the atmosphere at rates in excess of those occurring under natural conditions. The size of this sink, the potentially sequestered carbon (PSC), can be estimated by scaling the anthropogenic inorganic nutrients inputs according to the Redfield ratio (i.e. the atomic ratio in which C, N and P are consumed by phytophankton, namely 106:16:1). Estimates of total anthropogenic inorganic nitrate and phosphate inputs to the coastal zone are in the range of $(0.05 - 1.5) \times 10^{12}$ mol y^{-1}. The presumed coastal sink is therefore of the order of $(6 - 10) \times 10^{12}$ mol C y^{-1} or (0.07 - 0.12) Gt C y^{-1}. Organically bound nutrients which also enter the sea via sewage may - upon remineralisation - also increase coastal primary productivity. However, organically bound nutrients probably result in as much fixation as release of CO_2 when they are first remineralised, and therefore enhanced nutrient inputs may not result in an increase in the coastal sink of atmospheric CO_2.

This approach to the problem, however, does not provide information on when and how this PSC is fixed and where it eventually ends up. The PSC could be deposited in tidal flats, in estuaries, at deposition centres on the shelf or it could be transported over the shelf edge. Nor does this approach provide information on whether coastal seas are natural sources or sinks of CO_2.

The overall objective of this activity is to study, at regional and global scales, the disturbed and undisturbed carbon cycle of the coastal seas through the following:

- data collection designed to provide spatially referenced information on the synoptic distribution of key carbon variables,

- conduct of in-depth studies of the carbon dynamics of selected coastal systems,
- sensitivity studies designed to assess the reliability of transferring case study results from one area to another or from year to year,
- the use of spatially referenced data and model standardised reaction groups, to integrate the carbon budget for similar coastlines,
- develop global estimates on the basis of the regional budget values.

Task 3.1.1 Synoptic distribution of key carbon variables (data collection)

Rationale

Present data concerning carbon in the coastal ocean are both dispersed and inadequate in coverage, precluding the adequate assessment of the role of the coastal ocean in the disturbed and undisturbed carbon cycle.

Specific objectives

To assemble appropriate databases in georeferenced format, covering the following datasets:

- sediments: data on geographical extent and inorganic and organic carbon concentrations of shelf and inshore water sediments, on recent sedimentation rates and changes of sedimentation rates of Holocene deposition centres of coastal seas,
- particulate suspended matter: data on the concentration of total suspended matter, particulate organic carbon and inorganic carbon and chlorophyll in the water column covering both large areas and seasonality,
- dissolved carbon: data on the distribution and seasonality of dissolved organic carbon (DOC), the alkalinity, total inorganic dissolved carbon (TCO_2), pH and/or CO_2 - pressure (pCO_2),
- dissolved nutrients: phosphate, nitrate, ammonia and silicate,
- rates: data on remineralisation rates of organic carbon in surface sediments, respiration and photosynthesis in the water column, on air-sea CO_2 exchange and trace gas emissions and on the amount and the composition of the vertical particle flux.

Outputs

- distribution maps of various forms of carbon in the coastal environment, as input information for the calculation of simple mass balances (for example for air-sea exchange of CO_2),
- identification on a global scale of geographic sinks and sources for CO_2 in the coastal ocean,
- identification on a global scale of deposition centres of organic carbon in coastal sediments,
- in combination with some estimate of sediment deposition rate (isotope measurements, industrial markers, storm surge deposits etc.) an estimate of how much

organic carbon is deposited in relation to input of nearby rivers and/or average autotrophic fixation.

Task 3.1.2 Carbon dynamics of coastal systems investigated by case studies

Rationale

Measuring concentrations is only a first step in unravelling the carbon cycle. Therefore studies are needed which describe the dynamics of the system. This necessitates a thorough understanding of the hydrodynamics of the region under study as well as of the seasonality and long-term trends in climate and input forcing functions.

Specific objectives

- to initiate in-depth studies of the carbon cycle in coastal seas in representative areas determined on the basis of the coastal typology developed under Task F.2.1 and using high precision potentiometry, coulometry and infrared pCO_2 methods to acquire precise enough data fields to detect seasonal differences and possibly net fluxes
  - to improve the interpretation of remote sensing data by simultaneous ground truth measurements in a variety of river plumes differing in sediment, gelbstoff and chlorophyll contents,
  - to develop the capacity to model water exchange, biogeochemistry and sedimentation in coastal seas and validate it with measured concentration fields,
  - to resolve the dynamics of carbon export, carbon remineralisation, carbon sequestering, carbon degassing and carbon burial in estuaries of large pCO_2, DOC, POC, and/or nutrient gradients,
  - to unravel the fate of terrestrial organic carbon in the marine system by analysing the chemical and isotopic composition of the organic matter as it moves from the river source through the estuary and into the marine realm to its final site of deposition
  - to conduct studies with respect to the rate of deposition of organic carbon in near-shore waters, its rate of burial, its residence time and its remineralisation rate.

Outputs

- improved scientific understanding of the cycling of matter in the coastal zones,
- new analytical techniques to describe monitoring results by algorithms representing hydrodynamic, sedimentological, biogeochemical and biological processes.

Task 3.1.3 Sensitivity studies

Rationale

The coastal zone and shelf seas display high levels of heterogeneity in terms of their functioning in the carbon cycle; at present the importance of this spatial and temporal heterogeneity in determining the role of the coastal oceans in the global carbon cycle is inadequately documented.

Specific objectives

- to compare areas of similar settings and compare several seasonal cycles at the same area in order to assess how well the results of case studies can be transposed from one year to the next and from one area to another,
- to use statistics to discriminate groups of coastal sites with similar carbon cycle behaviour and establish averages of regional reactions in terms of characteristics of the seasonal cycle, turbidity and nutrient loading,
- to use statistics to discriminate assemblages of coastal settings according to similarity in the GIS data set, and establish groupings as to regional coastal characteristics.

Outputs

- monthly maps of the magnitude of the carbon sink/source function of certain oceanic areas,
- maps of the extent of certain ecosystem functions (for example magnitude of diatom versus flagellate blooms) may also be constructed leading to testable hypotheses about the size and seasonality of the carbon cycle in little-studied sections of the coastal zone, which then in turn can be tested in the field.

Task 3.1.4 Evaluation and validation of budget and system models

Rationale

The new models developed under Task 1.4.1 must be evaluated and validated if confidence in the results of global syntheses of model outputs is to be established and scenarios of future coastal systems under conditions of global change are to be developed.

Specific objectives

Short-term

- to strengthen and implement the modelling networks established under Task F.5.1,
- to evaluate and compare the results of budget and system models developed for similar coastal units,
- to evaluate and compare the results of budget and system models developed for contrasting coastal units,
- to explore the development of generic sub-models that can be transferred from one geographic area to another and used with confidence in building system models,
- to commence validation of regional system models using field data collected under Focus 3 Activities,
- to identify critical information gaps that should be addressed in extensive and intensive observational programmes.

Outputs

- a strong network of LOICZ numerical modellers,
- an improved empirical understanding of the functioning of coastal ecosystems,

- generic sub-models to be used in the construction of biogeochemical system models,
- validated numerical models of coastal systems for evaluating environmental change,
- a clearer understanding of the optimum way to develop fully predictive models of the coastal zone at regional scales,
- recommendations for new data collection programmes including process studies, time series observations at key coastal sites and remote sensing requirements for implementing coastal monitoring and predictive modelling of coastal environments.

Activity 3.2. Estimation of net fluxes of N_20 and CH_4 in the coastal zone

Activity 3.3. Estimation of global coastal emissions of dimethylsulphide

Both of these Activities were described as being of lower priority than Activity 3.1 and would be developed later.

Focus 4: Economic and social impacts of global change in coastal systems

The coastal zone occupies only about 8% of the planet's surface, but it accounts for approximately one fifth of total global primary production. Although only 5–10% of world food production is based on marine resources, as much as 85–90% of present world fisheries production comes from the waters of the Exclusive Economic Zones and approximately 60% of the population of developing countries derive between 40 and 90% of dietary protein from marine sources. Much of the world's productive agricultural land is found in low-lying coastal areas and in the flood plains of major rivers.

As a consequence, coastal areas have always been favoured locations for human habitation and present estimates suggest that about 60% of the world's population lives within 60km of the shoreline. Over two thirds of the world's cities with more than 2 million inhabitants are located in coastal areas, often in the vicinity of highly productive areas or wetlands. Coastal human populations in many countries are growing at approximately twice their natural growth rates due to migration and urban drift to coastal locations and cities. In addition, the bulk of world tourism centres on marine and coastal environments, resulting in extremely high transient population densities during peak seasons. The increase in coastal populations is of considerable concern given the high productivity of coastal areas, which is lost when land is occupied by infrastructure or housing or the sea is polluted with sewage and industrial effluents. It is suggested that the anthropogenically-induced transfer of nutrients from land to the coastal ocean now exceeds that of natural rates of transfer, a condition reached in the case of sediments much earlier in human history.

Thus many coastal areas are already stressed or degraded, or both, and the driving forces for short-term to medium-term change in these environments are anthropogenic. A conceptual framework for a research agenda proposed under Focus 4 of the LOICZ Implementation Plan is shown in Figure 2 while the Tasks and Subtasks within this Focus are given in relation to that framework in Figure 3. We do not propose to go through the details of the rationale, objectives, methodology and outputs for all the tasks and subtasks of Focus 4. This concludes want we want to say about the LOICZ Implementation Plan. We are not particularly interested in the details of the plan, but we are interested in the general questions that it addresses.

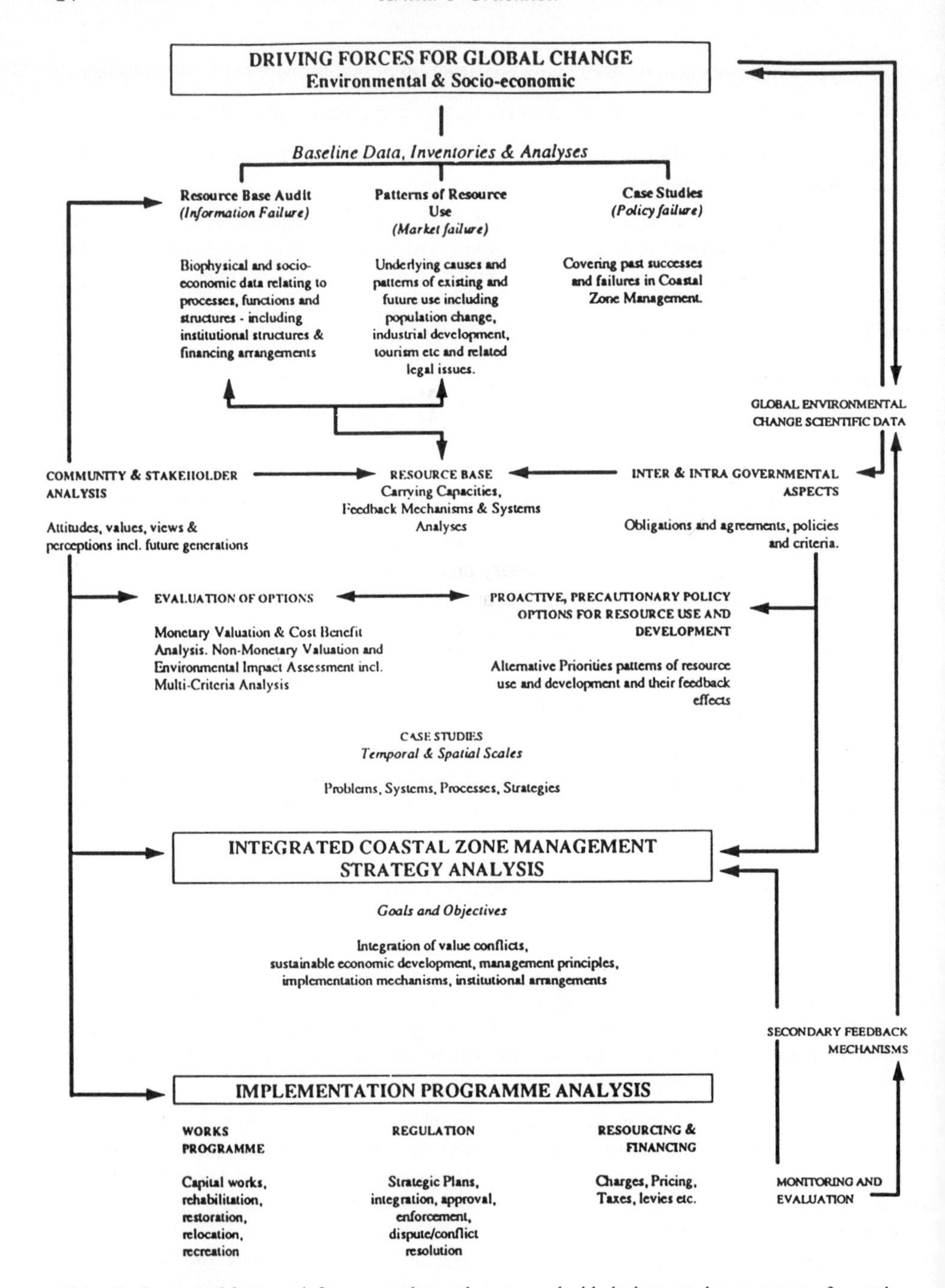

Figure 2. Conceptual framework for a research agenda concerned with the integrated management of coastal zone resource use and sustainable development. The LOICZ scientific research is focused on the components of the upper half of this diagram (Pernetta and Milliman 1995).

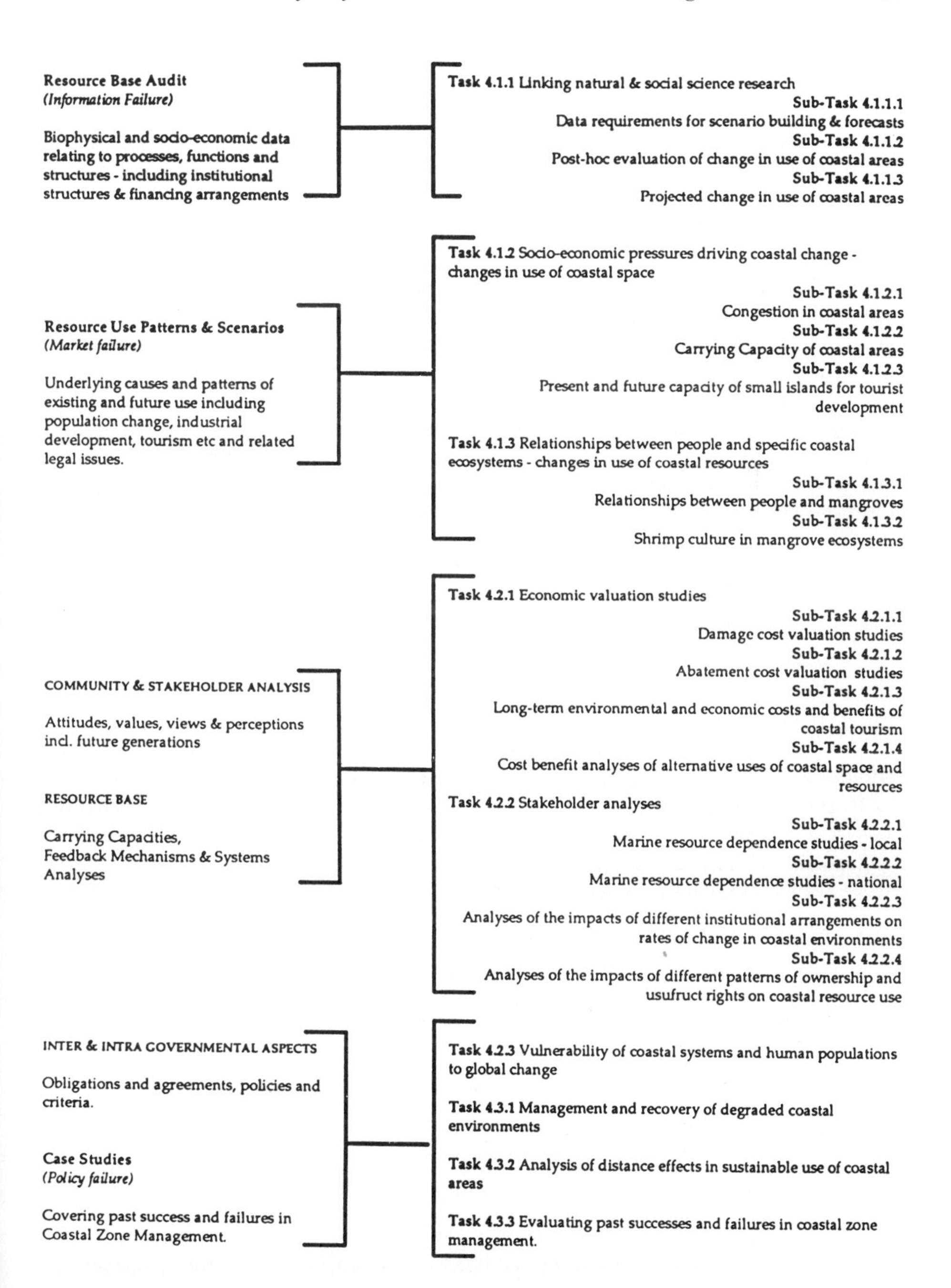

Figure 3. Relation between the LOICZ defined research Tasks and Sub-tasks and the conceptual research framework in Figure 3 (Pernetta and Milliman 1995).

4. The pressure-state-response (P-S-R) framework

The LOICZ Implementation Plan document (Pernetta and Milliman 1995) was published in 1995. Subsequently there have been a number of individual "LOICZ Reports and Studies" published on a variety of aspects of the LOICZ programme. I do not want to consider all of these in great detail, but would just like to note a few points. Especially I want to consider one major thought from the fourth of these Reports and Studies which is on "Coastal Zone Resources Assessment Guidelines" (Turner and Adger 1996). This particular document is concerned with the problem of assessing how the responses of coastal systems to global change will affect the habitation and use by humans of coastal areas and to develop further the socio-economic basis for the integrated management of the coastal environments. To do this requires the development of analysis and modelling approaches that can be applied in a number of situations that will produce comparable and consistent outputs. The document is thus concerned with providing general guidance on the application of socio-economic research methods and techniques in the context of coastal zone resource assessment and management. Therefore, while it is not primarily concerned with the details of the physical processes operating in the coastal zone, this scheme does necessarily have to take account of the physical processes involved in the coastal zone. A general pressure-state-response (P-S-R) framework, based on OECD environmental indicators format, is shown in Figure 4 and this has been adapted more specifically to coastal zones in Figure 5. In this volume we are concerned with the physical processes, with the gathering of physical environmental data (primarily by remote sensing) and with the modelling of physical processes; these tend to come at the top and on the left-hand side of Figure 5.

The coastal zone has been, and continues to be, affected both by natural events and processes and also by a wide range of human activities. The human activities may have a very direct effect, as for example in the case of dredging operations or the construction of coastal defences, or they may have a slightly less direct effect as in the case of regional-scale changes in land use, such as draining and clearing of wetlands for agriculture and residential development. There is no doubt that the coastal zone is subject to relatively high rates of change and therefore to subsequent pressures:

- the rate of population growth and economic development,
- the rate of degradation of natural resources,
- the rate of coastline modification resulting in dynamic changes, including barrier and nearshore islands,
- significant decline in biological productivity and biological diversity,
- increasing exposure of coastal populations to natural and anthropogenic hazards,
- increasing risk of over-utilisation of the sink assimilative capacity of the sea because of extensive links to 'upstream' human activities,
- declining management effectiveness resulting from complexities related to the problem of co-ordination between different management regimes for marine and land resources,
- vulnerability to potential climate change effects, including accelerated sea level rise.

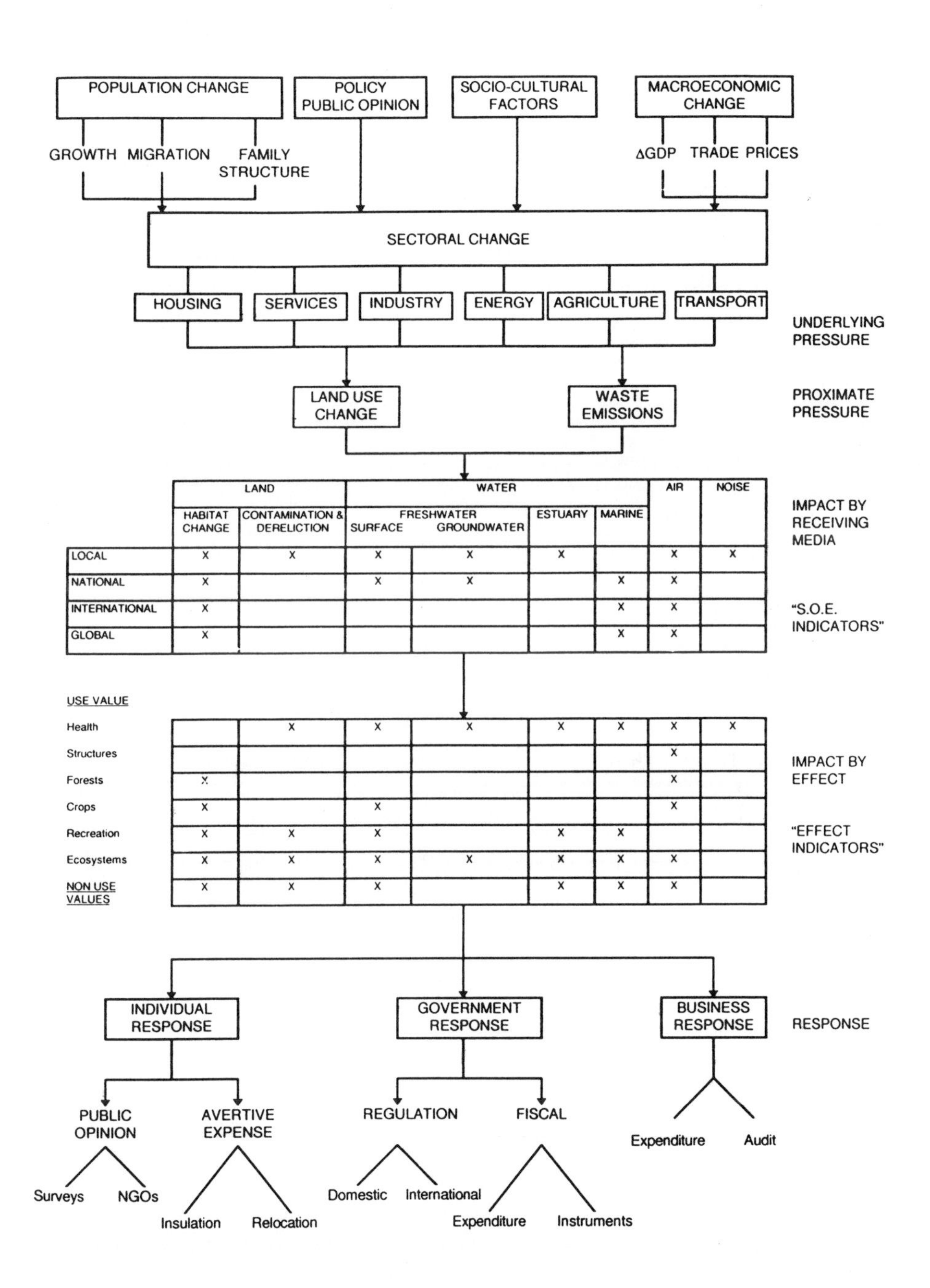

Figure 4. General Pressure-State-Response (P-S-R) framework (based on OECD environmental indicators format) (Turner and Adger 1996).

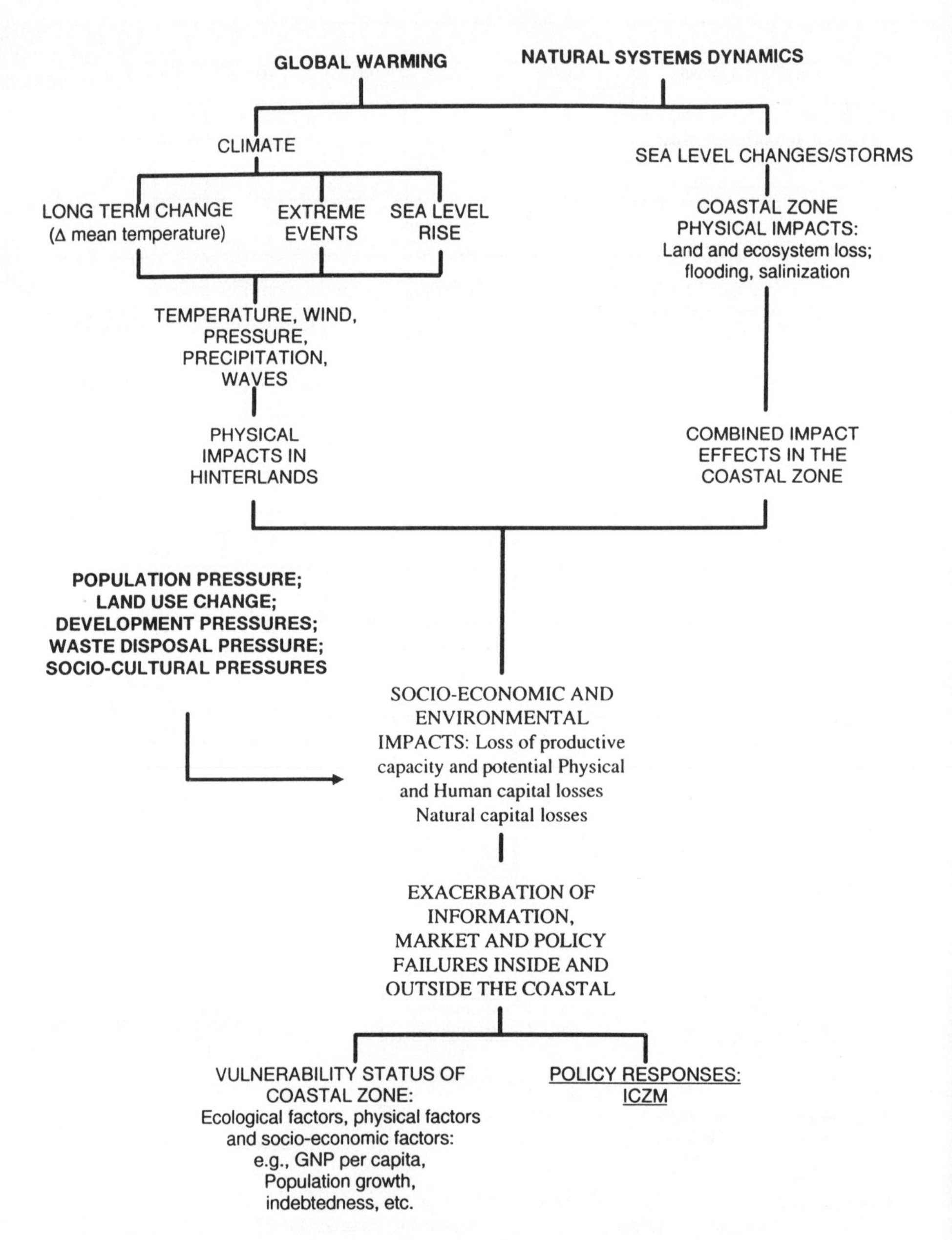

Figure 5. Coastal zone Pressure-State-Response (P-S-R) framework (Turner and Adger 1996).

'On-site' and 'off-site' linkages are usually found to be involved in the pressures on the coastal zone resources such as:

- urban sprawl and industrial and tourism development,
- pollution from riverine, airborne and marine sources,
- channelisation of the lowland sections of rivers and upstream diversion of rivers leading to beach loss and replenishment requirements,
- waste disposal in excess of assimilative capacities and posing human health risk,
- loss of coastal habitats such as coral reefs, wetlands and dune complexes,
- over-fishing,
- sand and gravel extraction,
- oil and gas exploitation and transport leading to shoreline loss and pollution.

In practice it is not always easy to separate changes arising as a result of human activities from changes that arise from natural causes

5. Conclusion

I have concentrated on outlining the LOICZ Programme in this chapter because

- it is, itself, very important,
- it provides a convenient framework within which to identify the main problems associated with the coastal zone,
- there is no substantial discussion of LOICZ in any of the later chapters.

This discussion should provide a general background to the material in this volume.

References

Holligan P M and de Boois H, 1993, Land-Ocean Interactions in the Coastal Zone: Science Plan. Global Change Report no. 25. (Stockholm: IGBP).

IGBP 1990, The International Geosphere-Biosphere Programme: A Study of Global Change (IGBP). The Initial Core Projects. Global Change Report no. 12 (Stockholm: IGBP)

Pernetta J C and Milliman J D, 1995, Land-Ocean Interactions in the Coastal Zone. Implementation Plan. Global Change Report no. 33. (Stockholm: IGBP).

Turner R K and Adger W N, 1996, Coastal Zone Resources Assessment Guidelines. LOICZ Reports & Studies (Texel, the Netherlands: LOICZ).

Viles H and Spencer T, 1995, Coastal problems: geomorphology, ecology and society at the coast (London: Arnold).

Fractals. How long is a coastline?

Gareth Rees

University of Cambridge

1. Introduction: How long is a coastline?

Let us start at the beginning. A coastline is the line separating the dry land from the ocean. We think we can define it very straightforwardly as a geometric thing. For an island, it is a line that closes on itself to form a loop. Inside the loop is dry land; outside is water. It is shown in atlases. It is shown on large-scale maps, although usually with some qualification as to the state of the tide to which it refers.

One obvious and important property of a coastline is its length. How could we determine this? We would need to compare the coastline against a standard length, a 'yardstick'. By laying copies of the yardstick end-to-end along the coastline, and then multiplying the number of yardsticks by the length of each, we would obtain a value for the length. Instead of a yardstick, let us use a pair of dividers with a gap δ between the points, and 'walk' the dividers round the coastline until we return to our starting point. The length of the coastline is then $N\delta$, where N is the number of steps in the walk.

Figure 1 shows the result of such an attempt. The 'island' (thick line) has been approximated by setting the dividers to a gap δ of 1cm, and the 'walk' of the dividers is shown by the thin line between the circles. Just over 17 steps of the dividers are needed, so the perimeter is between 17 and 18cm.

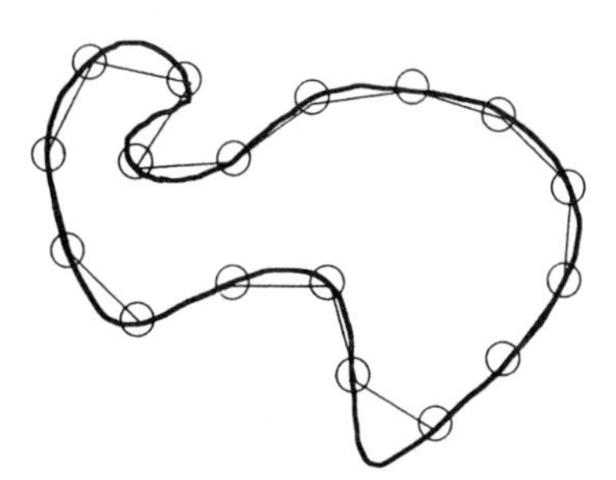

Figure 1. Measuring the length of a curve using dividers set at δ=1cm.

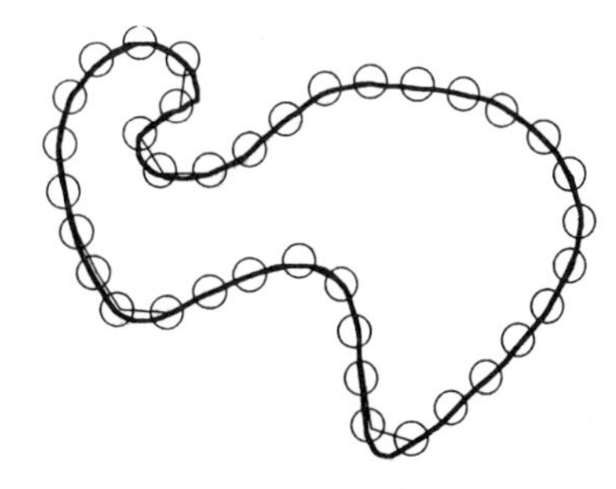

Figure 2. Repeating the exercise of the previous figure with δ = 0.5cm.

However, we might object that this result must be too small, because the 1-cm gap is too coarse to follow some of the tighter curves in the perimeter. Let us therefore try again with a gap of 0.5cm to obtain the result shown in Figure 2.

This time, just over 37 steps of the dividers are needed, giving a new estimate of between 18.5 and 19cm for the perimeter. We might expect that, as we continue this process of making δ smaller, our estimate of the perimeter will converge to a limit L, equal to the true length of the curve, i.e. that

$$N\delta \xrightarrow[\delta \to 0]{} L \quad (1)$$

This is indeed true for a smooth curve of the kind shown in Figures 1 and 2. In fact for such a curve, and for sufficiently small δ, it can be shown that

$$N\delta \approx L\left(1 - \frac{\delta^2}{24}\left\langle\frac{1}{R^2}\right\rangle\right) \quad (2)$$

where R is the radius of curvature and the average $\langle\ \rangle$ is calculated along the curve. This clearly tends to a limit L as δ tends to zero, and it is obviously meaningful to identify L as the length of the curve.

However, the smooth curve of Figures 1 and 2 is a very unrealistic-looking coastline. What happens when we apply the same approach to a real coastline? Table 1 shows data for the west coast of Britain, analysed by Mandelbrot (1982).

δ (km)	$N\delta$ (km)
1000	1000
500	1000
200	1200
100	1500
30	2100
10	3050

Table 1 : The length $N\delta$ of the west coast of Britain as a function of the divider opening δ

The value of $N\delta$ shows no sign of converging to a limit. In fact, if the data are plotted on a log-log graph as shown in Figure 3, we see that they closely fit a straight line of slope –0.25.

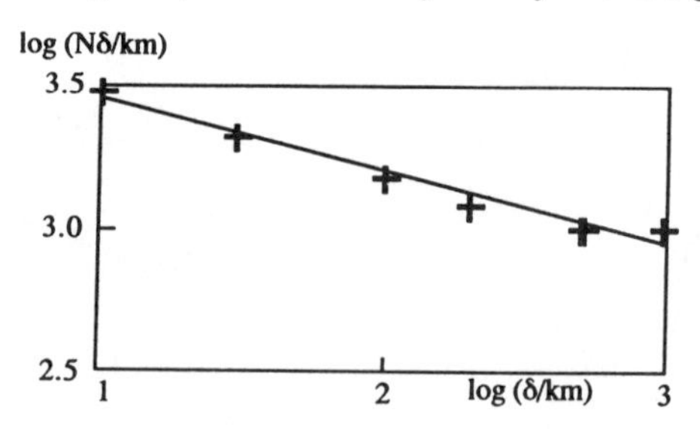

Figure 3. The data of Table 1 plotted as log ($N\delta$) against log (δ).

We can express this behaviour as a power-law:

$$N\delta \propto \delta^{1-D} \quad (3)$$

where $1-D = -0.25$ so $D = 1.25$ (the choice of $1-D$, rather than D, as the exponent will become clear later), at least over the range $10\text{km} \le \delta \le 1000\text{km}$. This type of power-law behaviour is very widely observed, although with various values of D. Table 2 summarises the results from some analyses of various coastlines. Note that a value of D (the **divider dimension**) close to 1 (such as is exhibited by the coast of South Africa) implies that the total length varies only slowly with the value of δ, whereas a large value of D implies a very marked variation. For example, the value of 1.52 given for the southern coast of Norway implies that the estimate of the total length obtained at $\delta = 0.6$km is $(80/0.6)^{0.52} \approx 13$ times longer than that obtained at $\delta = 80$km. To return to our introductory question – how long is a coastline? – we see that there is strong evidence to believe that the answer must depend on the spatial resolution with which it is defined.

Coastline	D	range of δ (km)	Source
Australia	1.13	100-2000	a
Britain (west)	1.25	10-1000	a
Co.Galway (Ireland)	1.32	0.2-10	c
Ireland	1.32	2-200	c
Norway (south)	1.52	0.6-80	b
South Africa	1.01	100-1000	a

Table 2. Some values of D and the range of d for which they have been obtained. Sources: a: Mandelbrot (1982); b: Feder (1988); c: this work.

A striking feature apparent from Table 2 is the range of length scales over which power-law behaviour, of the type represented by Equation (3), occurs. The two data sets from Ireland (entire coast, and a portion of the coast of County Galway) both yield a value of D of 1.32, suggesting that this value may be representative of the whole coast of Ireland, for values of δ between 200m and 200km. If so, this would imply that there is *no scale-length that is characteristic of the coastline*, at least over this range of scales, and that a small portion of coastline will, when magnified, appear similar in some sense to a much larger portion. This is illustrated by Figure 4, which shows two stretches of coastline from the west of Ireland, at scales differing by a factor of five.

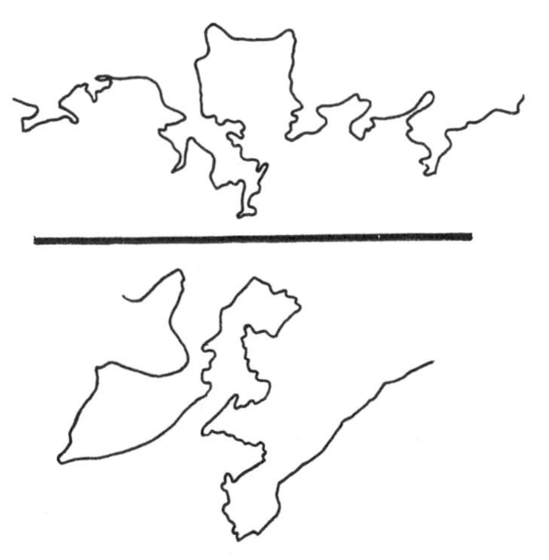

Figure 4. Two stretches of the west Irish coastline. Scale bar represents 25km (above) and 5km (below).

This property whereby any part of an object is in some sense similar to the whole object is one of the defining characteristics of a **fractal**, and will be explored in greater detail in the following sections.

2. The geometry of fractional dimensions

2.1 Box-counting length

An alternative approach to measuring the length of a curve is by counting boxes. We divide the plane containing the curve into a grid of square boxes of side δ, and count the number N of boxes through which the curve passes. For a smooth curve of the kind shown in Figure 1 the product $N\delta$ again tends to a limit as δ tends to zero. This is not surprising, since the box-counting method is actually very similar to the divider method of estimating the length of a curve. We can see this as follows.

Figure 5 shows a square box of side δ intersected by a straight line. The length l of the portion of the line covered by the box is obviously of the order of δ, although its precise value depends on the orientation and position of the line. If we consider all possible straight lines that intersect the box, we find that the average value $\langle l \rangle$ is given by

$$\langle l \rangle = \frac{2\sqrt{2}\delta}{\pi} \ln(1+\sqrt{2}) \approx 0.794\delta \quad (4)$$

so that our estimate for the length of the curve is $0.794N\delta$.

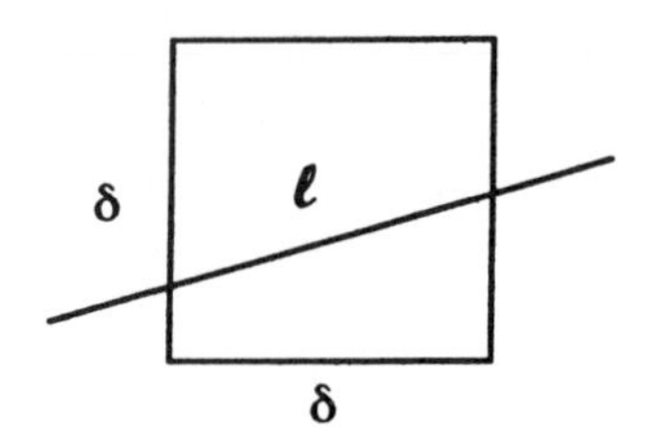

Figure 5. A square box of side δ cut by a straight line.
The length of the line enclosed by the box is l.

Figure 6 illustrates the box-counting method applied to the smooth curve of Figure 1 With a box size $\delta = 1$cm, the number of boxes N needed to cover the curve is 20, giving an estimate of 0.794 x 20 x 1cm ≈ 15.9cm for the length of the curve. For $\delta = 0.5$cm we find that $N = 46$, giving an estimate of 18.3cm for the length. Although these estimates do not agree exactly with those obtained using the divider method with corresponding values of δ, we will again find that, as d is made progressively smaller, the value of $N\delta$ will tend to a limit, and we can state that the length of the curve is 0.794 times this limit.

As we would expect, given the similarity of the divider and box-counting methods, when we apply the latter to real coastlines, rather than to smooth curves, we find that the value of $N\delta$ does *not* tend to a limit as δ tends to zero. Instead, the variation of N with δ

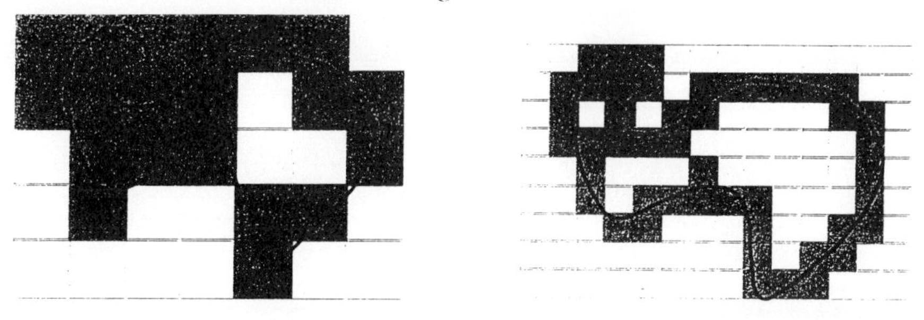

Figure 6. Box-counting approximations to the length of the curve shown in Figures 1 and 2, with box sizes of 1cm and 0.5cm.

follows the form of Equation (3). This is illustrated in Figure 7, where a portion of the west coast of Norway has been box-counted with values of δ corresponding to approximately 2.5, 5 and 10km. We find N = 258, 93 and 30 respectively, so that $N\delta$ = 645, 465 and 300km. These data conform reasonably well to equation (3) with a value of D of 1.55 ± 0.05, similar to the value given in Table 2 for the southern coast of Norway.

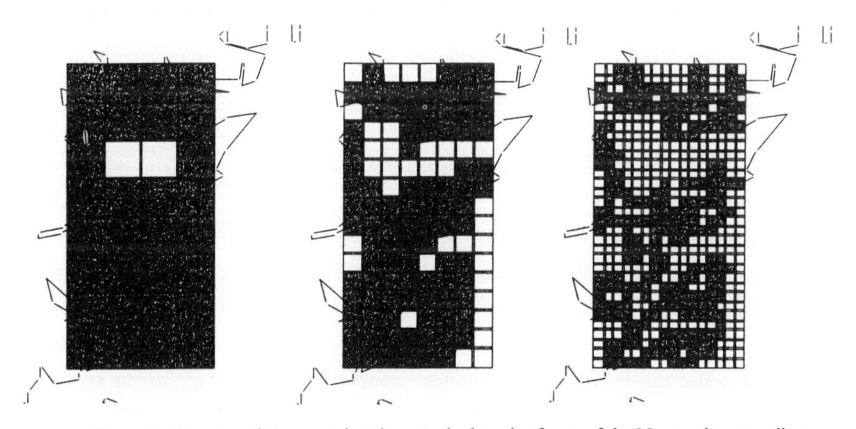

Figure 7. Box-counting approximations to the length of part of the Norwegian coastline.

2.2 The Hausdorff-Besicovitch dimension

The box-counting length leads us to a consideration of the concept of dimensionality, which is fundamental to the idea of fractals. In 'traditional' geometry, we are familiar with the idea that volumes are three-dimensional, areas have two dimensions, lengths have one dimension, and points are zero-dimensional. We can explore this concept in more detail using an extension of the box-counting approach.

We have seen how to associate a length with a curve by counting the number N of boxes of size δ that are needed to cover it: we defined the length as the limit, as δ tends to zero, of $N\delta$. Would it be meaningful to associate an *area* with a curve? By this, we do *not* mean the area *enclosed* by the curve: we mean the area of the set of points that make up the curve itself. Formally, we recognise that the area of each box is δ^2, so our estimate of the area of the curve must be $N\delta^2$. For the smooth curve of Figures 1 and 2, we know that $N\delta$

tends to a finite limit as δ tends to zero, so $N\delta^2$ must tend to zero as δ tends to zero. We therefore conclude that the area of the curve is zero.

Now let us consider the set of points *enclosed* by the curve. We can estimate the area of this set of points in a similar manner, by 'tiling' the plane with boxes of side δ and counting the number of boxes containing at least one point. Using our example curve, we obtain the results shown in Figure 8.

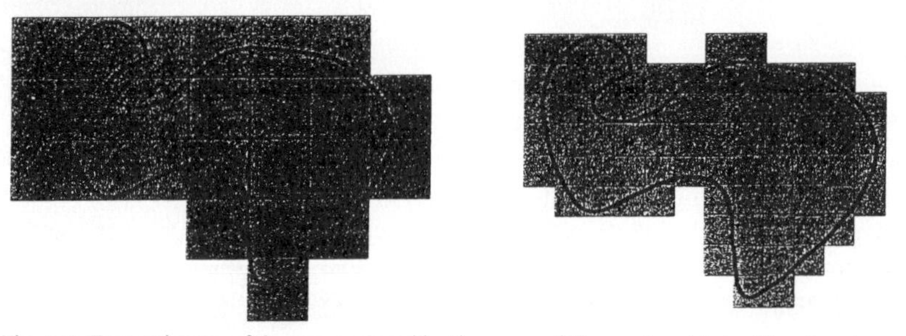

Figure 8. Two estimates of the area enclosed by the curve of Figures 1 and 2, by tiling the plane.

With a box size of 1cm, we find that $N = 24$ boxes are necessary to cover the set of points, so we estimate the area to be 24cm^2. With a box size of 0.5cm, 74 boxes are needed, giving an estimate of 18.5cm^2. Clearly as δ is reduced to zero the value of $N\delta^2$ tends to a limit which we call the area of this set of points.

We could attempt to estimate the *length* of this set of points, using the method we have already developed: count the number N of boxes of size δ needed to cover the set of points, and take the limit of $N\delta$ as δ tends to zero. Since we know that $N\delta^2$ tends to a finite limit, $N\delta$ must tend to infinity.

We could also attempt to estimate the *volume* of this set of points by filling space with an array of contiguous cubes of side δ and counting the number N of cubes containing at least one point. The volume estimate would be the limit as δ tends to zero of the value of $N\delta^3$, and this would clearly be zero.

In general, then, we have a prescription for finding the number of dimensions of an object regarded as a set of points. We divide space up into cubes of side length δ, and count the number N of boxes containing at least one point. Then we consider the limiting behaviour of $N\delta^d$ as δ tends to zero. If d is larger than the number of dimensions, the result will be zero. If d is smaller, the result will be infinite. We define the number of dimensions D as the critical value at which the limit switches from zero to infinity:

$$N\delta^d \xrightarrow[\delta \to 0]{} \begin{matrix} 0, d > D \\ \infty, d < D \end{matrix} \tag{5}$$

The value of D defined by this equation is called the **Hausdorff-Besicovitch dimension**.

Referring back to our attempts to measure the lengths of coastlines, we can now see that the value of D defined in equation (3) is just the Hausdorff-Besicovitch dimension, and that it is greater than unity. Thus we have the result that a coastline, which is topologically a one-dimensional object, behaves as though it has a non-integral number of dimensions that is greater than unity. This concept of 'fractional dimensionality' led Mandelbrot to coin the word 'fractal' (a contraction of 'fractionally dimensional'), and leads to another possible definition of a fractal: a set of points for which the Hausdorff-Besicovitch dimension is greater than the topological dimension.

2.3 The similarity dimension

We commented earlier on the fact that the fractal coastline tends to look similar to itself, in a statistical sense, at all scales. Next we consider the class of objects which are composed entirely of scaled-down copies of the object itself. The simplest example is a finite segment of a straight line. Let us suppose this segment has length L. If we scale this object by a factor r (<1) we produce a shorter line segment of length rL, and it is clear that a number N of these reduced copies can be fitted together to duplicate the original object, where $N = 1/r$.

Now we consider a rectangular object with sides a and i. Scaling by a factor r produces a smaller rectangle with sides ra and rb. Again, N of these can be fitted together to duplicate the original object, but this time $N = 1/r^2$.

Similarly, we find that for a cuboidal object, $N = 1/r^3$, and this leads us to propose the **similarity dimension**

$$D = -\frac{\log N}{\log r} \quad (6)$$

We can, of course, construct objects having non-integer values of the similarity dimension. The **Koch curve** (Figure 9) is a well-known example of such an object.

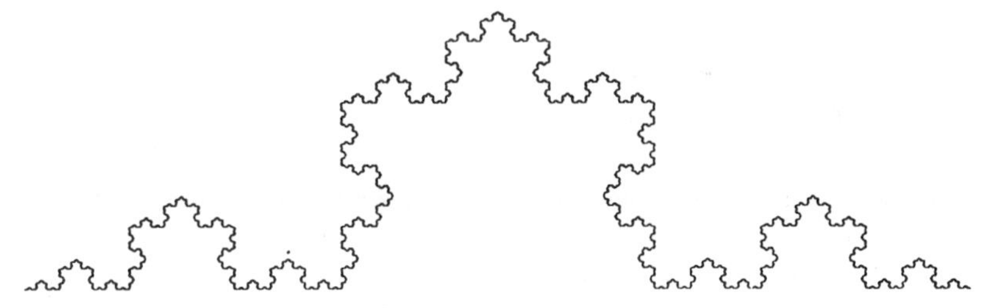

Figure 9. An approximation to the Koch curve.

The Koch curve can be constructed iteratively, beginning with a straight line segment of unit length. The middle third of the line is removed and replaced with two straight-line segments, each of the same length as the removed portion, inclined at angles of ± 60° to the remaining segments. This process results in four connected straight-line segments, each of length 1/3. The operation is then repeated on each of the four segments in turn, resulting in an object consisting of 16 connected straight-line segments each of length 1/9. This process is repeated an infinite number of times to generate the Koch curve. Clearly the Koch curve can broken into $N = 4$ identical parts each of which is $r = 1/3$ the size of the original object, giving a similarity dimension of (log 4)/(log 3) ≈ 1.26. The Hausdorff-Besicovitch dimension of the object is also equal to this value, while its topological dimension is clearly 1, so we conclude that it is indeed a fractal. Unlike our fractal coastlines, however, the self-similarity is exact rather than statistical. Other geometric fractals of this nature can clearly be constructed.

Synthetic fractal coastlines

Experimental data show that the western coastline of Britain has a fractal dimension of approximately 1.25, very similar to that of the Koch curve. It is immediately obvious that no real coastline would have the form of a Koch curve, and the reason for this is clearly its unnatural regularity. However, an iterative process similar to the one used to generate the Koch curve, but with appropriate randomisation, can yield realistic-looking synthetic coastlines.

Consider for example the following process of 'random midpoint displacement'. Begin with a line segment of unit length, and displace the mid-point a distance d in a direction perpendicular to the line, where d is drawn at random from a Gaussian distribution with mean zero and standard deviation σ. This will result in two connected line segments. Now repeat the process for each of these line segments, but this time drawing the displacement d at random from a Gaussian distribution with mean zero and standard deviation equal to σ multiplied by the length of the line segment. Figure 10 shows the first few steps of this process for $\sigma = 0.3$.

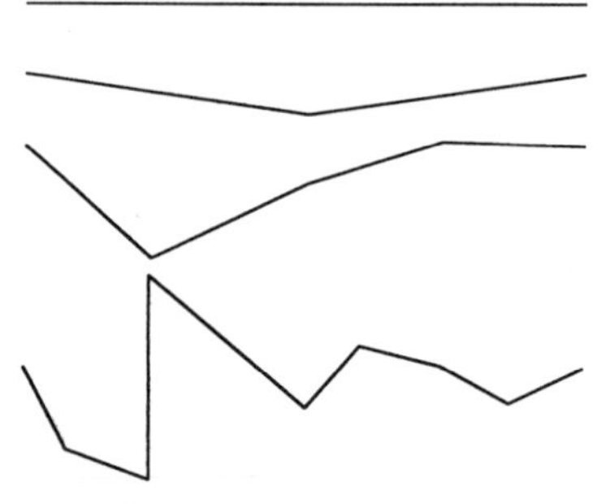

Figure 10. Three iterations of random midpoint displacement. At each iteration, the midpoint of a line segment is displaced by a distance d perpendicular to its length to give two new line segments.

We can see that this process should result in a fractal 'coastline' as follows. At some iteration, all the line segments will be of roughly the same length, say L. Displacement of the mid-point of one such segment by a distance d will generate two segments each of length $(L^2/4 + d^2)^{1/2}$. Since d is drawn from a Gaussian distribution with mean zero and standard deviation σL, this will be roughly $L(1+4\sigma^2)^{1/2}/2$. In other words, when the spatial resolution δ with which the coastline is measured is changed (decreased) by a constant factor of $(1+4\sigma^2)^{1/2}/2$, the number of steps needed to define the length is doubled. From equation (3), this implies that the relationship between σ and D is

$$1 + 4\sigma^2 = 2^{2-2/D},$$

so taking $\sigma = 0.3$ corresponds to a coastline with $D \approx 1.29$.

Figure 11 shows examples of reasonably natural-looking synthetic fractal coastlines constructed using the random midpoint displacement method.

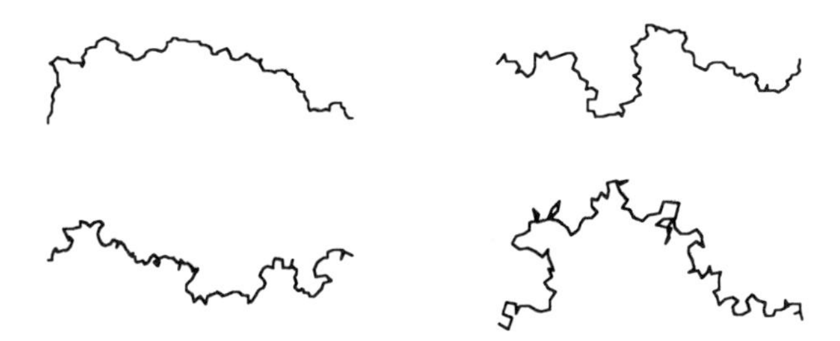

Figure 11. Synthetic fractal coastlines with fractal dimensions of 1.1 (top left), 1.2 (top right), 1.3 (bottom left) and 1.4 (bottom right).

L-systems

Another important type of self-similar fractal object is represented in the natural world by objects such as trees, such as Figure 12 which shows the branching structure of a small bush.

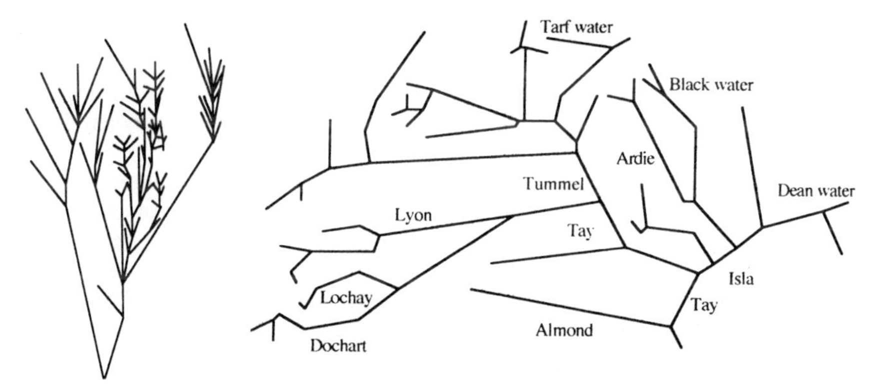

Figure 12. Branching structure of a bush.

Figure 13. Drainage network of the river Tay, Scotland (schematic).

The apparent complexity of this structure can be reduced by the following observations: (i) at each branch point, one twig gives rise to either two or three smaller twigs, with an average branching ratio of 2.43; (ii) at each branch point, the length of the smaller twigs is, on average, 82% of the length of the longer twig; (iii) branching angles are more or less constant. These characteristics are clearly consistent with a self-similar structure, and form the basis of a very simple set of rules for an algorithmic description (an 'L-system') of the structure. Such algorithmic systems have been used to generate remarkably natural-looking representations of plants, and may perhaps reflect the biological rules that govern the growth of real plants (see e.g. Prusinkiewicz and Lindenmayer 1990).

Figure 12 illustrates the simplest type of branching fractal 'tree', characterised by constant branching ratio r_N, branching angle and length ratio r_L. Clearly, the structure can be broken into r_N similar parts, each of which differs from the original by a factor of r_L in linear dimension, giving a similarity dimension (from equation 6) of

$$D = -\frac{\log r_N}{\log r_L}. \tag{7}$$

Similar structures are found when one considers river drainage patterns. Figure 13 shows schematically the major drainage network of the river Tay. This is characterised by a length ratio r_L of approximately 0.92 and a branching ratio r_N of approximately 1.17. It is notable, however, that the branching angle for this river network is not constant. River drainage networks are not, in fact, geometrically self-similar, since the shapes of the drainage basins change along the length of the river (e.g. Hack 1957). Although the fractal nature of river geometry is still not clear, it appears (e.g. Feder 1988) that the principle of self-similarity can nevertheless still be applied to it, and can provide a satisfactory explanation for the **length-area relation**. This empirical result states that the length L of the longest stream from a particular location to the upper boundary of the drainage basin supplying that location is related to the area A of the drainage basin through

$$L \propto A^n . \quad (8)$$

The value of n is related to the self-similarity dimension D of the drainage network through $n = 1/D$. For most river systems it lies in the range 0.6–0.7 (for the Tay it is 0.65 ± 0.02).

2.4 Fractal surfaces and fractional Brownian motion

We have seen that real coastlines can be well described as fractals. Since the coastline is the contour line corresponding to zero elevation, its spatial properties are determined by the spatial properties of the elevation. We might guess, therefore, that the surface profile of natural topography will also show fractal properties. A clue here is provided by the long-established observation (Venig-Meinesz 1951) that the spatial frequency spectra of topographic profiles follow a power law over a wide range of frequencies, thus implying self-affinity. Such profiles can be regarded as a manifestation of a phenomenon known as **fractional Brownian motion** (FBM).

FBM is, as its name implies, a generalisation of classical Brownian motion. Let us start by considering a one-dimensional random walk, in which a particle makes steps of length x along the x-axis at regular intervals t, with equal probability of moving forwards or backwards. The direction of any step is assumed to be uncorrelated with that of any other. Under these conditions, it is clear that

$$\langle x(t+t') - x(t) \rangle = 0$$

$$\langle [x(t+t') - x(t)]^2 \rangle = \frac{\xi^2 t'}{\tau}, \quad (9)$$

where the angle brackets < > denote the expectation value.

A more general statement of the essential property of the classical random walk is that the variance of the increment $x(t+t') - x(t)$ is proportional to $(t')^{2H}$, where $H = 1/2$. In a FBM, H can take any value between 0 and 1. This is equivalent to abandoning the requirement that increments are uncorrelated. If $H < 1/2$ the time-variation of x shows **antipersistence**, in the sense that an increasing trend in the past is likely to be followed by a decreasing trend in the future. Conversely for $H > 1/2$ the time-dependence shows **persistence**, with trends tending to remain of the same sign.

A function P of a single variable x (which can now denote either a spatial or a temporal coordinate) exhibits FBM if

$$\Pr\left(\frac{P(x+d)-P(x)}{d^H}<y\right)=f(y),$$

i.e. the cumulative probability distributions of $P(x+d)-P(x)$ are identical for different values of d apart from scaling by a factor of d^H, where H is a constant known as the **Hurst exponent**. If we write $f'(y)$ for df/dy, it follows that the probability distribution function of

$$Q=\frac{P(x+d)-P(x)}{d^H}$$

is $f'(Q)$, and hence that

$$\langle[P(x+d)-P(x)]^n\rangle=d^{nH}\int_{-\infty}^{\infty}f'(y)y^n\,dy.$$

Consider a statistically stationary variable P: it is clear that the first-order moment ($n=1$) is zero so the first non-zero integer moment will correspond to $n=2$, and be proportional to d^{2H}. Whether or not P is stationary we can define the **semivariance** $\gamma_P(d)$

$$\gamma_P(d)=\frac{1}{2}\langle[P(x+d)-P(x)]^2\rangle. \tag{10}$$

A plot of γ_P as a function of d is called a **semivariogram**, and it is clearly a condition of FBM behaviour that it should obey a power-law relationship $\gamma_P\ \alpha\ d^{2H}$. The corresponding fractal dimension is

$$D=N+1-H, \tag{11}$$

where N is the number of topological dimensions.

The use of semivariograms for the analysis of real spatial data is already well established (e.g. Woodcock *et al.* 1988a, 1988b); they are widely used in assessing the fractal characteristics of surface topography. We can illustrate the procedure with a simple example. Figure 14 shows a 16km north-south topographic transect in the south-west of Ireland, with a sampling interval of 0.25km. Figure 15 shows the corresponding log-log semivariogram, plotted for d = 0.25, 0.5, ... 5km. This figure shows that for small values of d the semivariogram is well described by a power law with an index of 1.44 ± 0.05, corresponding (Equation 11) to a fractal dimension D for the one-dimensional transect of 1.28 ± 0.03. Figure 15 also shows that the power-law behaviour breaks down at larger values of d – in other words, the topographic profile has a characteristic scale length of a few kilometres, as in Figure 14. In fact, this behaviour can also be described within the fractal framework, (see later), although it is necessary to introduce the concept of **multifractals**.

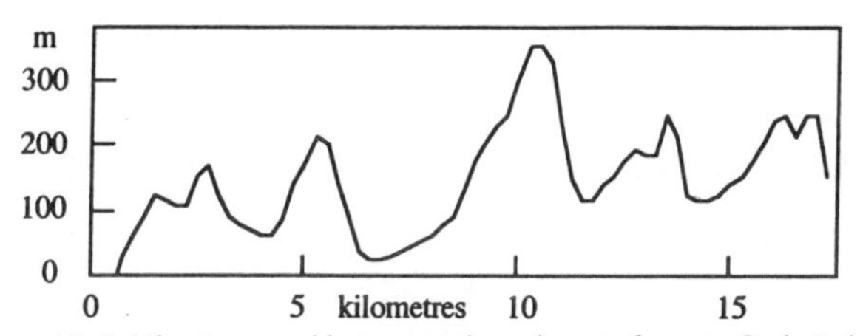

Figure 14. A 16km topographic transect through part of county Cork, Ireland.

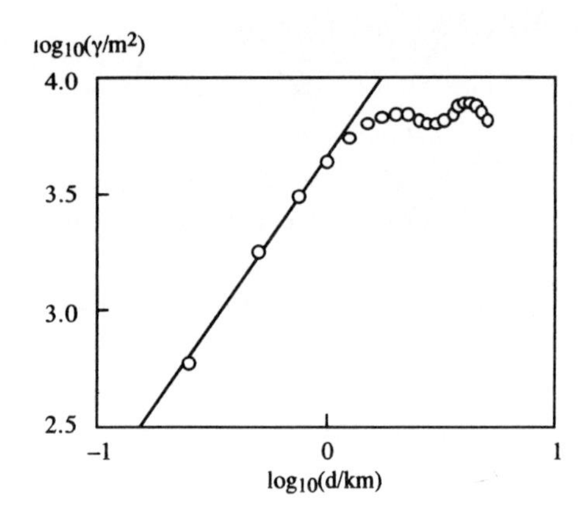

Figure 15. Log-log semivariogram of the data in Figure 14.

The semivariogram method can be applied directly to the problem of determining the fractal dimension of a topographic surface if a suitable terrain model (DTM or contour map) is available. In some circumstances, however, it is possible to infer the surface topography from variations in the radiance detected by an airborne or spaceborne imaging sensor. This method was applied by Rees (1992) to an analysis of the surface of an Arctic ice-cap, and used to show that the surface could be described by a fractal dimension $D = 1.12$ for scales between 0.3 and 10km. A fuller consideration of the conditions that must be satisfied in order that remotely-sensed imagery should represent the topographic properties of the surface, and the relationship between them, was presented by Rees in 1995.

Hurst's rescaled range analysis

A different approach to the analysis of time-series data was developed by Hurst *et al.* (1965), originally in order to assess the optimum storage capacity of a reservoir in order to ensure that it neither runs dry nor overflows. Writing x_i for the time-dependent variable (in Hurst's original case, the inflow to a reservoir in year i), Hurst needed to consider the behaviour of the function $X(i, j)$ defined as

$$X(i,j) = \sum_{k=1}^{i} \left(x_k - \langle x \rangle_j \right)$$

where $\langle x \rangle_j$ is just the mean value of x_i observed over the period j. The range R of this function is defined, for a particular value of j, as the difference between the maximum and minimum values of $X(i, j)$ observed for all i within the period j. Hurst observed that when the range R was scaled to the standard deviation S of the values of x_i observed over the period j, the results were well described by

$$\frac{R}{S} \approx \left(\frac{j}{2} \right)^H \tag{12}$$

with $H \approx 0.7$.

We can see that this result is consistent with the integrated function $X(i, j)$ showing fractional Brownian motion as follows. Suppose that $X(i, j)$ can be characterised by a power-law semivariogram ai^{2H}, so that over a period j we expect the range R to be of the order of $(aj^{2H})^{1/2}$, while the standard deviation S is just $a^{1/2}$. Thus R/S is of order j^H, and $H \approx 0.7$ corresponds to a fractal dimension $D \approx 1.3$ for the 'trace' (time-integral) of x.

Hurst found equation (12) to be obeyed for a wide range of natural phenomena, including sunspot numbers, rainfall, tree rings and atmospheric temperature. Feder (1988) shows that it is also a good model of ocean wave-height variations. In all cases the value of H was significantly greater than 0.5, implying persistent, rather than uncorrelated, behaviour of the variable in question.

As an example, Figure 16 shows the results of an R/S analysis performed on the 'Central England' temperature record over the period from 1700 to 1990, using values of t ranging from 19 years to 290 years. It can be seen that the data are well described by Equation (9), with a Hurst exponent H of 0.65±0.07. This compares very reasonably with the value of 0.68±0.09 reported by Hurst *et al.* (1965) for atmospheric temperature data over a range of 29 to 60 years.

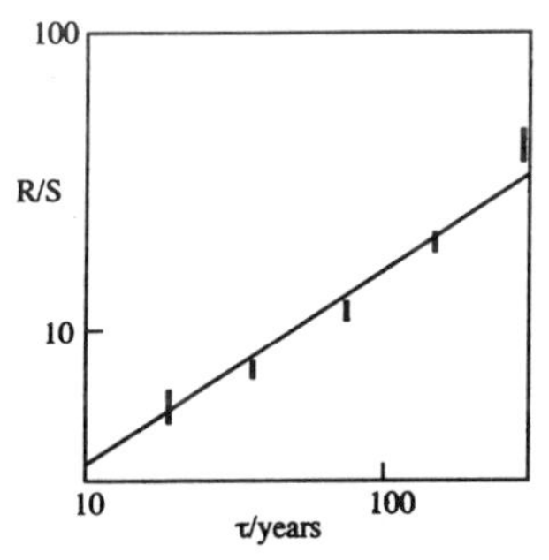

Figure 16. R/S analysis of 'Central England' temperatures between 1700 and 1990. The data are well described by a Hurst exponent of 0.65.

Synthetic fractal transects and surfaces

Using fractional Brownian motion as a model, we can easily construct transects through synthetic fractal surfaces. There are several available methods, but one of the simplest is that of 'successive random addition' due to Voss (1985). In a typical implementation of this algorithm, the value of the height $z(x)$ is first set to zero at x=0, 1/2 and 1. Each of these values of z is then changed by the addition of values drawn at random from a Gaussian distribution with mean zero and variance 1. Values of $z(x)$ are then calculated by linear interpolation for x=1/4 and 3/4. The next iteration involves random additions to all five values of z, this time from a Gaussian distribution with mean zero and variance 2^{-2H} (H being the Hurst exponent), followed by linear interpolation. At each successive iteration, the variance of the random addition is multiplied by a factor of 2^{-2H} and the spatial resolution is doubled by interpolation.

Figure 17 shows examples of FBM profiles calculated by this method for different values of H. Note that the profiles for H between about 1.0 (D = 1.0) and 0.6 (D = 1.4) have a reasonably natural appearance.

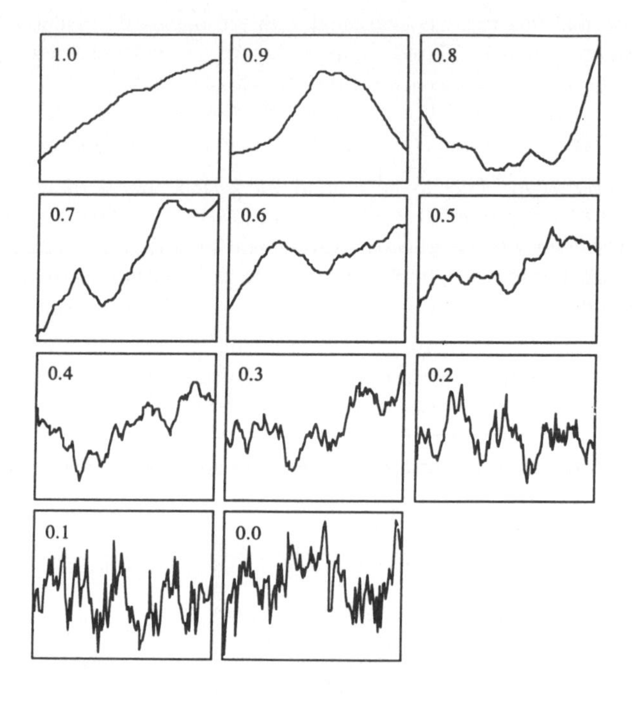

Figure 17. FBM profiles with different values of the Hurst exponent H.

A similar approach can be used to generate two-dimensional surfaces with FBM properties. Figure 18 shows a low-resolution (32 x 32 pixel) synthetic surface with $D = 2.3$, comparable to that of the western areas of Britain. The sea surface has been generated by truncating the profile at an arbitrary lower bound.

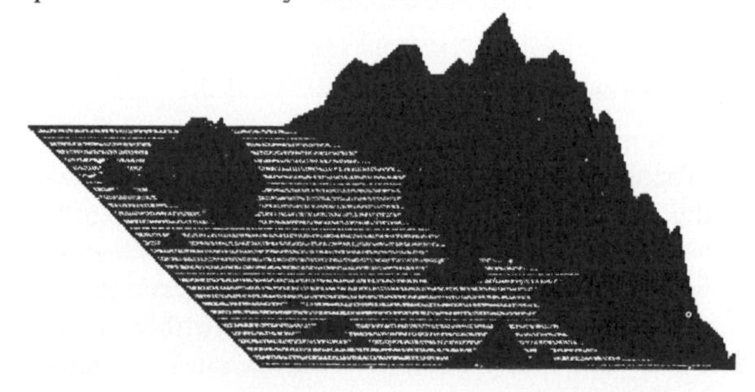

Figure 18. Low-resolution synthetic FBM topography with $D = 2.3$.

Again, this has a reasonably natural appearance within the limitations of the resolution. It shows a realistic topographic profile for the land surface, as well as a natural-looking coastline and distribution of islands.

Definitions of roughness

Examination of Figure 17 also suggests a connection between fractal dimension and the concept of 'roughness', since the profiles with lower H (higher D) are obviously more irregular, in some sense, than those with higher H (lower D). The simplest definition of the roughness of a height profile $h(x)$ is just the root mean square variation of h:

$$\sigma_h = \langle h^2 \rangle - \langle h \rangle^2,$$

where the angle brackets < > denote averaging over a suitable range of x. This definition is unsatisfactory for a number of reasons. Firstly, suppose the transect to be a straight line $h(x) = a + bx$, than which nothing could be smoother. Calculating σ_h over the range $x_0 < x < x_0 + L$ yields a value of $L|b|/\sqrt{12}$, which diverges as the length L of the transect tends to infinity. Obviously we could avoid this problem by first fitting a model to the transect, and then calculating the root mean square departure from the model. In our over-simple definition of roughness we have implicitly adopted the model h=0, whereas if we were to choose the best-fitting linear model we would, of course, retrieve the parameters a and b and hence deduce a roughness of zero.

This approach is, however, still unsatisfactory. We need to define a model of the surface, which is in itself an arbitrary step, and we need to choose a range of x over which to measure the departures from the model. For example, suppose we have a sinusoidal transect (which we would also tend to regard as fairly smooth): $h(x) = a + b\sin(kx)$ and fit a linear model for $x_0 < x < x_0 + L$. We find an RMS roughness that varies with L, being proportional to $k^2L^2|b|$ at small L and approximately equal to $|b|/\sqrt{2}$ for $L >> 2\pi/k$. The roughness in this case depends on the scale over which we examine the surface.

Perhaps unsurprisingly, our intuitive concept of roughness appears to be largely independent of scale, and Pentland (1984) has shown that the fractal dimension corresponds well to intuitive roughness.

2.5 Self-similarity and self-affinity

The Koch curve (Figure 9) and the fractal coastlines, both real (Figures 4 and 7) and synthetic (Figure 11), show self-similarity. The real (Figure 14) and synthetic (Figure 17) topographic profiles also show fractal properties, but it is clear that there are some basic differences between coastlines and topographic profiles. One important difference is that between self-similar fractals and self-affine fractals.

We have seen already that real coastlines (in a statistical sense) and the Koch curve (exactly) are self-similar, such that a magnified portion of such an object has the same properties of the object itself. However, this is not true for the FBM profiles of Figure 17, as can be seen from Figure 19.

This shows an FBM with H=0.2, and a portion of the FBM occupying one third of the original length, scaled up by a factor of 3. It can be seen that the scaled-up version is 'rougher' than the original. In order to preserve the statistical properties of the curve, it would be necessary to scale it vertically by *less* than a factor of 3 while scaling horizontally by a factor of 3. For a general FBM with Hurst exponent H, the scaling law is such that if we scale the horizontal coordinate by a factor of k, the vertical coordinate must be scaled by k^H. This kind of transformation, in which different coordinates are scaled by different factors, is called

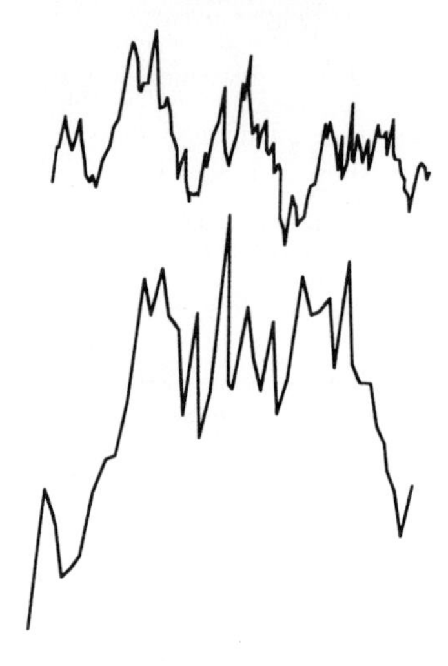

Figure 19. Brownian motion with H=0.2 (above) and a magnified portion of the same curve (below).

an **affine** transformation, and self-affine fractals are those for which a part reproduces the whole after a suitable affine transformation.

It is clear that the definitions of fractal dimension that we have elaborated for self-similar fractals cannot be applied without some modification to self-affine fractals, and in fact self-affine fractals have *no unique definition of fractal dimension*. Firstly, it is obvious that a self-affine fractal does not have a meaningful similarity dimension, since it is not self-similar. What about its box-counting dimension? Suppose we have a one (topological) dimensional FBM described by the semivariogram

$$\gamma_h(\delta) = \frac{1}{2} a^2 \delta^{2H}$$

where both h and δ are spatial coordinates (height and horizontal displacement respectively). In order to define the box-counting dimension of this FBM we choose a rectangular box of horizontal size w and vertical size kw, where k is a fixed aspect ratio, and count the number $N(w)$ of boxes needed to cover a horizontal range L of the FBM. Clearly this range is represented by L/w box-widths. Within a single box-width w, h varies over a range of the order of $(\gamma_h(w))^{1/2} = aw^H$ so, provided that $kw << aw^H$, the FBM will cross approximately aw^{H-1}/k boxes in the height dimension for each box-width in the horizontal dimension. Thus our estimate of $N(w)$ is

$$N(w) \approx \frac{aLw^{H-2}}{k}$$

From equation (3) we see that this implies a box-counting dimension of $2-H$, identical to the value given by equation (11). However, if $kw >> aw^H$, we find that the FBM crosses only a single box in the height dimension for each box-width, giving

$$N(w) \approx \frac{L}{w}$$

and hence a box-counting dimension of 1. For a fixed k, the ratio $(kw)/aw^H$ increases with increasing w (since $H < 1$), so we can see that the box-counting dimension of the FBM tends to $2-H$ at small scales and to unity at large scales – the FBM is **locally fractal**, but not globally.

We can also estimate the divider dimension of a FBM. Here we must suppose that the physical dimensions of $\sqrt{\gamma}$ and of its argument δ are the same (as in the example above). If they are not (for example, if we are considering a record of temperature as a function of time), the only way we can associate a length with the FBM is to represent it as a graph on paper.

We will again take as our model the FBM with semivariogram

$$\gamma_h(\delta) = \frac{1}{2} a^2 \delta^{2H}$$

If we consider two points on this FBM separated by a distance δ in the horizontal direction, they will have a vertical separation of typically $a\delta^H$, and the distance between them will therefore be

$$l \approx (\delta^2 + a^2 \delta^{2H})^{1/2}$$

We can take this value of l as the opening of our dividers. For sufficiently small openings we will have $l \approx a\delta^H$, whereas for large openings, $l \approx \delta$. To measure a stretch of the FBM subtending a horizontal distance L will require $N = L/\delta$ steps of the dividers, so we can write the number of steps as

$$N(l) \approx L\left(\frac{a}{l}\right)^{1/H}$$

for small openings of the dividers, and

$$N(l) \approx \frac{L}{l}$$

for large openings. Thus, as for the box-counting dimension, we find that the FBM has a fractal dimension of 1 at large scales. However, we find the divider dimension to be $1/H$ at small scales, in contrast to the value of $2-H$ that we found for the box-counting dimension.

3. Links between fractal properties of natural phenomena

3.1 Rough topography has a rough coastline

We remarked in Section 2.4 that the possession of a fractal coastline suggested the occurrence of fractal topography, on the basis that the coastline is the zero-altitude contour of the topography. The relationship is exact, although we do not prove it here. A surface with isotropic fractal properties, such that a one-dimensional transect has a fractal dimension D, will have a fractal coastline also with fractal dimension D. This certainly appears reasonable, since we would expect a rough surface topography to be reflected in a more jagged coastline, and the data presented here provided a limited quantitative demonstration. For example, Table 2 showed a fractal dimension of 1.25 for the western coastline of Ireland, and Figure 15 showed a fractal dimension of 1.28 for a transect through the surface.

The relationship between the fractal properties of a surface and those of its contour lines also leads to a characteristic property of the sizes of islands, known as the **Korcak distribution**. Consider a group of islands of various sizes, such that *N(A)* islands have an area greater than *A*. It is found empirically that *N(A)* is given approximately by the Korcak distribution

$$N(A) = aA^{-B} \tag{13}$$

where a and B are constants, with B lying typically between 0.5 and 0.75. We expect that the size distribution should be related to the fractal properties of the terrain, and hence to the fractal dimension D of the coastline. It can be shown that $B = D/2$, so that the observed range of values of B corresponds to a range of D between 1 and 1.5, consistent with the range of fractal dimensions observed for coastlines.

Figure 20 shows the size distribution of the world's 27 largest islands, from Australia to Sri Lanka, plotted as $\log_{10}$(rank) against $\log_{10}$(area/km^2). These data are fitted by a Korcak distribution with $B = 0.72 \pm 0.02$. When the list of islands is extended to include those much smaller than Sri Lanka, the 'global average' value of B is found to be about 0.65.

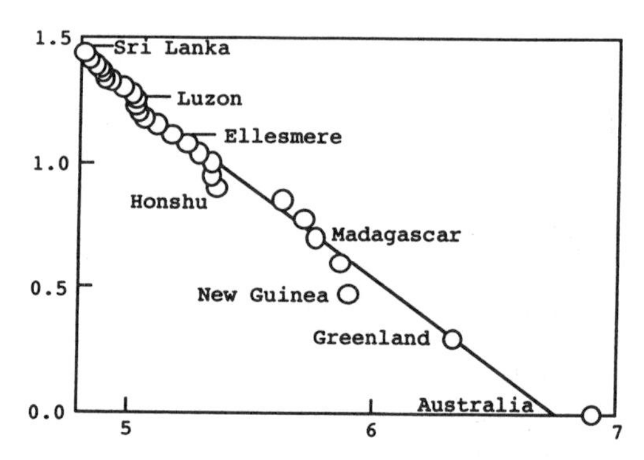

Figure 20. Log-log plot of the areas (in square kilometres) of the world's 27 largest islands (x-axis) against rank by area (y-axis). The data are well described by the Korcak distribution.

3.2 Perimeter-area relations

Returning to our simple (mono-) fractals, we can use them to characterise natural phenomena other than just surface topography and the properties of coastlines. We have already seen the link between the distribution of island area distributions and the fractal dimension of the coastline. This can be extended to a consideration of the perimeter-area relation for islands.

In traditional (non-fractal) geometry, a set of geometrically similar objects defined by their outlines drawn on a plane can be characterised by the constancy of the ratio of the perimeter to the square root of the area. For example, for circles this ratio is equal to $2\sqrt{\pi}$. Clearly, we do not expect this relationship to hold for the coastlines of real islands, since we cannot specify the perimeter without also specifying the spatial resolution at which it is measured. From equation (3), we expect the perimeter L_δ measured at resolution δ to be proportional to d^{l-D}, so we can put

$$L_\delta = C\delta^{1-D} A^{D/2} \tag{14}$$

where A is the island's area and C is a shape-dependent but size-invariant dimensionless constant. This is Mandelbrot's perimeter-area relation, which we expect to be obeyed by fractal islands provided that δ is sufficiently small.

Equation (14) has been used by Lovejoy (1982) to investigate the geometrical properties of cloud perimeters over a range of areas A from 1km^2 to 10^6km^2. He found the data to be well fitted by a single value of $D = 1.35\pm0.05$, a value satisfactorily accounted for by the turbulent diffusion model due to Hentschel and Procaccia (1984).

3.3 Problems with monofractals

Further consideration of the relationship between topography and the shapes of contour lines shows that the situation is not quite as simple as we have assumed. If we suppose that the surface of a terrain is a true fractal, then we must also assume that there is nothing special about the zero-altitude contour. A horizontal slice at any altitude should have a fractal outline with a fractal dimension independent of the altitude. This is manifestly not true for real topography: Figure 21 illustrates contours at 100m intervals for a 3km x 3km area of mountainous terrain in Wales.

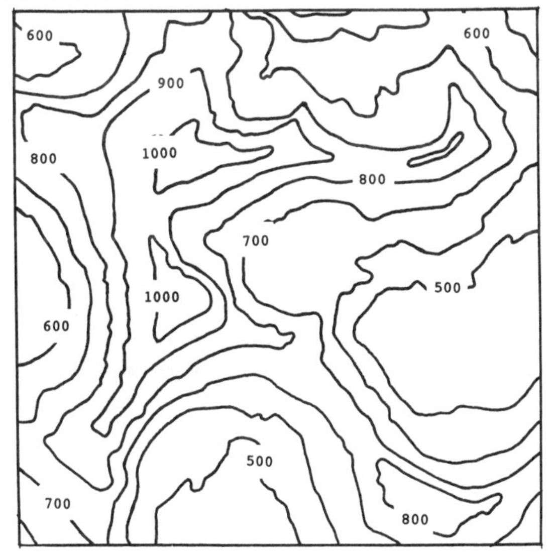

Figure 21. Contours of mountainous terrain. The area represented is 3km x 3km; contour interval is 100m.

It is clear that, at higher altitudes, the contours are smoother, corresponding to lower fractal dimension. Even after allowance is made for the effect of the spatial resolution of the data from which the contours are generated, this fact remains, and we are forced to conclude that the fractal dimension of natural surfaces decreases systematically with altitude. By performing simple divider-walks round each contour in turn, we find that for contours at about 700m and below the fractal dimension is close to 1.3, which is typical for west-coast Britain. At 800m the fractal dimension has fallen to about 1.2; by 900m it has reached 1.15 and by 1000m it is only 1.1. Thus we find that the correspondence between the fractal dimension of the coastline and that of the topography is only valid provided wc do not go too close to the maximum height of the terrain (in this case about 1085m). This behaviour is very common. Attempts to provide a more general framework in which to describe it have led to the topic of **multifractals** (essentially a distribution of monofractals), rather beyond the scope of these lectures (but see e.g. Lovejoy and Schertzer 1995 for an accessible discussion).

4. Where else can fractals be found?

The foregoing discussion has purposefully concentrated on geophysical fractals, since these are the most obviously relevant to considerations of the coastal zone. However, it would be wrong to conclude that the concept of fractals is only, or even principally, useful in this field. Following the mathematical exploration of non-Euclidean geometries in the late 19th century, Felix Hausdorff developed the study of fractional dimensionality in 1919, and although this work was extended by Besicovitch, it is probably fair to say that it was regarded as little more than a mathematical curiosity until Mandelbrot's pioneering work in the 1960s. Since then, however, the investigation and application of fractal geometries has invaded many areas of science, notably physics. A fairly diverse set of examples of fractal behaviour could include fluid turbulence, the distribution of galaxies in the universe (e.g. Peebles 1980), stock market fluctuations, viscous fingering, colloidal aggregation, critical phenomena in phase transitions (e.g. Nicolis 1989), and the temporal distributions of rainfall and of earthquake tremors.

A quite remarkable type of mathematical fractal is represented by objects such as the well-known Julia and Mandelbrot sets. These are objects of enormous complexity that are generated by extremely simple algorithms, a feature they share with the synthetic fractals we have already considered. The Mandelbrot set in its simplest form is the set of all complex numbers c_0 such that the iterative procedure

$$c_{i+1} = c_i^2 + c_0$$

does not cause $|c_i|$ to increase beyond 2 as i tends to infinity.

Another feature shared by all of the simple algorithmic descriptions of fractals that we have considered, including the Mandelbrot set, is that they are iterative, involving the repeated application of a rule that either does not take account of spatial scale at all, or varies with scale according to a power-law, i.e. without any natural scale length. It is not always obvious why the characteristic self-similarity or self-affinity of natural fractals occurs. In some cases, the scale-free algorithms that can be used to synthesise realistic-looking fractal structures undoubtedly mimic the natural processes fairly directly; in other cases, the link is harder to find. The branching structure of plants is probably due to a biological process that mimics the L-system algorithms used to 'grow' realistic-looking mathematical plants.

Fluid turbulence provides another illuminating example. Here the essential dynamical process is the transfer of energy from larger to smaller eddies, as a result of the non-linear form of the equation of motion (Navier-Stokes equation). This results in a power-law distribution of energy density as a function of wavenumber, the famous Kolmogorov 5/3 law. Fully-developed turbulence is in fact chaotic, in the sense that the trajectories of molecules of the fluid are practically unpredictable, being very sensitively dependent on the molecule's initial state (position and velocity). However, if the time-evolution of a smooth blob of the fluid is considered, we find that, at first, it retains its contiguity but becomes progressively rougher, i.e. its fractal dimension increases (see e.g. Tritton 1977 page 219), until it is eventually stirred uniformly through the whole fluid. The fractal form of the blob is thus indicative of a transition to chaotic dynamics.

5. Conclusion: implications and applications of fractals

5.1 Scale-dependence

The discovery that nature does not often use the classical Euclidean geometry, with which we are most familiar, has many ramifications which we are still in the process of exploring. We have seen that at least one simple geophysical question – how long is a coastline? – is in fact ill-posed as a consequence of the fractal nature of the coastline. The fact that so many natural phenomena display fractal characteristics alerts us to the danger that many essentially geometrical questions will be ill-posed unless we take careful account of their fractal nature and scale dependence. Two simple but important examples can be advanced to illustrate this point: what is the average area of the earth's surface covered by cloud? and what is the vegetated area within the boundary of a particular city? Both of these questions are accessible to investigation by remote sensing, and both are environmentally significant. Indeed the global cloud-cover is a fundamental parameter in understanding the global climate. Yet both of these questions yield scale-dependent answers, because the underlying phenomena appear to follow fractal geometry. As the spatial resolution of remote sensing systems has increased, estimates of global cloud cover have decreased (essentially, we are able to resolve smaller and smaller holes in the clouds). Similarly, one finds that the area of a city that can be classified as vegetation increases with the fineness of the resolution with which one can observe it (e.g. Vasiliev 1995). In this case one can probably come up with a meaningful single-parameter estimate of vegetation cover, by deducing the power-law corresponding to the observed fractal behaviour and extrapolating it to a scale corresponding to a single tree or a single plant, but it is not at all clear that a corresponding approach can be adopted in considering the global cloud cover.

These simple examples show the importance of **scale** in considering geophysical (and other) phenomena. If we do not understand the fractal behaviour of a phenomenon, we shall certainly fail to extrapolate observations from one spatial resolution to another. This point is clearly of fundamental importance in the application of remotely-sensed data. The monofractal concepts with which the bulk of these lectures have dealt are, fortunately, reasonably straightforward to recognise (through their power-law 'signature') and to apply, and they can often provide at least a 'first-order' understanding of the effects of scale. However, there is increasing evidence that many natural phenomena are in fact better described by multifractals, and these are considerably harder to grasp.

5.2 Description of complex phenomena

Whether a monofractal or multifractal model is appropriate, the fractal concept can often provide a simple language for the description of apparently complex phenomena. The models have predictive power, and the underlying mathematical ideas give the potential to link apparently unrelated physical phenomena. In some cases, the detection and characteristation of fractal behaviour can give a hint about the dominant dynamical processes responsible for the structure. However, even when this is not possible, a few simple tests performed on a set of data can indicate at least roughly the presence and significance of fractal behaviour, and hence provide an indication of where theories based essentially on Euclidean geometry are likely to be unreliable.

5.3 Other applications of fractals

Before finally abandoning this brief introduction to fractals, we should mention a few more applications that are of particular relevance in remote sensing and image processing. The first, a rather obvious application, is in the development of **scattering models**. 'Traditional' models of both surface and volume scattering normally describe the scatterer geometry in Euclidean terms, for example through the use of a surface roughness parameter and correlation length. It is becoming clear that such models are inadequate to describe scattering from real surfaces, and that fractal-based models are needed (e.g. Jaggard and Sun 1990). (Indeed, this is partly implicit in the recognition that the roughness parameters of a scattering model should be calculated only for a particular range of spatial scales, with longer-wavelength variations causing 'topographic' modulation of the scattering. We now recognise that a more general description of the scattering should not make this simple distinction between roughness and topography.) Similarly, fractal models of volume scattering processes, especially from vegetation canopies (e.g. Govaerts and Verstraete 1995), are beginning to find significant application. Partly for these reasons, fractal concepts are also finding increasing application in the quantification of texture in image classification procedures (e.g. Lévy-Véhel 1995).

The last application is to **image compression**. It is a commonplace remark that remote-sensing images can represent huge volumes of data, and image compression techniques have acquired increasing importance. Fractal image compression techniques (see e.g. Fisher 1992) involve identifying self-similar regions of the image at different scales. Once such regions have been found, they can be specified just by their positions and locations, rather than by the digital numbers of every pixel within them. Since for real images the self-similarity is likely to be statistical rather than exact, the data compression is irreversible in the sense that information is lost. Nevertheless, very large compression ratios can be achieved with comparatively little degradation of the image.

Acknowledgements

The excellent book by Feder (1988) forms the basis of the discussion presented in various parts of this chapter, and the book can be recommended to any reader wishing to gain a deeper understanding of this subject. I am also indebted to the organisers of, and my fellow participants at, the expert meeting on 'Fractals in geoscience and remote sensing', held at the Joint Research Centre of the European Commission at Ispra, Italy, in April 1994.

References

Feder J, 1988, Fractals , Plenum Press (New York)

Fisher Y, 1992, A discussion of fractal image compression (in Chaos and Fractals, by Peitgen H-O, Jürgens H and Saupe D, Springer-Verlag (New York) pp 903-919

Govaerts Y and Verstraete M M, 1995, Applications of the L-systems to canopy reflectance modelling in a Monte Carlo ray tracing technique (in Fractals in Geoscience and Remote Sensing. JRC Image Understanding Research Series, volume 1, edited by Wilkinson G G, Kanellopoulos I and Mégier J, Office for Official Publications of the European Communities (Luxembourg), pp 211-236

Hack J T, 1957, Studies of longitudinal stream profiles in Virginia and Maryland, US Geological Survey Professional Paper 294-B

Hentschel H G E and Procaccia I, 1984, Relative diffusion in turbulent media: the fractal dimension of clouds, Physical Review A, **29**, 1461-1470

Hurst H E, Black R P and Simaika Y M, 1965, Long-term storage: an experimental study, Constable (London)

Jaggard D L and Sun X, 1990, Scattering from fractally corrugated surfaces, Journal of the Optical Society of America A, **7**, nnn-nnn

Lévy-Véhel J, 1995, Multifractal analysis of remotely sensed images (in Fractals in Geoscience and Remote Sensing. JRC Image Understanding Research Series, volume 1, edited by Wilkinson G G, Kanellopoulos I and Mégier J, Office for Official Publications of the European Communities (Luxembourg), pp 85-101

Lovejoy S, 1982, Area-perimeter relation for rain and cloud areas, Science, **216**, 185--187

Lovejoy S and Schertzer D, 1995, How bright is the coast of Brittany? (in Fractals in Geoscience and Remote Sensing. JRC Image Understanding Research Series, volume 1, edited by Wilkinson G G, Kanellopoulos I and Mégier J, Office for Official Publications of the EC (Luxembourg), pp 102-151

Mandelbrot B B, 1982, The fractal geometry of nature, W H Freeman (New York)

Nicolis G, 1989, Physics of far-from equilibrium systems and self-organisation (in The New Physics, edited by Paul Davis, Cambridge University Press (Cambridge), pp 316-347

Peebles P B E, 1980, The large-scale structure of the Universe, Princeton University Press (Princeton)

Pentland A P, 1984, Fractal-based description of natural scenes, IEEE Transactions on Pattern Analysis and Machine Intelligence, **6**, 661-674

Prusinkiewicz P and Lindenmayer A, 1990, The algorithmic beauty of plants, Springer-Verlag (New York)

Rees W G, 1992, Measurement of the fractal dimension of ice-sheet surfaces using Landsat data, International Journal of Remote Sensing, **13**, 663-671

Rees W G, 1995, Characterisation and imaging of fractal topography (in Fractals in Geoscience and Remote Sensing. JRC Image Understanding Research Series, volume 1, edited by Wilkinson G G, Kanellopoulos I and Mégier J, Office for Official Publications of the EC (Luxembourg), pp 298-324

Tritton D J, 1977, Physical fluid dynamics, Van Nostrand Reinhold (New York)

Vasiliev L, 1995, Fractals in Geosystems and Implications for Remote Sensing (in Fractals in Geoscience and Remote Sensing. JRC Image Understanding Research Series, volume 1, edited by Wilkinson G G, Kanellopoulos I and Mégier J, Office for Official Publications of the European Communities (Luxembourg), pp 252-276

Venig-Meinesz F A, 1951, A remarkable feature of the Earth's topography, Proceedings of the koninklijke nederlandse akademie van wetenschappen, series B (Physical Science), **54**, 212-228

Voss R F, 1985, Random fractal forgeries (in Fundamental Algorithms in Computer Graphics, edited by R. A. Earnshaw), Springer-Verlag (Berlin)

Woodcock C E, Strahler A H and Jupp D L, 1988a, The use of semivariograms in remote sensing. I. Scene models and simulated images, Remote Sensing of Environment, **25**, 323-348

Woodcock C E, Strahler A H and Jupp D L, 1988b, The use of semivariograms in remote sensing. II. Real digital images, Remote Sensing of Environment, **25**, 349-379

Fluid dynamics in the coastal zone

Andrew M Folkard

University of Strathclyde

1 Introduction

The aim of this chapter is to provide an introduction to the physical nature of the fluid processes which are dominant in determining the behaviour of coastal seas. As we have seen in the introductory article of these proceedings (Cracknell), the definition of the term "coastal zone" can mean many different things to different people. At one end of the spectrum, it is used to refer to the coastal zone of the land, that is the strip of land next to the shore in which the ecology is partly determined by the presence of the sea. At the other end of the spectrum, physical oceanographers use the term to describe the region of the oceans in which the circulation is affected by the presence of the coastal boundary. This can extend to hundreds of kilometres from the shore in some cases. An attempt has been made to cover this whole spectrum to some extent, concentrating on areas related to my own particular experience, and perhaps glossing over some others.

The approach I took in structuring this article was to try and picture a casual observer of the sea standing on the shoreline. The first thing that he or she will notice is the action of the waves, so the first major section investigates these, initially from the point of view of fundamental dynamics, and then by looking at how waves behave in the coastal zone. After a while, the casual observer will start to discern the tidal rising and falling of the sea surface, so an investigation of tides, along the same lines as that carried out for waves, follows in the next section. Finally, in the last section I cover phenomena that a casual observer may never become aware of, but which nevertheless are of fundamental importance in the coastal zone.

2. Waves

2.1 Introduction

Waves at the seashore are short period waves, passing fixed points in a few seconds, and existing purely on the surface of a water body. However, the term wave is used to refer to any periodic oscillation in any physical parameter. The great variation in depth and density of water found in the world's oceans, leads to an enormous menagerie of different motions which can all be described as waves.

Within the coastal zone, waves are very important for transporting energy, sediment and chemical and biological matter. In coastal engineering terms, they are of the utmost importance because of the forces they apply to structures, and because of the coastal erosion which they cause.

We shall begin our discussion of waves by looking at a simple, idealised model of surface waves in a homogeneous fluid, and develop this to examine some of the complexities of wave behaviour when they interact with shallow water, finishing with an investigation of some other types of waves important in the coastal zone.

2.2 Linear wave theory

Waves result from any perturbation to a fluid system - the fluid cannot sustain any displacement applied to a part of it, so the perturbation is propagated away as an oscillating motion. Idealised waves have three defining properties: they are periodic, they propagate energy and they cause no net transport of the fluid. As described below, in reality waves do transport material, and this is one of their most important properties in the coastal zone. We begin our investigation of wave motion by looking at a simple, idealised model. See for example Gill, 1982.

Consider a homogeneous body of water of infinite horizontal extent and uniform depth H. The pressure at any depth z is given by $p_o(z) = -g\rho z$. Assume now that the system is perturbed so that the pressure changes by an amount $p'(x,y,z,t)$ and the position of the free surface is $z = \eta(x,y,t)$. What will the motion of the fluid look like?

To answer this question, we consider the equations of motion, namely the equations of mass and momentum conservation. Mass conservation follows from the assumption of fluid incompressibility, and states mathematically that one cannot increase or decrease the amount of fluid in a fixed volume, i.e. you can't squash (or stretch) water. If we consider a small cube of fluid of dimensions δx by δy by δz, incompressibility means that the amount of fluid going into the cube over any fixed time must equal what goes out in the same time. This can also be thought of in terms of speeds of entry and exit of fluid: the speed of entry minus the speed of exit (or the change in speed over the length of the cube) must equal zero when all three dimensions are considered. This is written mathematically as

$$\nabla \cdot \underline{u} = \frac{\partial u}{\partial x} + \frac{\partial v}{\partial y} + \frac{\partial w}{\partial z} = 0. \qquad (1)$$

The equation of momentum conservation, which can be derived by considering the net forces acting on an fluid element, is just Newton's Second Law of Motion ($F = ma$). The forces here are due to horizontal differences in pressure, the mass can be substituted by density in the case of unit volume and acceleration is the derivative of speed with respect to time, i.e.

$$\rho\frac{\partial u}{\partial t} = -\frac{\partial p'}{\partial x} \;;\; \rho\frac{\partial v}{\partial t} = -\frac{\partial p'}{\partial y} \;;\; \rho\frac{\partial w}{\partial t} = -\frac{\partial p'}{\partial z} \;. \qquad (2)$$

Combining these leads to Laplace's equation:

$$\nabla^2 p' = \frac{\partial^2 p'}{\partial x^2} + \frac{\partial^2 p'}{\partial y^2} + \frac{\partial^2 p'}{\partial z^2} = 0. \qquad (3)$$

We must now consider the boundary conditions which dictate the form of the solution to this equation. At the surface, where $z = \eta$, the pressure must become zero, so that $p = -g\rho z + p' = 0$, which implies that

$$p' = g\rho\eta \quad at \quad z = \eta. \qquad (4)$$

Furthermore, any particle in the free surface must remain there. This means that the vertical speed of a particle must always equal the rate of change of the surface height, that is

$$w = \frac{\partial \eta}{\partial t} \quad \text{at} \quad z = \eta . \tag{5}$$

Finally, there is no vertical speed at the bottom of the tank, so

$$w = 0 \quad \text{at} \quad z = -H . \tag{6}$$

In finding the solution to these, the first step is to take a guess at the answer. We know that disturbing a fluid results in a periodic motion, so we look at sinusoidal solutions from which we can construct a Fourier series to give us any shape we like. The surface elevation will then be of the form

$$\eta = \eta_0 \cos(kx + ly - \omega t) . \tag{7}$$

If we assume that the pressure perturbation is proportional to the free-surface height everywhere (this is essentially the hydrostatic approximation), and therefore substitute (4) and (7) into (3), we reach

$$\frac{\partial^2 p'}{\partial z^2} - \kappa^2 p' = 0 \tag{8}$$

where $\kappa = (k^2 + l^2)^{1/2}$ is the magnitude of the wavenumber (the number of waves per unit distance). The general solution of this equation is

$$p' = A\cosh(\kappa z) + B\sinh(\kappa z) \tag{9}$$

for which

$$\frac{\partial p'}{\partial z} = A\kappa \sinh(\kappa z) + B\kappa \cosh(\kappa z) . \tag{10}$$

From (2) and (6) we have $\partial p' / \partial z = 0$ at $z = -H$ therefore

$$0 = -A\kappa \sinh(\kappa H) + B\kappa \cosh(\kappa H) . \tag{11}$$

Since the value of η is fairly small, we can assume (4) holds at $z = 0$. Thus equation (9) gives immediately $A = \rho g \eta$. Using this result alongwith equation (11) gives

$$p' = g\rho\eta\left\{\cosh \kappa z + \frac{\sinh(\kappa H)}{\cosh(\kappa H)} \sinh(\kappa H)\right\} . \tag{12}$$

which can be written as

$$p' = \frac{\rho g \eta \cosh \kappa(z + H)}{\cosh(\kappa H)} = \frac{\rho g \eta_0 \cos(kx + ly - \omega t) \cosh \kappa(z + H)}{\cosh(\kappa H)} \tag{13}$$

So we have our equation for p'. We still have to consider the effect of boundary condition (5), for which we need to know the vertical velocity, w. Using the third equation in (2), we can quickly translate (13) into an equation for w:

$$w = \frac{\kappa g \eta_0 \sin(kx + ly - \omega t)\sinh \kappa(z + H)}{\omega \cosh(\kappa H)} \tag{14}$$

If, again, we assume that (5) holds at $z = 0$ as well as $z = \eta$, then we have

$$\frac{\kappa g \eta_0 \sin(kx + ly - \omega t)\sinh(\kappa H)}{\omega \cosh(\kappa H)} = \eta_0 \omega \sin(kx + ly - \omega t) \tag{15}$$

which implies that

$$\omega^2 = g\kappa \tanh(\kappa H). \tag{16}$$

Now ω is the frequency of the waves, that is the number passing a fixed point per unit time. Hence this relation tells us how the wavelength (i.e. the reciprocal of the wavenumber) is related to its frequency. Such an equation is called a **dispersion relation**. It also leads us directly to the speed of propagation of the wave, sometimes called the phase speed or the celerity. Since, by definition, the wave speed $c = \omega\ \kappa$ (this is the same idea as the familiar $c = f\lambda$), we get

$$c = \left(\frac{g \tanh \kappa H}{\kappa} \right)^{1/2} \tag{17}$$

This relationship is shown graphically in Figure 1. κH, the variable on the horizontal axis, is, physically, the number of waves in the horizontal distance equal to the depth of the water. The figure shows that when $\kappa H > 0.1$, the speed of each wave depends on its wavenumber. So if we start off a set of disturbances from the same point, they will travel away at different speeds and disperse (hence the name: dispersion relation).

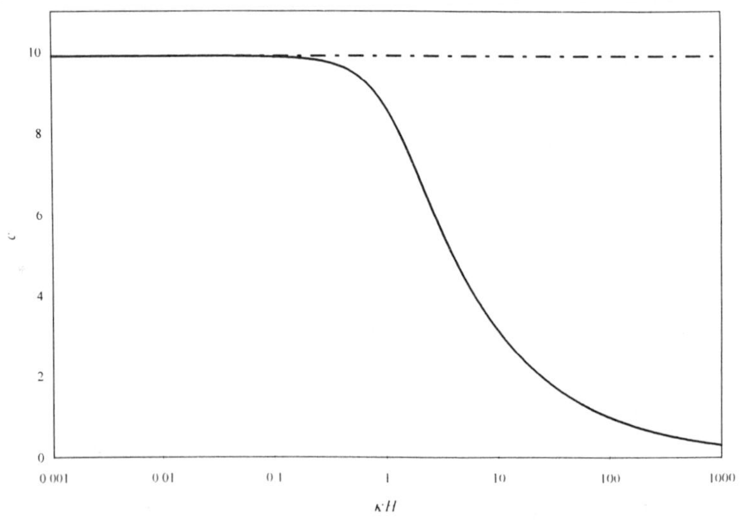

Figure 1 The variation of wave speed c with κH. The dotted line represents the shallow-water limit.

At the coast, when the water becomes shallow, κH will become very small. At small values of x, $\tanh(x) \sim x$, so (17) becomes

$$c = \left(\frac{g\kappa H}{\kappa} \right)^{1/2} = (gH)^{1/2} \tag{18}$$

and we arrive at the familiar expression for shallow water waves. Note that the speed now no longer depends on the wavenumber, so the waves are now non-dispersive. This is illustrated in Figure 1, where for $\kappa H < 0.1$, c is virtually constant. Note also, that in deep water, where κH is very large, $\tanh(\kappa H) \sim 1$, so $c = (g/\kappa)^{1/2}$. In this limit, the shortest waves (those with the highest wavenumber) move more slowly than the larger waves. But this is a diversion, and we will focus now on the behaviour of waves in shallow water.

The shallow water wave speed equation, given in (18) implies that c decreases as H does. That is, waves slow down as they move into shallower water. The obvious way to explain this is to say that they experience the dragging effect of the bottom more as the water gets shallower, so that holds them back. But this does not really give the whole picture because there is no term for bottom drag in our equations. The only effect of the bottom considered was in the boundary condition, where we stated that there could be no vertical motion at the bottom. Drag would effect horizontal velocities, not vertical ones. In order to understand why waves behave like this in shallow water, we must consider the motions of particles.

2.3 Particle motion in waves

Let us for simplicity look at the motion of waves in two dimensions (x, z). The perturbation pressure given in (13) then becomes

$$p' = \frac{\rho g \eta_0 \cos(kx - \omega t) \cosh k(z + H)}{\cosh(kH)}. \tag{19}$$

The expression for the vertical velocity w in (14) is then

$$w = \frac{kg\eta_0 \sin(kx - \omega t) \sinh k(z + H)}{\omega \cosh(kH)}. \tag{20}$$

We can use the first equation in (2) to get a similar equation for u, the horizontal speed

$$u = \frac{-kg\eta_0 \cos(kx - \omega t) \cosh k(z + H)}{\omega \cosh(kH)}. \tag{21}$$

For chosen values of k, η_0 and ω, we can then calculate the velocity at any height. An example is shown in Figure 2. The overall result is that particles are forced into elliptical orbits. If H is very large (deep water) these elliptical orbits tend towards a circular limit, whereas for shallower water, the orbits become flatter. Furthermore in shallow water, since we have $w = 0$ at $z = -H$, the orbit is completely flat, that is the particles move back and forth horizontally. Note that the particles never actually travel - they just go round and round. This is consistent with the idea of waves not causing any material movement in the fluid.

So how does this relate to the way in which waves slow down as H decreases? To help us see this more clearly, let us look at the situation at z = 0, and take the shallow water approximation $kH << 1$, so that $\tanh(kH) \sim kH$. This gives us expressions for u and w as

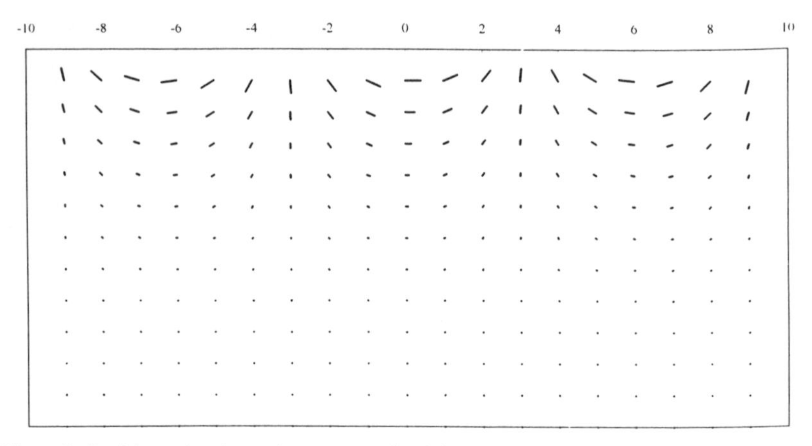

Figure 2. Particle motions beneath a wave at a fixed time. As time proceeds, this pattern moves from left to right, resulting in elliptical particle orbits.

$$w = \omega^{-1} k^2 g \eta_0 H \sin(kx - \omega t);$$

$$u = -\omega^{-1} k g \eta_0 \cos(kx - \omega t) \qquad (22)$$

So although the horizontal speed at the surface is not related to the depth, the vertical speed does depend on depth – the shallower the water, the slower the particles move up and down. This goes right back to the equations in (2): vertical motions are caused by vertical variations in the perturbation pressure. If the depth is only very small, these will not be very large, and thus will not cause much vertical motion. As a result, the particles will not bob up and down quickly and the wave will not be able to progress quickly. So the relation between wave speed and water depth in the shallow water limit is caused not by bottom drag, but because the bottom limits the amount of vertical pressure variation and thereby constrains the speed of the wave oscillations.

2.4 Energy propagation

As we have already stated, waves transport not material, but energy. Therefore, having covered what they are, let us look now look at what they do.

So far, we have looked at a perfectly sinusoidal wave and have derived its speed in shallow water as $c = (gH)^{1/2}$. It is very rare, however, for perfectly sinusoidal waves to be found in the ocean, since a random perturbation, such as a gust of wind, will tend to excite waves of numerous different wavenumbers and frequencies. Returning to the dispersion relation in (16), which in the deep water limit (where the waves originate) became $c=(g/\kappa)^{1/2}$, we recall that as they move away from their origin, waves spread out according to their wavenumber - long waves moving more quickly than the short ones. Hence waves of approximately equal wavenumber tend to stay together. Let us therefore consider the effect of two waves of similar ω and k superimposed on one another. The overall result will be:

$$\eta = \cos[(k+\delta k)x - (\omega+\delta\omega)t] + \cos[(k-\delta k)x - (\omega-\delta\omega)t] = 2\cos(\delta kx - \delta\omega t)\cos(kx - \omega t) \qquad (23)$$

Physically, this results in a wave of the average wavenumber and frequency of the two original waves, enveloped by a wave which has a speed equal to

$$C_g = \frac{\delta\omega}{\delta k} \approx \frac{d\omega}{dk}. \tag{24}$$

This speed depends on the derivative of the frequency because the change in the shape of the large envelope wave depends on the change in phase difference between the two waves.

Any number of waves of similar frequency and wavenumber can combine in this way, and form a group of waves travelling with this velocity. It is therefore called the **group velocity** and represents the rate at which energy is carried by waves. There is a partial analogy with a *peloton* of cyclists: cyclists within the group may be travelling at different speeds (some moving to the front, some dropping to the back), but the group can also be assigned an overall speed with which it is moving. Within each group of waves, Lamb (1932) describes the action of each particular wave as follows:

> *"If attention be fixed in a particular wave, it is seen to advance through a group, gradually dying out as it approaches the front, while its former place in the group is occupied in succession by other waves which have come forward from the rear."*

Thus, when energy propagation by waves is being considered, we must always think in terms of group velocity, which is nothing more than the combined effects of several waves of similar wavenumber and frequency which have become grouped together because of the process of dispersion described above.

Having calculated how waves transport energy, we now ask how much energy a given wave carries. The energy of a wave is, of course, made up of two kinds of energy: kinetic and potential. Still considering our wave in the x and z dimensions only, but considering a portion of this wave of unit length in the y-direction, we can write the kinetic energy as

$$E_K = \int_0^L \int_{-H}^{\eta} \rho \frac{u^2 + w^2}{2} dzdx \tag{25}$$

where $L = 1/k$ is the wavelength. By substituting our expression for u and w from (20) and (21) (the working is left as an exercise - assume $\eta \sim 0$), we arrive at

$$E_K = \rho g \frac{\eta_0^2 L}{4}. \tag{26}$$

Similarly, the potential energy relative to the stationary state of the water is given by

$$E_P = \int_0^L \int_0^{\eta} \rho g z dz dx \tag{27}$$

which gives, using the two-dimensional form of (7)

$$E_P = \rho g \frac{\eta_0^2 L}{4}. \tag{28}$$

So the total energy, which is made up of equal parts of kinetic and potential energy is given by

$$E_T = \rho g \frac{\eta_0{}^2 L}{2}. \tag{29}$$

Clearly the longer and higher the wave, the more energy it carries. If the integrations in (25) and (27) are carried out with different limits, it turns out that 99% of the wave's energy is carried in a depth of less than $L/2$ below the water's surface. Thus although waves must be thought of as passing through the whole water column, rather than just being surface oscillations, they have very little effect on the sea-bottom unless they are either extremely energetic, or in very shallow water.

2.5 Stokes drift and material transport by waves

The statement made earlier about waves not transporting material is true only for idealised waves such as the ones considered mathematically above. In the previous considerations, the height of the wave η was considered to be approximately equal to zero in order to make some of the boundary conditions easier to apply. Attempting to correct this approximation leads to an equation (first derived by George Stokes) that superimposes a mean transport on the oscillatory motion already described. This type of motion is known as Stokes Drift. The flow of momentum due to these effects is known as radiation stress, particularly when the coastal zone is being considered. The waves predicted by this theory tend to have sharper crests and longer flatter troughs than those predicted by our simplified theory. This non-ideal form of behaviour is the basis of wave-induced currents and sediment transport by wave-induced motions. These topics will be touched upon later.

2.6 Behaviour of waves in shallow water

So far, we have looked at the effect of shallow water on waves purely in terms of what happens to their velocity as a result of pressure variations. There are, however, numerous other phenomena relating to waves in shallow water, and we will study some of those now.

Wave damping by bottom friction

We have already determined that bottom friction is not responsible for the slowing down of waves as they approach the shoreline. Friction, however, does have an effect on waves, namely that it dissipates their energy and causes a reduction in wave height. It is perhaps of more importance to consider what the drag on the bottom sediments caused by waves is like, rather than *vice versa*. This is a very complex subject, and has been approached historically mainly through experimental work.

In general, the **shear stress** τ (i.e. the shear force per unit area) at the sea bottom is found to be related to the flow velocity at the bed u by an equation of the form

$$\tau \propto \tfrac{1}{2} \rho u^2. \tag{30}$$

Jonsson (1975) found that the effect of waves could be incorporated into this sort of equation by adding a **wave friction factor** f_w such that

$$\tau = \tfrac{1}{2} f_W \rho u^2 \tag{31}$$

where f_w is given by

$$\ln(f_{\mathrm{W}}) = -5.977 + 5.213\left(\frac{\xi_0}{k_{\mathrm{N}}}\right)^{-0.914} \quad (32)$$

where ξ_0 is the amplitude of the horizontal, wave-induced motions at the bed and k_N is the roughness parameter of the seabed.

Reflection

Reflection is the simplest of the three classical wave processes (refraction and diffraction, the other two, being covered below) to consider. Let us consider our surface wave of the form given in (7) (ignoring the y-component) hitting a vertical wall at $x=0$. Part of the wave's energy will be dissipated, but partial reflection will occur, resulting in a returning wave of the form

$$\eta^* = \eta_{\mathrm{R}} \cos(kx + \omega t) \quad (33)$$

So the resulting disturbance will be

$$\eta_{\mathrm{T}} = \eta + \eta^* = (\eta_0 + \eta_{\mathrm{R}})\cos(\omega t) \quad (34)$$

at particular points where $x = 2\pi k$, $4\pi k$ and so on. This is a sinusoidal oscillation at these fixed points which has an amplitude equal to the sum of the amplitudes of η and η^*. If reflection is strong enough, this can lead to a violent collision such as is often seen near sea-walls during storms. Usually, waves will approach a boundary at an angle, so this collision effect will not be so marked as in the head-on case.

Refraction

Refraction of any type of wave occurs when one part of the wave slows down or speeds up before the rest of the wave, and thereby causes a change in direction. Surface water waves behave in this way, just like any other type of wave. When waves approach the coast obliquely from deep water, the parts of the waves which reach shallow water first tend to slow down. The result of this process is that the waves eventually take up the shape of the bottom contours and arrive at the shore line moving approximately perpendicular to it.

If the angle between the direction of motion of a train of waves and a sudden change in depth is large enough (the latter being the analogy of a change in refractive index in an optical system), total internal reflection can occur, and the wave train is reflected back rather than continuing onwards.

That waves tend to follow depth contours implies that topographic underwater features play an important role in determining where waves are concentrated and rarefied. This is illustrated in Figure 3, where the effects of undersea canyons and ridges are illustrated. Waves will tend to be diverged away from canyons, and converged towards headlands. This process tends to **coastal uniformity**: headlands will tend to receive more wave action than the surrounding coast, and hence be worn down more rapidly, and vice versa.

Diffraction

Diffraction occurs when part of a wave becomes blocked by a solid object, leaving the rest of the wave to propagate in a different fashion from that which would have occurred had the solid object not been there. An example is shown in Figure 4. This commonly occurs in the coastal zone, where waves encounter both natural (bars, spits and headlands) and man-made (breakwaters, harbour walls and piers) obstacles. It is clear from Figure 4 that three distinct regions can be identified in a diffracted wave system. The first consists of purely diffracted waves, the second of unaffected waves, and the third of reflected waves.

The heights of the diffracted waves will be lower than those of the unaffected waves, and a significant loss of energy will occur because of the process of diffraction.

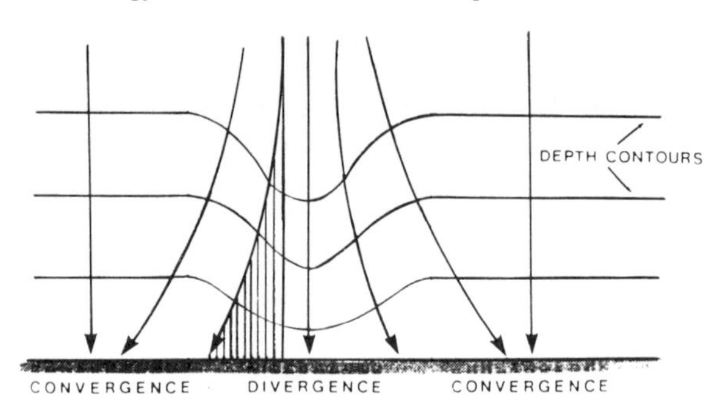

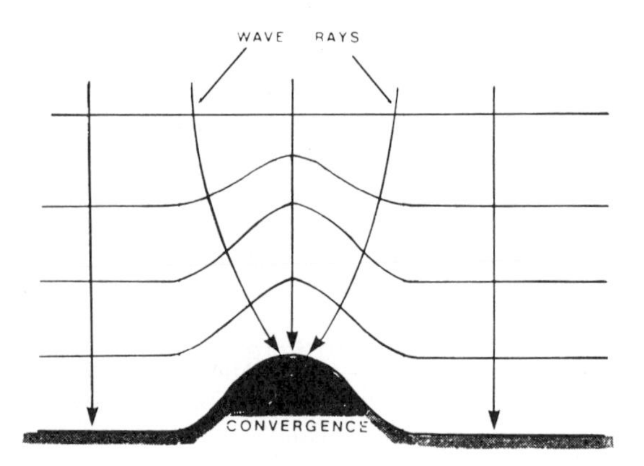

Figure 3 The effect of undersea canyons and ridges on waves via refraction. Wave-fronts travel parallel to depth contours in shallow water, so wave rays travel perpendicular to them. This leads to wave energy being rarefied at near-shore canyons and concentrated at headlands (after Beer, 1983).

Wave breaking

Waves release their energy at the coastline through the process of breaking. This is a process hydrodynamically-similar to the production of a **tidal bore** or a **hydraulic jump** and is once again dependent on the fact that waves travel faster in deep water. When a train of waves approaches a coastline, it begins to slow down. However, the frequency of waves passing through two different points in a direction perpendicular to the coastline must remain constant. If this were not the case, the number of waves within some region as the train approached the shore would either keep increasing until there existed an infinite number of waves in this finite region, or would decrease to zero; two situations which are physically impossible. If the speed decreases but the frequency remains constant, this implies that the wavelength must decrease also i.e. the waves become *steeper* as they approach the shore.

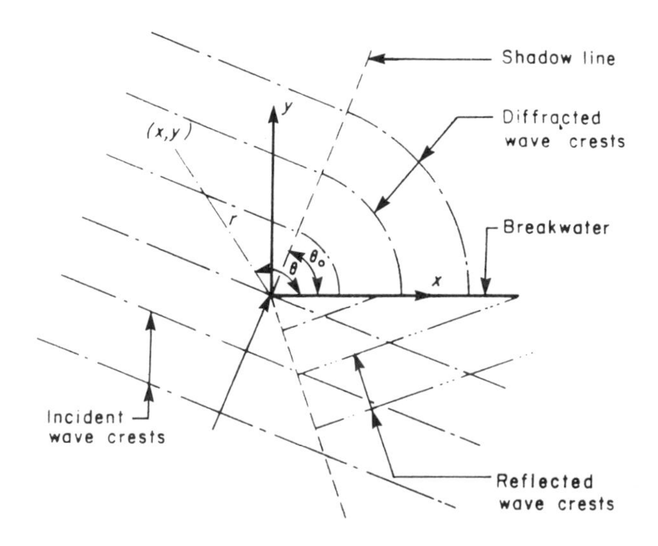

Figure 4 Waves diffracting around a breakwater. Note three distinct areas: one of purely diffracted waves, the second of unaffected waves and the third of superimposed reflected and incident rays (after Muir-Wood and Fleming 1981).

The second effect that leads to the breaking of waves depends on speed variations within each wave. Crests are travelling over deeper water than troughs, so they move faster. As they do so, the waves start to become *asymmetric*, and lean forward. Eventually, the crests try to overtake the troughs, and the waves become unstable. They start to topple over and break up, entraining air and forming surf. Finally, the waves crash to the floor, their oscillatory energy having been converted to random turbulent motions.

Longshore currents

When waves approach a coastline at an oblique angle, because of the material transport described in Section 2.5, they produce currents which flow parallel to the shore. This current is found to be strongest in the region close to the shore where the waves have broken and given up their oscillatory motion. The theory that describes this process is somewhat complex and outwith the scope of this chapter. The classical exposition is given by Longuet-Higgins (1970). Velocity profiles of longshore currents are shown in Figure 5. The parameter P represents the importance of horizontal mixing ($P = 0$ corresponds to no mixing). The offshore distance (X) and current speed (V) have been non-dimensionalised using the offshore width of the breaker zone and the maximum current speed in the case of no mixing respectively. Most field data can be fitted to curves for which $P \sim 0.1$–0.4. These currents are very important for sediment transport along coastlines.

2.7 Trapped waves

So far we have considered surface gravity waves than propagate freely in open waters. There are, however, a great many types of waves peculiar to certain environments or representing periodic oscillations in parameters other than the height of the water surface. We shall now complete our look at waves by considering two of these special classes of waves. In this section we will look at coastally-trapped waves, and in Section 2.8 we shall consider internal waves.

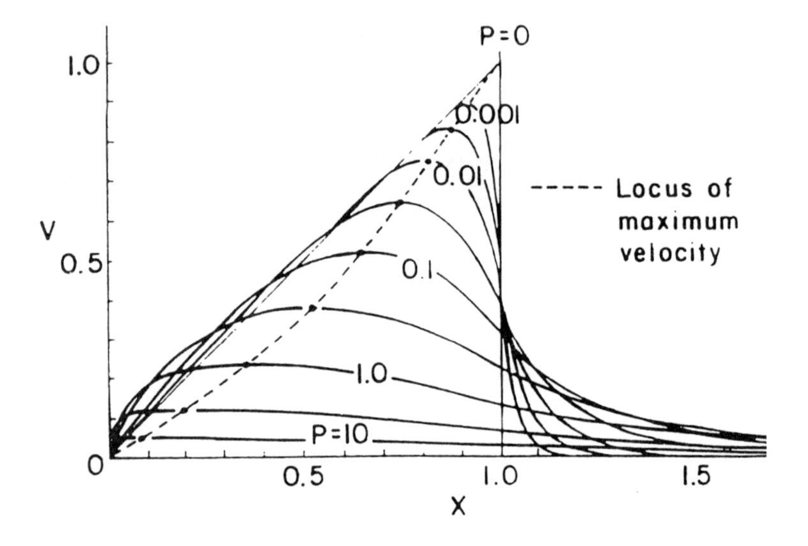

Figure 5 Longshore current profiles, after Longuet-Higgins (1970). Note the majority of the current's position within the breaker region close to the coast.

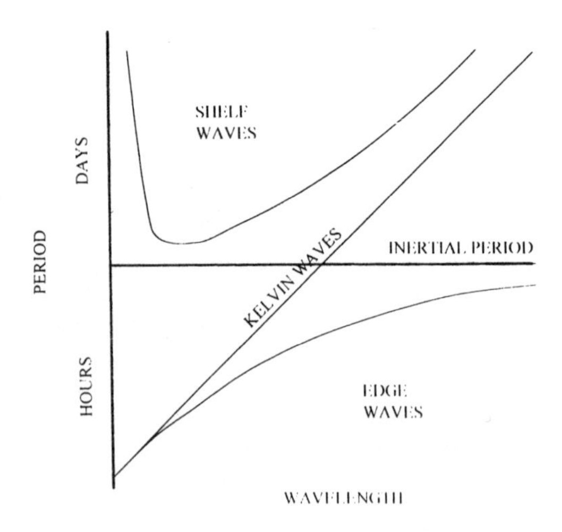

Figure 6 Dispersion diagram for coastally-trapped waves.

Trapped waves are those which owe their existence to the presence of physical or dynamical boundaries. These include coastlines, sudden changes in depth, and the equator. The last of these is a dynamical boundary since the direction of the Earth's rotation changes as it is crossed (in the sense that the Earth is rotating anti-clockwise when viewed from above the North Pole, but clockwise when viewed from above the South Pole). We will consider only waves which owe their existence to, and are therefore "trapped" next to, coastlines.

Coastally-trapped waves can be divided into three categories: edge waves, Kelvin waves and shelf waves. Their respective wavelengths and periods are shown in Figure 6. They will be discussed in turn.

Edge waves are short period waves, that is they take very much less than a day to pass by. At this time scale, the effects of the Earth's rotation are relatively small and edge waves are trapped by the processes of refraction and reflection. Consider a wave train moving at a slight angle away from the coast. The seaward parts of these waves will be moving faster than the onshore parts, so the wave will be steered around through refraction by the topography so that they are moving back in towards the shore. They will be then become reflected back out to sea, and the process will repeat itself, thus maintaining the progress of the wave along the coastline. The overall effect is a wave shaped like a surface gravity wave (that is, sinusoidally), but with an amplitude that decays exponentially away from the coast.

Kelvin waves are a very wide category of waves; as shown in Figure 6, their periods can vary from hours to days. They are the dominant response of coastal waters to any major disturbance, and result from a combination of rotational (Coriolis) and pressure forces. The effect of the boundary is to ensure that in a Kelvin wave there are no motions perpendicular to the coastline. If we define x, y and z as the along-shore, across-shore and vertical directions respectively, and u, v and w as the corresponding velocity components, then this boundary effect can be written as v 0. In the x-direction, we can apply Newton's Second Law of Motion to a unit volume, taking the force as that due to the hydrostatic pressure gradient, and arrive at

$$\frac{\partial u}{\partial t} = -g\frac{\partial \eta}{\partial x} \tag{35}$$

where η is once again the height of the surface above its undisturbed level. The presence of rotation tells us that for steady motion, geostrophic balance must apply, that is

$$fu = -g\frac{\partial \eta}{\partial y} \tag{36}$$

Finally, we have the continuity equation, which states that in a given interval δx, the amount of fluid leaving minus the amount of fluid arriving in a given time δt is equal to the change in height of water within δx. This is written

$$\frac{\partial \eta}{\partial t} + H\frac{\partial u}{\partial x} = 0 \tag{37}$$

Combining (35) and (37) leads to

$$\frac{\partial^2 u}{\partial t^2} = gH\frac{\partial^2 u}{\partial x^2} \tag{38}$$

which is the standard wave equation, with a general solution

$$\eta = F(x + ct, y) + G(x - ct, y) \tag{39}$$

$$u = -\left(\frac{g}{H}\right)^{1/2} \{F(x+ct),y) - G(x-ct,y)\}. \tag{40}$$

Substituting these into (36) we get

$$\frac{\partial F}{\partial y} = \frac{f}{(gH)^{1/2}} F, \qquad \frac{\partial G}{\partial y} = \frac{-f}{(gH)^{1/2}} G, \tag{41}$$

which lead to general solutions

$$F = \exp\left\{\frac{f}{(gH)^{1/2}} y\right\} F'(x+ct), \qquad G = \exp\left\{\frac{-f}{(gH)^{1/2}} y\right\} G'(x-ct). \tag{42}$$

Rejecting the first as unphysical and choosing G′ as peroiodic (with Fourier Series in mind), we take our final solution for the Kelvin wave as

$$\eta = \eta_0 \exp\left\{\frac{-f}{(gH)^{1/2}} y\right\} \cos(kx - \omega t) \tag{43}$$

$$u = \left(\frac{g}{H}\right)^{1/2} \eta_0 \exp\left\{\frac{-f}{(gH)^{1/2}} y\right\} \cos(kx - \omega t). \tag{44}$$

These waves thus have the same shape as gravity waves, but like edge waves have amplitudes which decay exponentially away from the coast. They travel with the coast to their right in the Northern Hemisphere, and to their left in the Southern Hemisphere. An illustration of them is shown in Figure 7.

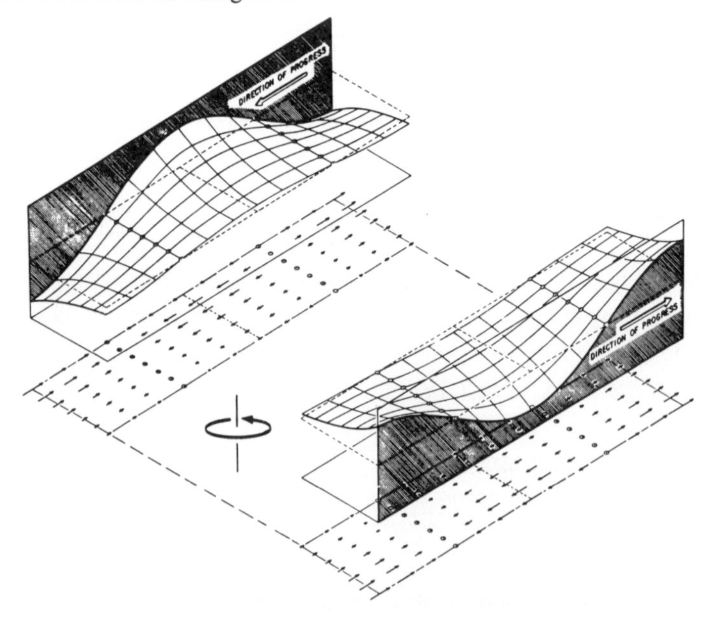

Figure 7 Elevation and velocity structure of a Kelvin wave, after Gill (1982).

Finally, we consider **shelf waves**. These are coastally-trapped waves which are very slowly oscillating, having periods of longer than a day. As a result their behaviour is strongly determined by the effects of the Earth's rotation. Just as Kelvin waves are driven by a combination of the Coriolis force and cross-shore pressure gradient forces (resulting from the increase in wave amplitude at the coastal side of the wave), shelf waves are driven by a combination of Coriolis forces and the effects of bottom topography which slopes away from the coastline. Unit volume columns of water are initially disturbed on (off) shore. Because of the sloping topography, this causes them to contract (extend) vertically. Conservation of angular momentum then dictates that they will gain cyclonic (anticyclonic) vorticity. The combined effect of these two processes results in the progression of the line of particles shown in Figure 8. Thus these waves can only exist where topography slopes away from the coastline on a large scale.

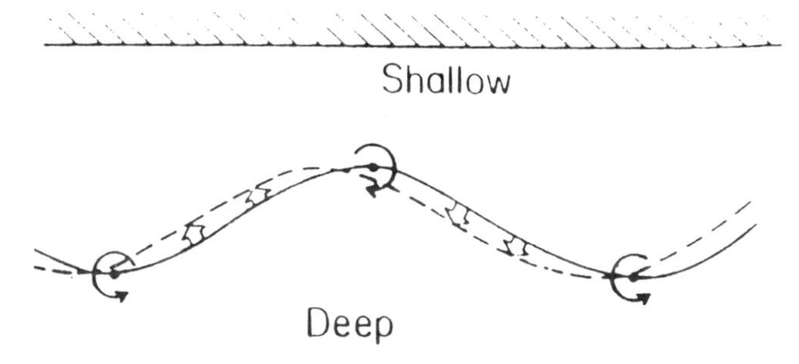

Figure 8 The process by which shelf waves progress along the coast., after Gill (1982).

2.8 Interfacial and internal waves

Thus far, we have considered only waves in homogeneous fluids, that is, fluids which have uniform density. The oceans, however, are far from homogeneous. In some places (the thermocline, for instance), there are strong, quasi-discontinuous changes in density. In other places, the density increases, usually approximately-linearly, with depth. Waves can propagate in both these situations. In the former situation, the boundary between the fluids of different densities is very much like the boundary between the water and the air at the surface, and the waves propagated here behave similarly to surface waves, with phase velocity and group velocity moving in the same direction. These are called interfacial waves. In a continuously-stratified fluid, the waves behave rather differently, and are called internal waves.

Both of these types of waves play an important role in determining the energy budgets in the coastal zone, and react in complex and not yet fully understood ways with boundaries that they impinge upon. We will now take an overview of the nature of each of these waves in turn.

Firstly, we look at **interfacial waves**. Consider a two-layer fluid, with the upper and lower layer densities ρ_1 and ρ_2 respectively, where $\rho_1 < \rho_2$. If there are oscillations on the surface, we would expect these to be mirrored by oscillations on the underwater interface. It is not unreasonable to picture these happening so that the interfacial oscillations are either in phase or 180° out of phase with the surface oscillations. Mathematical consideration of the situation (e.g. Gill 1982) shows that this is exactly what happens. When the two interfaces are in phase, the wave moves along with a phase speed equal to $(gH)^{1/2}$ where H is the total depth of the water, i.e. in exactly the same way as a surface gravity wave does. This is called the **barotropic** mode. When they are out of phase, the amplitude of the surface oscillations is very small, and that of the interfacial oscillations is much larger. These waves move with a phase speed of $(g'H)^{1/2}$ where

$$g' = \frac{(\rho_2 - \rho_1)}{\bar{\rho}} g \tag{45}$$

$\bar{\rho}$ being the mean density of the two layers. g' is called the **reduced gravity**, and because the difference between ρ_1 and ρ_2 is small, this results in a phase speed much less than that of the surface waves. Thus interfacial waves are larger and slower than their surface counterparts. This is called the **baroclinic** mode. These waves are illustrated in Figure 9.

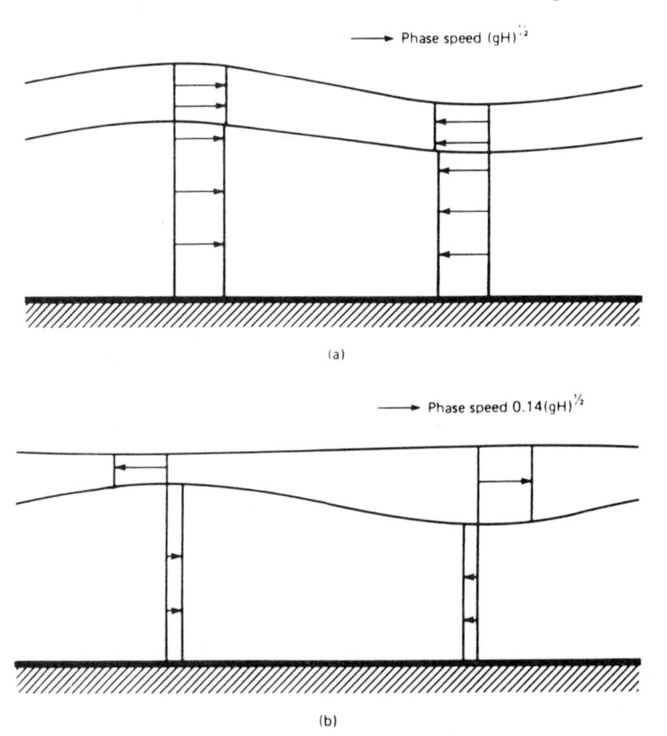

Figure 9 The two modes of solution for oscillations in a two-layer system: (a) barotropic waves and (b) baroclinic waves, after Gill (1982).

Interfacial waves can be easily observed by filling a jar with layers of oil and water. Swinging the jar should cause the surface to remain relatively steady, but in fact causes great agitation at the interface between the two fluids. Examples showing the variations of the depth of interfacial waves from the coastal zone off San Diego (US) are shown in Figure 10.

We move now to the more general situation where the density varies continuously with depth. Let us consider the situation where $\partial\rho/\partial z$ is a constant, that is, the density varies linearly with depth. This is often referred to as **linear stratification**.

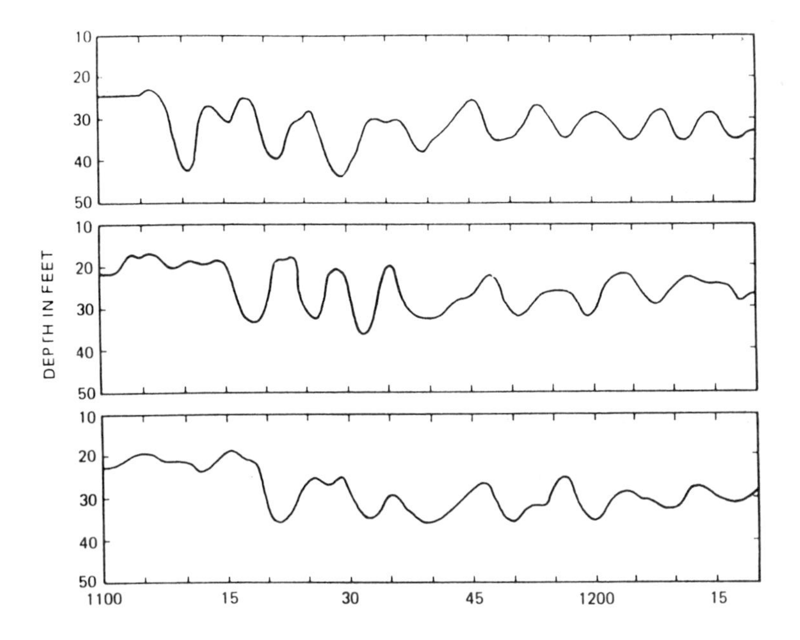

Figure 10 Field observations of interfacial waves recorded off the Californian coast, after Beer (1983).

If a particle of water of density ρ_0 is displaced from its equilibrium position by a vertical distance z, the water surrounding our particle will have density

$$\rho_1 = \rho_0 + z\frac{\partial \rho}{\partial z} \tag{46}$$

so, by Archimedes Principle, assuming the particle is of unit volume, there will be a downward buoyancy force on the particle resulting in an acceleration of

$$z'' = \frac{(\rho_0 - \rho_1)}{\rho_0} g = \frac{-gz}{\rho_0}\frac{\partial \rho}{\partial z} \quad \Rightarrow \quad z'' + \frac{g}{\rho_0}\frac{\partial \rho}{\partial z} z = 0 \tag{47}$$

which is the equation of simple harmonic motion with oscillation frequency

$$N = \left(\frac{g}{\rho_0}\frac{\partial \rho}{\partial z}\right)^{1/2} \tag{48}$$

N is called the **buoyancy frequency** or, rather less concisely, the Brunt-Vaisala frequency. Thus wave motion can be induced in a linearly-stratified fluid (and, for that matter, any stratified fluid) by oscillating it vertically. It also turns out (see Tritton 1988, for a full exposition) that oscillations can be set up in a stratified fluid by oscillating fluid particles at any angle. However, unlike surface waves where the frequency of the waves depends on their speed and wavenumber, here the frequency depends only on the angle at which the particles are oscillated. That is,

$$\omega = N\cos\theta \tag{49}$$

where θ is the angle that the particle oscillation makes with the vertical. Internal waves, therefore, have a maximum upper limit to their frequency, namely N. If some sort of forcing oscillation occurs in a linearly-stratified fluid at a frequency greater than N, no internal waves will result because the fluid cannot support oscillations at this frequency. If the forcing oscillations are at a frequency less than N, then internal waves will result and will radiate out at an angle $\theta = \cos^{-1}(\omega/N)$ from the point at which they are being forced. Unlike surface waves, the particle motions in internal waves are perpendicular to the phase velocity (the direction in which crests and troughs move). The group velocity - the direction in which energy moves - is directed parallel to the particle motions and perpendicular to the phase velocity.

Internal waves are very important in transporting energy to, from and within the coastal zone. The Kelvin waves and edge waves described previously can form within a stratified fluid and become **internal trapped waves**. Tidal waves can also occur internally, leading to **internal tides** (see, for example, Beer 1983). Current research is being carried out on the way in which internal waves reflect and/or break when they impinge on the coastline - i.e. on what happens to their energy when they reach the coastline (e.g. Wu and Lin 1994; Thorpe 1994 and Taylor 1993).

3. Tidal motion

3.1 Introduction

The action of the tides will be familiar to anyone who has spent any time at a coastline. If the casual observer I invoked in Section 1 stayed a while on the shore watching the waves, he or she would soon notice the rising and falling of the tide over an approximately 12-hour cycle. If this observer returned to the shoreline regularly, he or she would come to be aware of the fortnightly variation in the height of high tide, and the existence of spring (fortnightly-high) and neap (fortnightly-low) tides. After several years of visiting the shoreline, one would notice that the highest tides of each year happened at the same time each year, indicating some sort of annual time scale variation. If the shoreline was in certain places, anomalous situations would be observed. At Southampton in the south of England, for example, there are four distinct high tides each day instead of the normal two. Finally, if the observer travelled the shorelines of the world, very marked differences would be noticed in the range of high tides: in the Bay of Fundy in North-Eastern USA, the tides vary by several metres, for example, whereas in the harbour at Marseilles the height of the water hardly changes at all. This section will aim to explain these and many other features of tidal motion, and to illustrate its role in the coastal zone.

3.2 Mathematical models od tidal motion

The simplest model

The dominant cause of tidal motion is, of course, the gravitational pull of the Moon. Let us therefore begin with a situation where we have a perfectly-spherical Earth, covered entirely with water of uniform density, salinity, temperature and depth. The Earth is not rotating about any star, but it rotates on its own axis once every 24 hours. At a fixed distance from this idealised Earth the Moon sits in the equatorial plane of the Earth. The situation is shown in plan view in Figure 11.

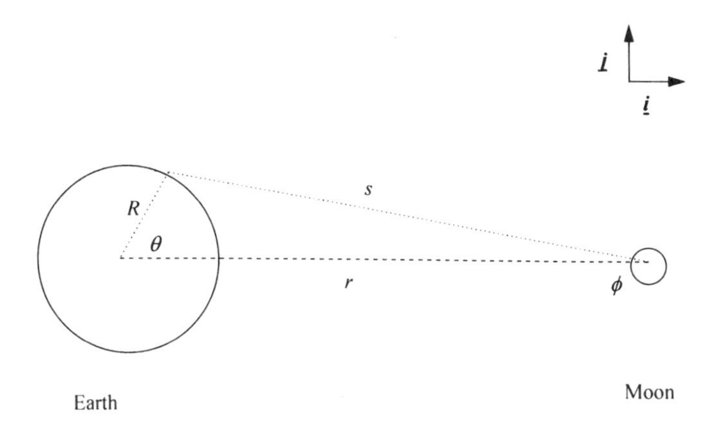

Figure 11 Diagram of the idealised Earth-Moon system. The Earth is rotating about its own axis once every 24 hours, but the Moon is stationary.

We wish to find the effect of the Moon's gravitational pull on the water, relative to the Earth. We therefore need to subtract the effect of the Moon's gravitational pull on the Earth from that on the water. Consider therefore the action of the gravitational pull of the Moon. Using the co-ordinate system in Figure 11, the acceleration of the Earth (which we are assuming is a non-deformable body) caused by the gravitational force of the Moon is simply

$$\mathbf{a}_{\mathrm{EM}} = \frac{GM_{\mathbf{M}}}{r^2}\mathbf{i} \tag{50}$$

where G is the universal gravitational constant (which has a value of 6.67×10^{-11} Nm^{2}kg^{-2}), M_{M} is the mass of the Moon, and r is the distance from the centre of mass of the Moon to the centre of mass of the Earth.

Similarly, the acceleration of a particle of water on the surface of the Earth due to the gravitational force of the Moon is given by

$$\mathbf{a}_{\mathrm{WM}} = \frac{GM_M}{s^2}\cos\phi\,\mathbf{i} + \frac{GM_M}{s^2}\sin\phi\,\mathbf{j} \tag{51}$$

where s is the distance from the centre of mass of the Moon to the particle and ϕ is the angle between the line of action of the Moon's gravitational force on the particle and the line connecting the centres of mass of the Earth and the Moon. To make things easier to visualise, we should convert this expression for $\mathbf{a}_{\mathrm{WM}}$ to one in terms of θ, rather than ϕ. From Figure 17, it is clear that

$$\phi = -\sin^{-1}\left(\frac{R\sin\theta}{s}\right) \tag{52}$$

hence

$$\mathbf{a}_{\mathrm{WM}} = \frac{GM_{\mathrm{M}}}{s^2}\left\{\frac{\left(s^2 - R^2\sin^2\theta\right)^{1/2}}{s}\right\}\mathbf{i} - \frac{GM_M}{s^2}\left\{\frac{r\sin\theta}{s}\right\}\mathbf{j} \tag{53}$$

which leads to

$$\mathbf{a}_{\mathrm{WM}} = \frac{GM_{\mathrm{M}}}{s^3}\left[(r - R\cos\theta)\,\mathbf{i} - R\sin\theta\,\mathbf{j}\right]. \tag{54}$$

Subtracting the expression for $\mathbf{a}_{\mathrm{EM}}$ given in (50) gives if we assume that s~r,

$$\mathbf{a}_{\mathrm{WE}} = \frac{GM_{\mathrm{M}}R}{r^3}\left[\cos\theta\,\mathbf{i} - \sin\theta\,\mathbf{j}\right]. \tag{55}$$

Note that this is $\mathbf{a}_{\mathrm{WE}}$ not $\mathbf{a}_{\mathrm{WM}}$ i.e. it is the acceleration of the water particle relative to the Earth due to the gravitational pull of the Moon.

We are interested at the moment not so much in the size of this acceleration (the $GM_{\mathrm{M}}R/r^3$ part of (55)), but in its direction. This is illustrated in Figure 12. For example, when $\theta = 0°$, the term in the brackets in (55) is equal to **i**, that is the gravitational pull is directly away from the centre of the Earth. This is also the case at $\theta = 180°$, whereas at $\theta = 90°$ and 270°, the acceleration is directly towards the centre of the Earth. The acceleration arrows at angles intermediate to these can be drawn in by interpolation.

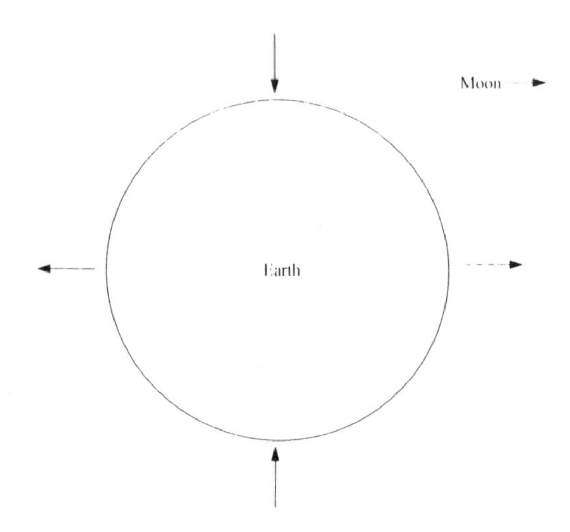

Figure 12 Directions of gravitational pull of the Moon on water particles on the Earth.

The resultant effect can be clearly seen to cause a bulge of water on either side of the Earth along the line which connects the centres of mass of the Earth and the Moon, and a thinning of the sheet of water at the furthest points on the Earth's surface from that line. So far, in Figures 11 and 12, the idealised Earth has been drawn in plan, but the arguments presented work equally well if the Earth is viewed from the side, so this thinning of the water surface occurs in a plane perpendicular to the line connecting the centres of mass, passing through the North and South poles. Taking a slice through the Earth at any latitude, the bulges will still appear at the points closest to and furthest from the Moon, although they will be less marked the higher the latitude one takes the slice.

Since the Earth is rotating exactly one every 24 hours, each point on the solid surface of the Earth (below the sheet of water) passes through a bulge once every 12 hours, therefore experiencing two high waters and two low waters every day.

But the Moon goes round the Earth!

Let us now take a step closer to our actual situation from the highly-idealised state described in Section 3.2, by assuming that the Moon rotates around the Earth, always staying in the Equatorial plane, once every 28 days (this is an approximate figure). The Moon is travelling in the same direction as the rotation of the Earth, and it takes with it the bulges of water. This means that each time a point on the solid Earth comes round to where the bulge was on the previous rotation, the bulge has moved on through 1/28th of a full rotation. It will therefore be a little more than 12 hours now between consecutive high waters. We can work out how long this will be. Allow time $t = 0$ to be the time at which a given point on the solid surface, A, is at a high water point. It is travelling at an angular speed of 360 degrees/day. The bulges are travelling at 360/28 degrees/day. In the time that the other high water point passes through $x°$, point A must pass through $180+x°$ to catch it up. So

$$t = \frac{x}{360/28} = \frac{x+180}{360} \quad \Rightarrow \quad x = 180/27 = 6.67^{\circ}$$

$$\Rightarrow \quad t = 0.519 \text{ days} = 12 \text{ hrs } 27\text{mins.}$$

Thus high tides occur a little more than 12 hours apart, hence the change in the time of high tides by about a little less than one hour every day (two tidal cycles).

The tidal motion we have discussed so far is the principal lunar constituent of the tide and is semi-diurnal, occurring roughly twice per day. The rate at which it recurs is most usually quoted in degrees/hour, the precise value being 28.98°/hr. Since it is related to the Moon and occurs twice per day, it is referred to as the $\mathbf{M_2}$ constituent of the tide.

What about the Sun?

Let us now add another piece of reality to our picture, by considering the action of the gravitational force of the Sun. Recall that the approximate size of the relative acceleration produced by the Moon was

$$|\mathbf{a}_{\mathrm{WE}}| = \frac{GM_{\mathrm{M}}R}{r^3} = \frac{6.67\times10^{-11}\times7.36\times10^{22}\times6.38\times10^{6}}{\left(3.8\times10^{8}\right)^{3}} = 5.7\times10^{-7}\ \mathrm{ms}^{-2} \qquad (56)$$

The equivalent figure for the Sun is

$$|\mathbf{a}_{\mathrm{WE}}| = \frac{GM_{\mathrm{S}}R}{r_{\mathrm{SE}}^3} = \frac{6.67\times10^{-11}\times1.99\times10^{30}\times6.38\times10^{6}}{\left(1.49\times10^{11}\right)^{3}} = 2.6\times10^{-7}\ \mathrm{ms}^{-2} \qquad (57)$$

where M_S is the mass of the Sun and r_{SE} is the mean Earth-Sun distance (all figures taken from Sears *et al.* 1982). Thus the Sun will produce tidal variations of a comparable magnitude to those produced by the Moon. They will also occur approximately twice per day, and hence are referred to as the $\mathbf{S_2}$ constituent of the tide. In the same way as described above for the Moon, their speed can be calculated as 30.00°/hr.

How do the Solar and Lunar tides interact?

Because of their different periods, the S_2 and M_2 tidal constituents move in and out of phase, a process referred to as beating. If we redefine the magnitude of the M_2 tide as 1 (in units of "M_2 tidal magnitudes"), the magnitude of the S_2 tide becomes 0.46. If we also think of the tidal frequencies in terms of cycles per hour, rather than degrees per hour, we find values of 0.0805 hr^{-1} and 0.0833 hr^{-1} for the M_2 and S_2 tides respectively. Hence we can represent these two tides mathematically as

$$H_{M2} = \cos(2\pi \times 0.0805 \times t) \quad \text{and} \quad H_{S2} = 0.46\cos(2\pi \times 0.0833 \times t) \tag{58}$$

The combined effect of these two tides is, by the principal of superposition, simply H_{M2} + H_{S2}. By using the trigonometric rule cos2A + cos2B = 2cos(A+B)cos(A-B), this is given by

$$H_{M2} + H_{S2} = 0.54\cos(2\pi \times 0.0805 \times t) + 0.92\cos(2\pi \times 0.0819 \times t)\cos(2\pi \times 0.0014 \times t) \tag{59}$$

Thus there is a constituent that has the same frequency as M_2 but only half its magnitude, and another constituent that has the average frequency of M_2 and S_2, is the magnitude of the S_2 constituent, but which is modified by a very low frequency variation that causes it to come and go at a frequency of 0.0014 cycles hr^{-1}. The whole pattern is shown in Figure 19. 0.0014 cycles hr^{-1} is equivalent to approximately 29.75 days per cycle. From Figure 13 it is clear that this causes a modulation of twice this frequency or approximately 14.8 days per cycle. This tidal constituent is called the "luni-solar fortnightly tide" and is denoted M_{sf}. It is one of the two main reasons for the phenomenon of spring and neap tides - the approximately fortnightly variation in tidal amplitude. The other will be described shortly.

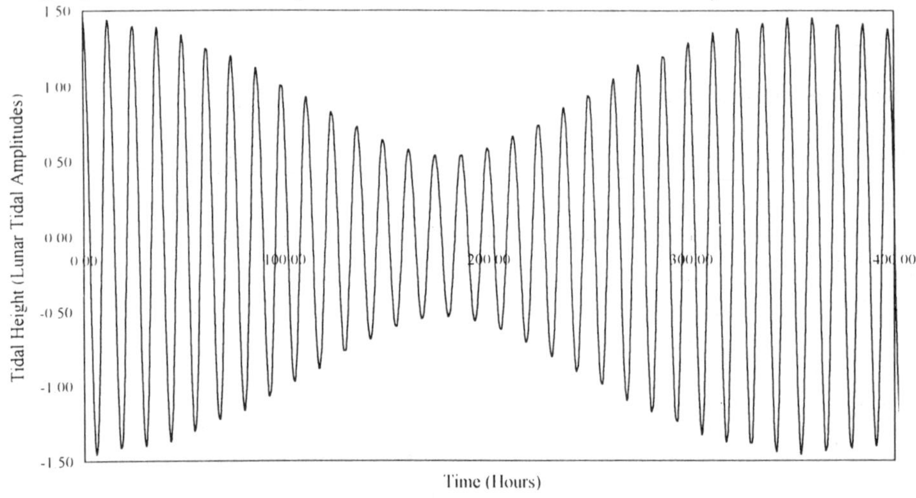

Figure 13 Luni-solar fortnightly tidal pattern. Note the steady component with an amplitude of 0.54, and the varying component with an amplitude of 0.92.

The Moon's orbit is not circular

Like all bodies in the Solar System, the Moon has an orbit which is elliptical, not circular, which can be most easily described in terms of Kepler's Laws of Planetary Motion (see any basic physics book for these). Thus we must adjust our idealised (but gradually improving) picture to incorporate this fact and determine its effect on terrestrial tides.

The ellipticity of the Moon's orbit means that sometimes it is closer to the Earth, so the tidal forces are stronger, and at other times it is further from the Earth, so the tidal forces are weaker. The mean Earth-Moon distance is 3.84×10^8km, where as the perigee (closest approach) is 3.56×10^8km and the apogee is 4.07×10^8km. Recalling that the tidal acceleration of water relative to the Earth is 5.7×10^{-7}ms^{-2} for the mean distance, we get extreme values of

$$|\mathbf{a}_{\mathrm{WE}}| = \frac{GM_{\mathrm{M}}R}{r^3} = \frac{6.67 \times 10^{-11} \times 7.36 \times 10^{22} \times 6.38 \times 10^6}{\left(3.56 \times 10^8\right)^3} = 6.94 \times 10^{-7}\,\mathrm{ms}^{-2} \qquad (60)$$

and

$$|\mathbf{a}_{\mathrm{WE}}| = \frac{GM_{\mathrm{M}}R}{r^3} = \frac{6.67 \times 10^{-11} \times 7.36 \times 10^{22} \times 6.38 \times 10^6}{\left(4.07 \times 10^8\right)^3} = 4.64 \times 10^{-7}\,\mathrm{ms}^{-2} \qquad (61)$$

This gives a periodic variation in tidal strength due to the ellipticity of the Moon's orbit of amplitude 2.30×10^{-7}ms^{-2}. The amplitude of the main semi-diurnal lunar tidal constituent is twice the 5.7×10^{-7}ms^{-2} value given, because at high tide the water is being pulled up with this acceleration, and at low tide the water is being pushed down with this acceleration (recall Figure 18), so this variation due to the Moon's orbital ellipticity gives a tidal constituent whose amplitude is approximately $(2.3\times100)/(2\times5.7) = 20\%$ of that of the major M_2 constituent.

This variation takes place over the month that it takes the Moon to orbit, and produces the monthly lunar tidal constituent M_m. There is also another effect of this tide, caused by its interaction with the M_2 tidal constituent. This interaction arises from some details we discarded when we assumed that s ~ r in deriving equation (55). This can be dealt with by defining a scalar tidal potential which is the gradient of the tidal force vector, and expanding it in terms of Legendre polynomials. This is covered in detail by Godin (1972), but is outwith the scope of these lectures. We simply note that this monthly anomaly adds constituents of two new frequencies which are the sum and difference of the M_2 and the M_m constituents. That is, since the M_2 frequency is 28.98°hr^{-1} and the M_m frequency 0.54°hr^{-1}, these two new constituents have frequencies of 28.98+0.54 = 29.52°hr^{-1} and 28.98-0.54 = 28.44°hr^{-1}. The new constituents are called L_2 and N_2 respectively (because they lie either side of M_2). N_2 has an amplitude of 0.20 that of M_2 for the reasons laid out above, but L_2 is much weaker.

Of course, the Earth's orbit around the Sun is also elliptical, and this in turn produces variations in the tidal height. The most basic one is the **solar annual constituent,** S_a, which is the direct analogy of M_m. The other major constituent is the **solar semi-annual tide,** S_{Sa}, which is produced by the interaction between the annual constituent and the semi-diurnal solar constituent.

The Moon does not stay in the Equatorial Plane and the Earth's axis is tilted

The final adjustment to our model is the recognition of the fact that the Earth's axis of rotation is tilted at ~23.5° to the plane of its orbit around the Sun, and that the Moon undergoes some "vertical" motion (i.e. out of the plane of its orbit around the Earth).

The effect that this has on the tides at a particular place is that each point on the Earth will tend to pass through a larger part of one of the two tidal bulges it experiences each day, and through a smaller part of the other. This leads to a cycle that, at least partly, has a period of approximately one day. The same effect is found in relation to the position of the

Sun relative to the point on the Earth where the tides are being measured. There are found to be three major constituents of this diurnal (daily) tide. These are called K_1, O_1 and P_1.

Anything else?

The tidal constituents we have talked about so far interact with each other *ad infinitum* to produce finer and finer structure in the spectrum of tidal cycles. However, these new hybrid constituents have very small amplitudes in general, and we can be satisfied with our model at the level of complexity to which we have built it so far. The major tidal constituents which we have derived are listed in the following table.

Name	Description	Speed (deg hr^{-1})	Relative Magnitude
	Semi-diurnal constituents		
M_2	Principle Lunar constituent caused by rotation of Earth about its own axis relative to Moon	28.98	1.00
S_2	Principle Solar constituent caused by rotation of Earth about its own axis relative to Sun	30.00	0.46
N_2	Caused by interactions between M_2 and M_m constituents	28.44	0.20
L_2	as for N_2, but weaker	29.53	0.03
	Diurnal constituents		
K_1	Made up of combination of parts of "unequal bulge" effect due to position of Sun and Moon	15.04	0.58
O_1	Other part (with K_1) of "unequal bulge" effect due to position of Moon	13.94	0.42
P_1	Other part (with K_1) of "unequal bulge effect" due to position of Sun	14.96	0.20
	Long period constituents		
M_{sf}	Luni-solar fortnightly - the cause of spring/ neap variations	1.02	0.06
M_m	Lunar monthly - due to ellipticity of the Moon's orbit	0.54	0.04
S_{sa}	Solar semi-annual - caused by interaction of S_a with S_2	0.08	0.03
S_a	Solar annual constituent - caused by ellipticity of Earth's orbit around Sun	0.04	0.03

3.3 Terrestrial effects

So we have satisfied ourselves that we have covered all the effects of the Moon, the Sun and for that matter any other celestial bodies. We can therefore now concentrate on the situation on the ground. So far, we have been considering an Earth entirely covered with water. Furthermore we have not considered the effect of the rotation of the Earth on the movement of the water. Since we are particularly concerned here with the coastal zone, it would be useful to look at these effects.

In order to study these effects, it is most convenient to think of the tides as very long waves which take several hours or days to rise and fall, rather than the couple of seconds which more familiar waves take to do this. We will refer to the periodic tidal motions, therefore, as tidal waves. These are not to be confused with the sort of tidal waves beloved of disaster movie producers, which are usually called "tsunamis" and which have nothing to do with tides.

Tidal waves have very long wavelengths. More importantly, they have wavelengths which are long compared to the depth of water over which they are travelling. In Section 2, this was the definition of a shallow-water wave, the speed of propagation of which is

$$c = (gH)^{1/2} \tag{62}$$

where H is the depth of the water. Taking a ballpark figure for the depth of the oceans as 4000m gives a wave speed of $c \sim 200 \mathrm{ms}^{-1}$. A wave travelling at this speed would take about two days to travel around the circumference of the Earth. Another way of looking at this is that it takes two days for the effect of the astronomical tidal forces to reach all around the Earth. Hence the ocean never takes up the idealised double bulge shape that we discussed in the previous section, because it can not respond quickly enough to the changing forces. Furthermore, the propagation of the tidal wave is hindered by the presence of continental boundaries. Finally, the propagation speed is of the same magnitude as the rotation speed of the Earth itself (which varies from zero at the poles to $\sim 458 \mathrm{ms}^{-1}$ at the equator), so we must consider the effects of rotation on the tidal waves. Clearly the situation is very complex. Having come from astronomical effects in the last section to terrestrial effects, we shall continue our down-scaling process by starting with gross global scale effects and working our way towards the local or coastal scale.

The characteristic pattern of tidal motions at the global ocean scale

The Earth's oceans can be thought of as a series of deep basins connected by relatively shallow and narrow channels and separated by the continents. We can therefore get a general idea of the way in which the tides behave at a global scale by thinking of water in an enclosed basin. The water is being forced, by the astronomical tidal forces, into the wave motion we have already discussed. The basin is also being rotated. The rotation causes the tidal wave always to be forced to the right in the northern hemisphere and to the left in the southern hemisphere; this gives rise to a rotation of the high tide about a central point where there is little tidal variation. This central point is called an **amphidrome**.

Extending out from the amphidrome we can draw lines where the state of the tide (that is, the time before or after high tide) is the same. These are called **co-tidal lines**. Perpendicular to these, we can draw contours along which the amplitude (the height at high water) of the tide is the same. These are called **co-range lines**. Figure 14 shows the measured pattern of tides around the United Kingdom. Note how, as the sea gets less like a uniform basin (i.e. as the coastline gets more complicated), so the tidal pattern becomes more complex, and multiple amphidromes appear.

Many attempts have been made to model these tidal patterns - one example is shown in Figure 15, which is a model of the global pattern of M_2 tides as calculated by Accad and Pekeris (1978). These models are successful at simulating some observed features, such as the very small semi-diurnal tides in Western Australia due to the proximity of an amphidrome to the coast. They also, however, badly misrepresent some other features. The tides along the Californian coast, for example, are shown moving southward, whereas in fact they are observed to move in the opposite direction.

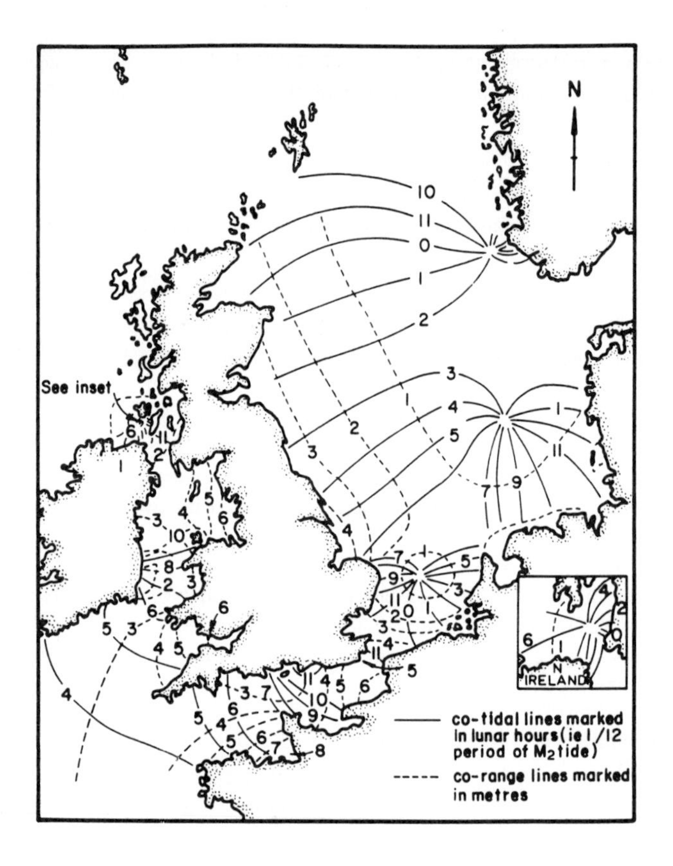

Figure 14 System of amphidromes, co-tidal lines and co-range lines around the U.K.. Note the existence of multiple amphidromes in the North Sea, which is essentially a broad gulf (see Section 3.3 and Figure 23), after Gill (1982).

Tides in narrow gulfs and channels

Channels and gulfs are usually the areas of the coast where the most complex hydrodynamics is occurring. Channels link two open basin areas, and the tides in them must therefore act to match the tides at either end. A good example of this occurs in the Strait of Gibraltar, where the relatively large tides of the North Atlantic and the much smaller tides of the Mediterranean meet. A thorough study of this is described by Candela *et al.* (1990) and Candela (1991). In both channels and gulfs (which are closed at one end), the tides also adapt to match the boundary conditions caused by the surrounding land.

If a channel or gulf is sufficiently narrow, there will be insufficient space for rotational forces to take effect, and the motion of tidal waves can be considered in a non-rotational context. This is valid, at least as a first approximation, for areas like the Bay of Fundy, the Bristol Channel and the Gulf of California. Let us consider this mathematically, following Gill (1982):

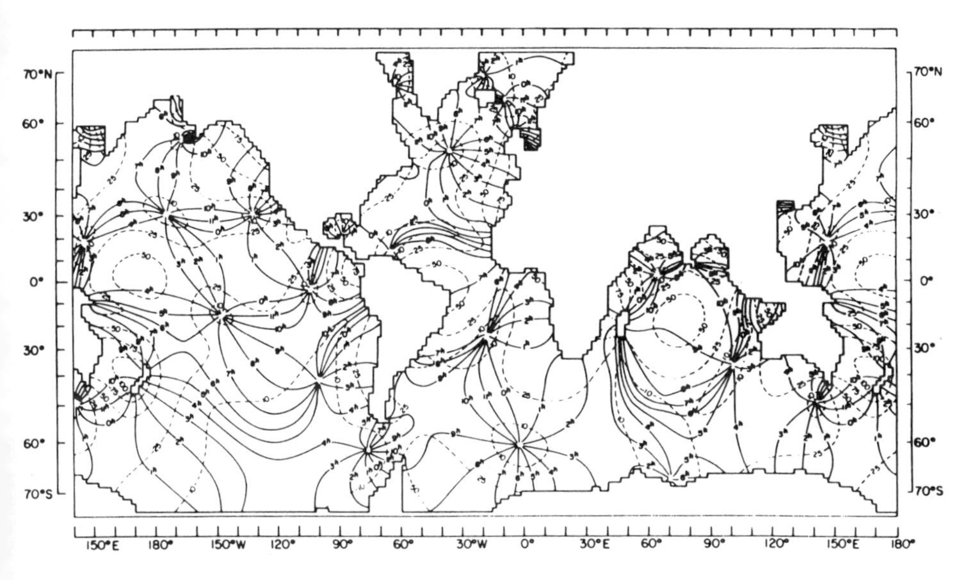

Figure 15 Model results (from Accad and Pekeris 1978) of tidal patterns using M_2 and S_2 modes only. Note the accurate prediction of an amphidrome near Western Australia, and the erroneous prediction of southward-bound tidal waves on the coast of California.

Consider a section of the idealised gulf in Figure 16 of length δx. Because of its narrowness, we concern ourselves only with motions for which the surface elevation does not vary across the gulf. The rate of net transfer of fluid mass into this section is given by

$$\frac{\partial m}{\partial t} = \delta x \frac{\partial(\rho A)}{\partial t} = -\delta x \frac{\partial(\rho A u)}{\partial x} \tag{63}$$

Taking the limit $\delta x \to 0$, leads to the continuity equation

$$\frac{\partial A}{\partial t} + \frac{\partial(Au)}{\partial x} = 0\,. \tag{64}$$

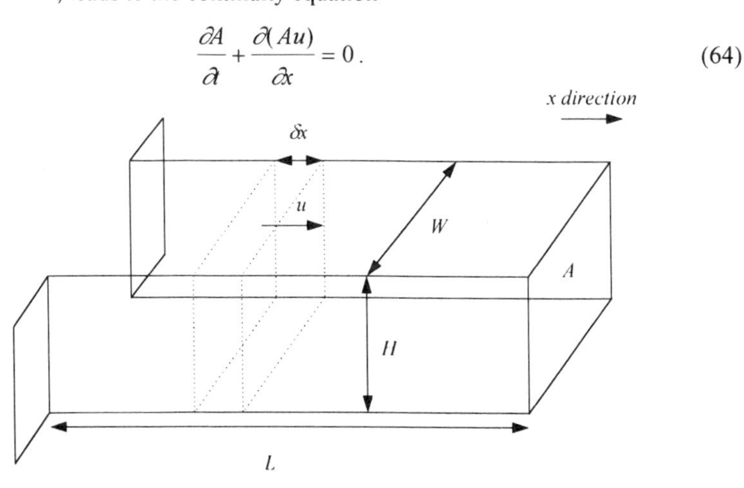

Figure 16 Idealised narrow gulf investigated in Section 3.3.

We can re-write A as $W\eta$, where W is channel width and η surface elevation. This gives

$$W\frac{\partial\eta}{\partial t}+\frac{\partial(Au)}{\partial x}=0. \tag{65}$$

For simplicity, we now consider a gulf of uniform depth and width. This allows us to write

$$\frac{\partial\eta}{\partial t}+H\frac{\partial u}{\partial x}=0. \tag{66}$$

As well as this continuity of mass equation, we also need to consider the continuity of momentum. This can be written

$$\rho\frac{\partial u}{\partial t}+\frac{\partial p'}{\partial x}=0 \tag{67}$$

or, using the hydrostatic approximation $p' = \rho g\eta$

$$\frac{\partial u}{\partial t}+g\frac{\partial\eta}{\partial x}=0. \tag{68}$$

Substituting (68) into (66) gives

$$\frac{\partial^2\eta}{\partial t^2}=c^2\frac{\partial^2\eta}{\partial x^2} \tag{69}$$

where c is given by (62). This clearly has the form of a wave-equation. Choosing $x = 0$ at the closed end of the gulf (where $u = 0$, but oscillations in η are possible) gives the solution

$$\eta=\eta_0\cos kx\cos\omega t \tag{70}$$

and the corresponding equation for the velocity

$$u=\frac{C\eta_0}{H}\sin kx\sin\omega t \tag{71}$$

(the working is left as an exercise). In order to calculate values for k and ω, we must consider the situation at the open end of the gulf at $x = L$, say. Here, we must match what is happening on in the gulf with the situation in the open sea. This matching must cover two quantities: the pressure ($\rho g\eta$), and the mass flux (ρAu). Hence the ratio of these two, $g\eta Au$ must also match. This quantity is called the **impedance** and is usually denoted Z. Using the equations (21) and (22), we obtain an expression for Z:

$$Z=\left(\frac{g}{W^2H}\right)\cot kL\cot\omega t \tag{72}$$

In the open sea, W and H both effectively tend to infinity, so $Z = 0$. Thus Z in the channel must also be zero. This gives solutions

$$k=\frac{\left(n+\frac{1}{2}\right)}{L}\pi; \qquad \omega=\frac{\left(n+\frac{1}{2}\right)}{L}\pi c \tag{73}$$

These are the natural frequencies of oscillation in a closed narrow gulf of uniform width and depth.

This situation is analogous to the problem of deriving wave motions in a pipe open at one end that will be familiar from secondary school physics classes. The analogy to a pipe closed at both ends, namely a lake or reservoir, and that to an open pipe, namely a channel, can be treated in the same way. See Section 5.8 of Gill (1982).

Tides in broad channels and gulfs

In broad channels, such as the North Sea, the effects of rotation must be considered, since the across-channel effects will now be significant. This will lead to the sort of rotation around an amphidrome seen in the basins of the open ocean. If the gulf or channel is sufficiently small, its entire length will be traversed by one tidal wave during a tidal cycle. Often, however, tidal waves will take longer than one period to move round a gulf, and this will lead to two or more amphidromes within a single gulf, see Figure 17. (This is similar to the situation in the North Sea shown in Figure 14). An excellent exposition of the formulation and a solution of problems such as these is given in Chapter 10 of Gill (1982).

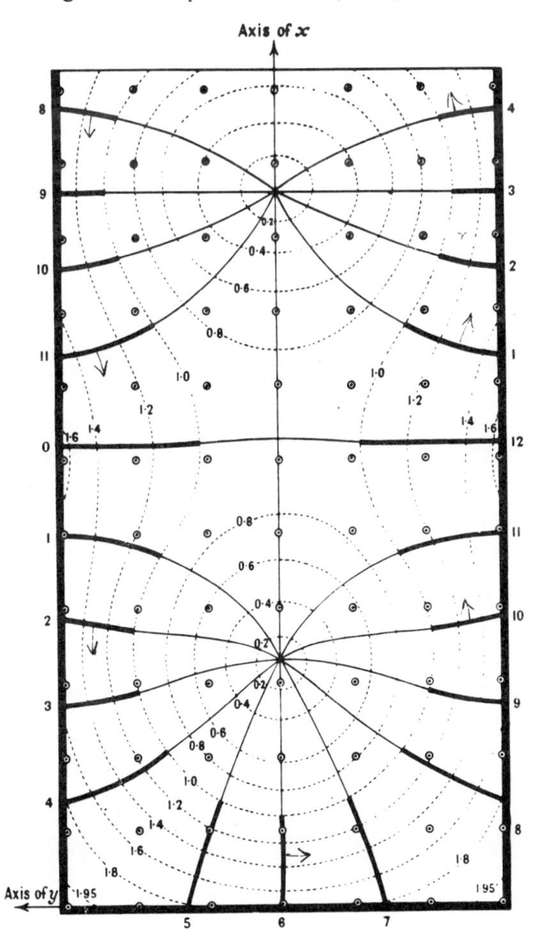

Figure 17 A tidal wave travelling around an idealised wide gulf.

Shallow water effects

Having moved from astronomical scales to global scales and then down to the scale of a local sea such as the North Sea or the Adriatic, we can now move right up to the coast and look at what happens to the tide when it reaches the shallow coastal waters. The following derivation is adapted from that given by Muir-Wood and Fleming (1981).

Consider the tidal wave in Figure 18. It is travelling at a speed c relative to the sea bottom in the x-direction. It also causes the sea level to be changed by an amount η, where η is a function of x only and the integral of η over all x is zero (i.e. there is no net change in the sea level). The tidal wave causes fluid particles beneath it to move at a speed $u = u(x)$. We assume that u is not a function of depth, and that we can draw **streamlines** which have some component in the x-direction.

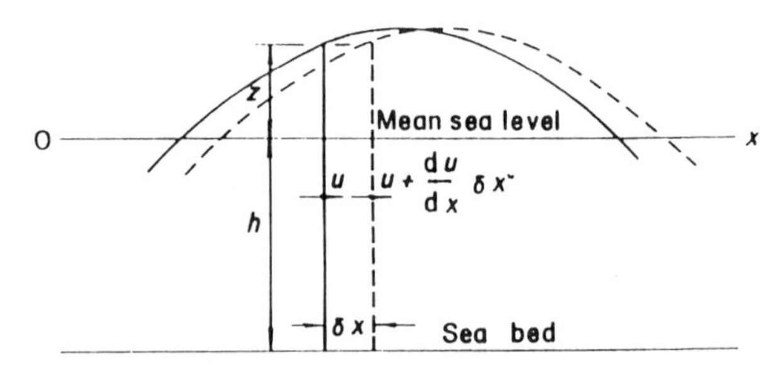

Figure 18 Schematic diagram of tidal wave, after Muir-Wood and Fleming (1981).

Consider a streamline joining two points, one at x and the other at $x+\delta x$. Because these are on the same streamline, we can apply **Bernoulli's equation** which we write as

$$\frac{2p}{\rho}+u^2+2g\eta=\frac{2\left(p+\frac{dp}{dx}\delta x\right)}{\rho}+\left(u+\frac{du}{dx}\delta x\right)^2+2g\left(\eta+\frac{d\eta}{dx}\delta x\right) \qquad (74)$$

If we assume that there is no energy loss between the two points, no work will have been done between them, so we can ignore the p terms (which are simply the work per unit volume, multiplied by $2/\rho$). Furthermore, let us consider the problem from the frame of reference of the part of the tidal wave for which $z = 0$. This is moving at c, so we must change the speed u to u-c:

$$(u-c)^2+2g\eta=\left(u-c+\frac{du}{dx}\delta x\right)^2+2g\left(\eta+\frac{d\eta}{dx}\delta x\right) \qquad (75)$$

which reduces to

$$-2(u-c)\frac{du}{dx}=2g\frac{d\eta}{dx} \qquad (76)$$

if we neglect (δx^2) terms. We also have the continuity equation, which we can write as

$$(u-c)(h+\eta)=\left(u-c+\frac{du}{dx}\delta x\right)\left(h+\eta+\frac{d\eta}{dx}\delta x\right) \qquad (77)$$

which can be re-written as

$$-(u-c)\frac{dz}{dx}=\left(h+\eta\right)\frac{du}{dx}. \tag{78}$$

Combining (76) and (78) gives

$$c=u+\left\{gh\left(1+\eta/h\right)\right\}^{1/2}. \tag{79}$$

Now in the frame of reference of the sea bed, there is no net transport. So in our frame of reference, the mean transport per unit width is *-ch*. This means that

$$-ch=(u-c)(h+\eta). \tag{80}$$

Substituting for u in (79) then gives

$$c=\left\{gh\left(1+\frac{\eta}{h}\right)\right\}^{1/2}\left(1+\frac{\eta}{h}\right). \tag{81}$$

If $\eta \ll h$, as is usually the case even in shallow water then we can apply the binomial theorem, and write (81) as

$$c=\left(1+\frac{3}{2}\frac{\eta}{h}\right)\left[gh\right]^{1/2}. \tag{82}$$

The presence of the shallow water changes the speed of the tidal wave from $(gh)^{1/2}$ to the form shown in (82). This implies that the parts of the wave which have heights above the mean level (η>0) travel faster in shallow water, and those below the mean level (η<0) travel more slowly. So the peaks begin to catch up with the troughs, the flood tides become shorter, and the ebbs become longer. This is the tidal wave equivalent of the process which leads to surface wave-breaking described in Section 2.

If we analyse this process in detail, we find that the distorted form of the wave is the same as its original sinusoidal form, modified by harmonics of shorter and shorter periods. This can lead to anomalous situations where **more than two high or low waters** are observed each day. A good example of this occurs, off Southampton in Southern England.

Tidal Bores

The pattern of steepening tidal waves caused by shallow water covered in the previous section has a particularly notable effect when it occurs in estuaries and rivers. Given favourable topographic conditions, this can lead to the presence of a tidal bore – a sudden change in the water level travelling upstream. Bores are simply extreme cases of the condition in Figure 13, and produce instant tidal flood phases. The hydrodynamics of bores are identical to those of breaking waves, or hydraulic jumps. Noteworthy examples of these spectacular phenomena occur in the River Severn (England), and the Yellow River (China).

Resonance

So far we have discussed forcing oscillations driven by gravitational forces. Resonance can also occur in the case of tides, and results in very high tidal excursions. The primary example of this is in the Bay of Fundy, on the eastern seaboard of North America. The bay here is precisely the right length to hold an integer number of tidal waves caused by the lunar gravitational forces, and as a result some of the biggest tides in the world are found there.

Resonance is also important when considering wave motion, and its effects on structures and partially-enclosed areas such as harbours. The harbour will have a natural oscillation period associated with it. For a rectangular harbour with dimension L perpendicular to the coastline the period is

$$T = \frac{4L}{\sqrt{gH}}. \tag{83}$$

The resonance of this oscillation can be excited by wind, or by currents passing along the top of a harbour (analogous to blowing across the top of a bottle). Avoiding resonances of this type is a major factor in harbour design.

Atmospheric tides and storm surges

As well as being changed by gravitational forces, the sea level can also change in response to variations of pressure and wind speed in the atmosphere. These are not often noticeable unless there is very little astronomical tide, or the atmospheric effect is particularly strong (during a storm for example). Changes in pressure alone cause only minor changes in sea level, about 10mm for 1mb, so for the usual maximum variation of atmospheric pressure from ~960–1030mb, this is a maximum variation of 700mm.

More marked effects occur in storm surges, where the pressure effect is exacerbated by strong winds which pile water up in particular areas, inducing wave motions. The amplitudes of these waves are generally small in deep water, but become magnified in shallow regions.

4 Flows and fronts

4.1 Introduction

Although waves and tides are the two most obvious features of the coastal sea to the casual observer, there are many other processes occurring that may never become obvious to them, or would only become obvious if the observer were to start taking measurements of density, temperature, salinity or velocity.

These phenomena can be split into two main categories, namely flows and fronts. The dominant forms of each of these will be discussed in this section.

4.2 Flows in the coastal zone

Currents in the coastal zone can arise from a large variety of difference sources and occur at a large number of different scales. Tidal currents, arising from the gradual procession of high water up or down a coast as it rotates around its amphidrome (see Section 2) are perhaps the most immediately obvious, occurring on timescales of hours and oscillating back and forth. Non-oscillatory coastal currents can arise from outflows of rivers. These can be dominant close to the river mouth, but will eventually be swallowed up in the general oceanic flow. If the river flow is large enough, it will be forced to the right (in the Northern Hemisphere) by the Coriolis force and flow along the coastline. At larger scales, the Coriolis force acts on density-driven flows and forms boundary currents. The Algerian Coastal Current in the Western Mediterranean and the Leeuwin Current around Western Australia are good examples of such a flow. At the largest scales are the major coastal currents - the Agulhas Current around South Africa, the Kuroshio Current around Japan, the Gulf Stream up the east coast of America and the California Current. These all appear to be formed by the response of the ocean basin to wind forcing and are just one illustration of the very strong coupling between the atmosphere and ocean that determines both their behaviours.

Although the last of these types of flows fall within what oceanographers refer to as the coastal zone (i.e. that part of the world's oceans where the flow is affected by the presence of a boundary), we are concerned here with rather smaller scales. Therefore, we shall investigate no further the behaviour of these global scale flows. Tidal flows have already been covered in Section 2. We shall focus, therefore, on **density-driven boundary currents** found as a result either of riverine outflows, or the meeting of waters from two different seas.

Density-driven boundary currents

The nature of these currents is determined by a combination of geostrophic balance (i.e. the balance of Coriolis and pressure forces), boundary conditions and the continuity of volume. Griffiths (1986) gives a thorough review of the theory of these features. The currents can be divided into two categories: in the first, the water in the current is lighter than that surrounding it, leading to a surface current, whereas in the second the water in the current is heavier and flows as a bottom current. The former is the most relevant to coastal zone studies and is found to have a depth profile of

$$h \quad H_0\left[1-\frac{\cosh(y \quad \lambda_0)}{\cosh(L \quad \lambda_0)}\right] \tag{84}$$

and a velocity profile of

$$u \quad \sqrt{g' H_0}\,\frac{\sinh(y \quad \lambda_0)}{\cosh(L \quad \lambda_0)} \tag{85}$$

where the current is flowing along a vertical wall, as shown in Figure 19. Here h=0 at $y \quad L$, $h \quad H_0$ at $y = 0$ and $\lambda_0 \quad (g'H_0)^{1/2}/f$.

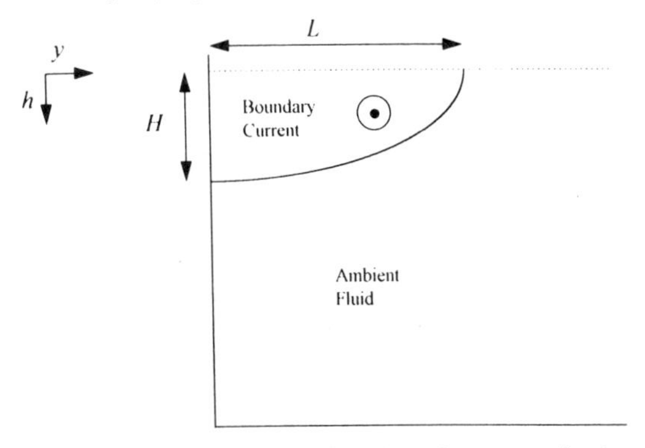

Figure 19 Schematic diagram of a density-driven boundary current flowing against a vertical wall.

As hinted above, these flows are important for carrying water from one sea to another or from a freshwater system to the marine system. They often occur where there is only a restricted passage between two open basins. In the Greenland Current, they carry cold Arctic water and icebergs into the Atlantic. In the Algerian Current, they carry Atlantic Water into the interior of the Mediterranean. As such, they are very important for the mixing and circulation processes of the oceans and the physical, chemical and biological matter which they transport. Often, these currents become unstable, in as much as their boundaries become convoluted and enhance the mixing between the boundary current water and that in the ambient flow - indeed this is often the main form of mixing between the two water bodies. (see Griffiths and Linden, 1981).

Flow around headlands and in bays

All flows which occur in the coastal zone must interact with the coastline itself to some extent. We have considered the behaviour of flows thus far mainly on perfectly flat coastlines. Let us look now at what effect headlands (protrusions from the coast) and embayments (indentations in the coast) have on coastal flows.

Some of these effects are illustrated qualitatively in Figure 20. It is clear that the width of bays and the positions of headlands and bays with respect to the coastal flow will have a significant effect in determining the extent to which flow is maintained within the bay, and therefore how well pollution is flushed from it.

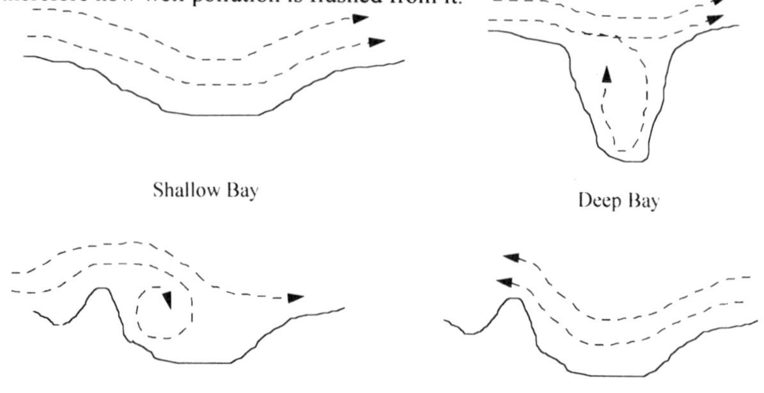

Figure 20 The effects of various configurations of headland and embayments on along-shore currents.

Many studies have been made of flows around headlands, because of their importance in the coastal zone (e.g. Davies *et al.* 1995 and Verron *et al.* 1991). The most obvious feature of flow around a headland is the formation of **eddies**. Signell and Geyer (1991) give the following clear explanation of why this should occur.

Consider a headland, similar to the 'Bay with Headland Upstream' case in Figure 20 but with no bay. As the flow approaches from the left, it speeds up due to the constricting effect of the headland. Within the boundary layer, close to the coastline, there is associated with this increase in speed a decrease in pressure (the Bernoulli effect). Downstream of the headland, the flow decreases again, and the pressure in the boundary layer increases. From the tip of the headland downstream, therefore there is within the boundary layer an adverse pressure gradient, which slows the flow down and then reverses it. The point at which the flow changes direction is called the point of separation since at this point a streamline separates from the boundary layer and extends into the interior of the flow. Continuity considerations, which match the flow in the boundary layer with the main flow then lead to an eddy in the lee of the headland.

In the common cases where tidal flows are dominant, an eddy may be formed on either side of the headland, depending on whether the tide is ebbing or flowing. In between, as the tidal flow gradually slackens off, the point of separation moves up towards the tip of the headland (decreased flow speed means that the adverse pressure gradient can reverse it more quickly). As the change in direction of the tide approaches, the eddy will be very close to the tip of the headland. Fishermen often report that the tides turn first at the tip of headlands, and this process appears to be the reason for this observation.

Flow in estuaries

One of the most important aspects of the coastal zone is the presence of estuaries. These are areas of both intense human activity, and great ecological diversity and fragility. As a result, there has been a huge amount of research concerning estuaries over many years. Excellent introductions are given by Dyer (1973) and McDowell and O'Connor (1977). Since we are concerned here only with fluid dynamics (rather than sediment transport or chemical or biological fluxes), we will look here at the dynamic balance in estuaries.

What we mean by the dynamic balance of estuaries is essentially the equations of motion in longitudinal, lateral and vertical directions. The general equation of motion relates the acceleration of a fluid particle to the forces which cause it. These are inertial, rotational, pressure gradient and frictional forces. The overall equation can be written as

$$\frac{\partial \mathbf{u}}{\partial t} + \mathbf{u} \cdot \nabla \mathbf{u} = -\frac{\nabla p}{\rho} - (\mathbf{f} \times \mathbf{u} + \nu \nabla^2 \mathbf{u} + \mathbf{F} \qquad (86)$$

where f is the Coriolis parameter, ν is the coefficient of viscosity, and F represents body forces. In the vertical, (86) reduces to

$$\frac{\partial w}{\partial t} + u\frac{\partial w}{\partial x} + v\frac{\partial w}{\partial y} + w\frac{\partial w}{\partial z} = -\frac{1}{\rho}\frac{\partial p}{\partial z} + \nu\left(\frac{\partial^2 w}{\partial x^2} + \frac{\partial^2 w}{\partial y^2} + \frac{\partial^2 w}{\partial z^2}\right) + g \qquad (87)$$

Usually, averaged vertical accelerations are relatively very small, and vertical stresses are also considered negligible, so this reduces to the hydrostatic approximation

$$\frac{1}{\rho}\frac{\partial p}{\partial z} = g. \qquad (88)$$

To determine the longitudinal dynamic balance, we break the velocities down into three constituents: the mean over a tidal cycle ($\bar{u}$) a sinusoidally changing fluctuation over a tidal cycle (U) and a turbulent fluctuation (u') as shown in Figure 21. The longitudinal equation of motion then becomes

$$\begin{aligned}\frac{\partial \bar{u}}{\partial t} + \bar{u}\frac{\partial \bar{u}}{\partial x} + \bar{v}\frac{\partial \bar{u}}{\partial y} + \bar{w}\frac{\partial \bar{u}}{\partial z} + \frac{\partial}{\partial x}(\overline{UU}) + \frac{\partial}{\partial y}(\overline{UV}) + \frac{\partial}{\partial z}(\overline{UW}) = \\ -\overline{\frac{1}{\rho}\frac{\partial p}{\partial x}} + f\bar{v} - \frac{\partial}{\partial x}(\overline{u'u'}) - \frac{\partial}{\partial y}(\overline{u'v'}) - \frac{\partial}{\partial z}(\overline{u'w'})\end{aligned} \qquad (89)$$

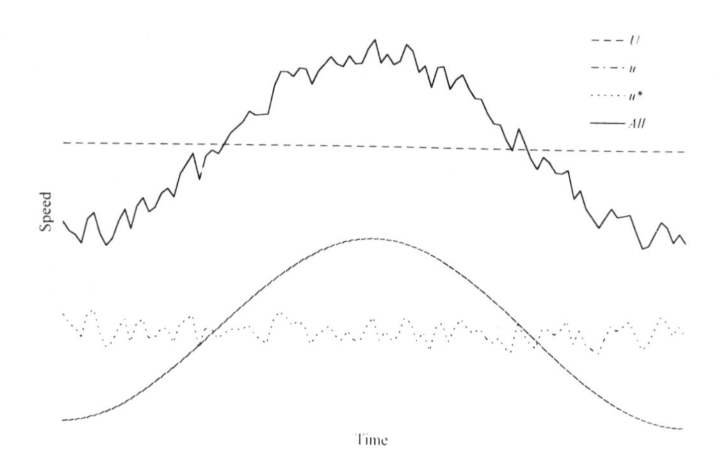

Figure 21. Definition and role of each component of velocity in estuarine flow fields.

Not every possible cross-product appears here, because the equation has been averaged over a tidal cycle, and therefore where terms average to zero over this time (for example U, V, and W) or there is no reason for multiplied and averaged terms to correlate (in which case their product is assumed to have no net effect, and therefore to average out to zero) they have been left out.

The first four terms on the left hand side give the acceleration of a particle of fluid. The next three terms represent net effects of the tidal flow. The last three terms on the right hand side represent turbulent stress, which will be much larger than molecular stress.

From here on, use of this equation becomes a case of deciding what can and cannot be assumed negligible. For example, assumed lateral homogeneity and no lateral motion (a reasonable assumption in a long, thin estuary) eliminates all the terms with any kind of v in them. This process usually depends on the specific state of the estuary which is being considered. Dyer (1973) gives several examples.

Similarly, the lateral balance can be written as

$$\begin{aligned}\frac{\partial \bar{v}}{\partial t} + \bar{u}\frac{\partial \bar{v}}{\partial x} + \bar{v}\frac{\partial \bar{v}}{\partial y} + \bar{w}\frac{\partial \bar{v}}{\partial z} + \frac{\partial}{\partial x}(\overline{VU}) + \frac{\partial}{\partial y}(\overline{VV}) + \frac{\partial}{\partial z}(\overline{VW}) = \\ -\overline{\frac{1}{\rho}\frac{\partial p}{\partial y}} + f\bar{u} - \frac{\partial}{\partial x}(\overline{v'u'}) - \frac{\partial}{\partial y}(\overline{v'v'}) - \frac{\partial}{\partial z}(\overline{v'w'})\end{aligned} \tag{90}$$

Coastal upwelling

We have considered in quite some detail motions which travel along the coastline and perpendicular to it, but have said little about vertical motions in the coastal zone. One form of vertical motion - coastal upwelling - is often a very important aspect of coastal circulation, as well as driving the economic well-being of the coastal community.

Coastal upwelling occurs when surface water is driven away from the coast, and is replaced (by continuity) by the water from below it. This deeper water tends to be cooler, less polluted but nutrient-rich because of the flux of dead organic material from the surface. By bringing these nutrients up into the eutrophic zone, the marine food chain is maintained and important fisheries are found in these upwelled waters. In fact, Ryther (1969) estimates that upwelling regions, which comprise ~0.1% of the world's oceans, produce about half of the world's fish supply.

The phenomenon of upwelling occurs when the prevailing wind blows along a coastline with the coast to the left (or to the right in the southern hemisphere, although for clarity in the rest of this argument, we shall assume we are in the northern hemisphere). A good example of this situation may be found on the Atlantic coast of Portugal, where the Azores High Pressure System maintains southerly winds. The wind stress on the surface layer of the ocean forces it initially into a southerly motion, which is modified by the Coriolis force into a motion directed partially away from the coast. This motion now acts on the next layer of ocean down. The viscous stresses only transmit part of the momentum downwards, so this second layer travels slightly slower than the surface one. The Coriolis force also acts on this motion, pushing it around further to the west.

This process continues, layer by layer, with each layer travelling slightly slower and slightly further to the right than the one above it, until the velocity becomes negligible. The overall result is a spiral of velocity, known as the **Ekman Spiral**, as shown in Figure 22. Tritton (1988) gives a clear exposition of the mathematical derivation of this formation. The overall effect is a driving of the surface waters at the coast out to sea, leading to the upwelling motion described above.

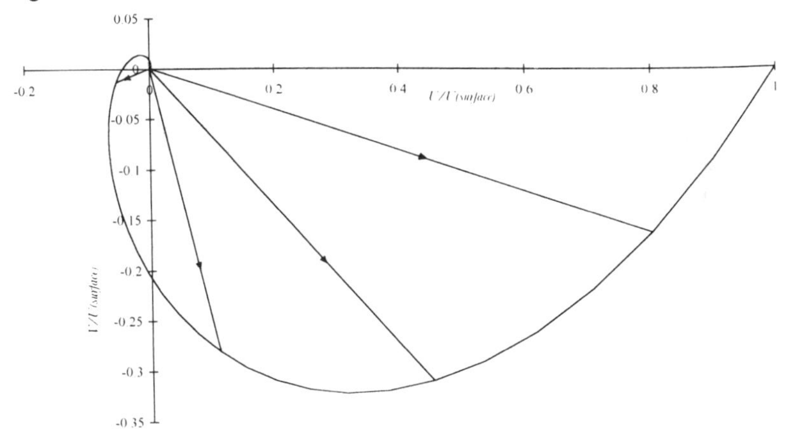

Figure 22 The Ekman Spiral of Velocities in the surface layer of the ocean, demonstrating the combined effects of rotation and friction. Note how the overall effect will force water away from the coast when the surface velocity has the coast to its left.

4.3 Fronts in the Coastal Zone

Our final topic is fronts - regions where two water masses of different physical properties meet and form a sharp gradient in those properties. Fronts are regions where intense mixing takes place, and strong vertical velocities exist. They are also often zones of convergence, and as such can be clearly demarcated by the collection of debris along their length. We shall consider three types of fronts important in the coastal zone: geostrophic fronts, coastal mixing fronts and tidal fronts.

Geostrophic Fronts

These are found at the edge of density-driven boundary currents, which were discussed in the previous subsection. Most of the flow associated with them is directed parallel to the front, but there are also secondary flows causing upwelling on the less dense side of the fronts, and down-welling on the other side, as illustrated in Figure 23. Approaching from the left will induce cyclonic rotation, causing stretching of fluid columns and thereby causing down-welling. The opposite effect on the other side of the front will cause upwelling. These two processes combined will lead to a secondary, cross-frontal circulation at the front. The upwelling brings nutrients to the surface, and these regions are often found to support a high level of biological activity from plankton up to large fish, sea birds and marine mammals.

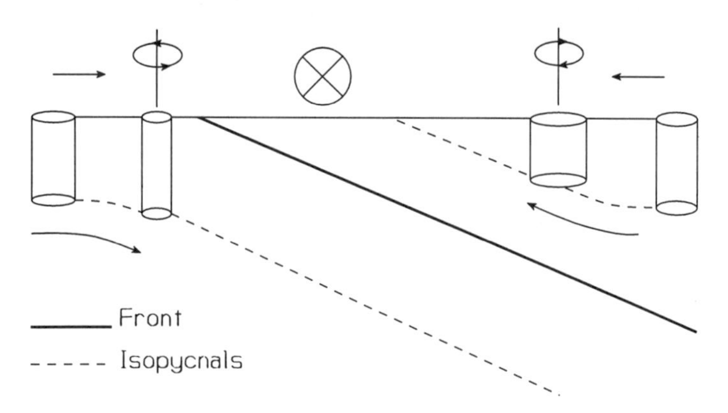

Figure 23 Schematic diagram of a cross-section of a geostrophic front.

Coastal Fronts

These form the boundary between the region close to the coast which is shallow enough to be well-mixed by tidal forces throughout its depth, and the deeper regions of the open sea. Well-mixed water will have a colder surface, and hence these fronts can be easily identified on satellite infra-red imagery. These fronts generally have a structure of well-mixed water close to the coast, which has a density of approximately the mid-depth water in the unmixed region. Thus the isopycnals (line of equal density) are turned upwards (and downwards) and form a sharp front at the surface. Planktonic matter (that which cannot propel itself) often tends to flow along isopycnals, so this type of front again causes nutrient upwelling which leads to a high concentration of macrofauna.

These fronts can be accurately identified using the **stratification parameter**

$$S_H = \log_{10}\left(\frac{H}{U^3}\right) \tag{91}$$

where H is the water depth (in metres) and U is the surface tidal speed at the peak flow of a mean spring tide in metres per second. This is best thought of as a ratio of the potential energy of a well-mixed column of water (which is proportional to H) and the energy dissipated by tidal flow (which is proportional to U^3). Low values of S_H correspond to well-mixed water. Fronts appear to form where $S_H \sim 2$.

Tidal Fronts

Tidal fronts form in estuaries or other restricted areas where salt and freshwater meet. They represent the boundary between the landward-flowing saltwater, and the seaward-flowing freshwater. In the absence of rotation, this boundary would be horizontal and would be entirely below the surface, between the denser saltwater below and the freshwater above. However, where estuaries are wide enough to allow rotation to have a significant effect, the front is tilted and often intersects the surface. Like the other types of fronts, they are zones of convergence and a line of debris, called a **foamline** or a **tideway**, can often be observed.

Since these fronts only form when there is a significant landward flow of saltwater, they are most usually observed on flood tides. The formation of the fronts is often very complex. Often two fronts from either side of the estuary can come together to form a V-shaped intrusion. Very good examples of these fronts are easily visible in the Tay Estuary at Dundee from either the road or the rail bridge - examples of their locations are shown in Figure 32. The location of the foamlines changes with stage of the tide. Recent studies of the tidal front system in the Tay are described in Ferrier and Anderson (1997).

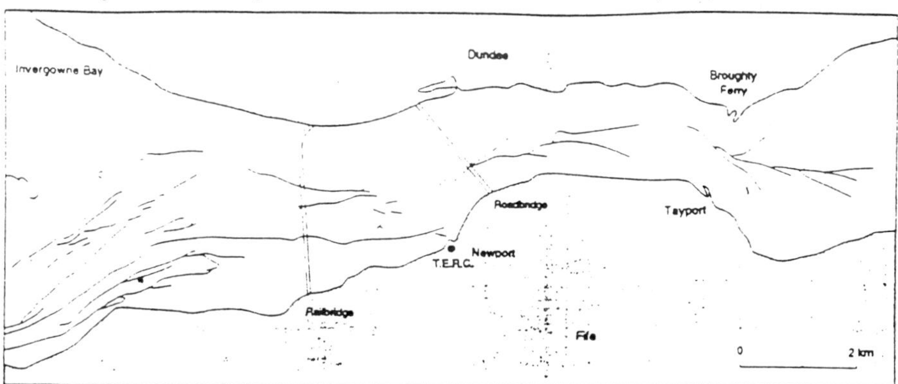

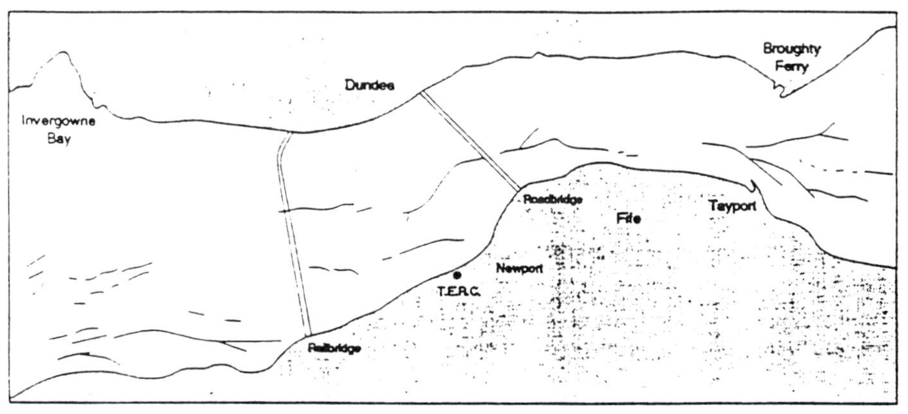

Figure 24 Typical foamlines (delineating tidal fronts) in the Tay Estuary. Note the complexity of the system, and the tendency for fronts to join together to form V-shaped intrusions. The two pictures were taken at different stages of the flood tide.

References

Accad Y and Pekeris C P, 1978, Solution of the tidal equations for the M_2 and S_2 tides in the world oceans from a knowledge of tidal potential alone, Philosophical Transactions of the Royal Society of London, Series A, **290**, 235-266

Beer T, 1983, Environmental Oceanography, Pergamon Press, Oxford

Candela J, 1991, The Gibraltar Strait and its role in the dynamics of the Mediterranean Sea, Dynamics of Atmospheres and Oceans, **15**, 267-299

Candela J, Winant C and Ruiz A, 1990, Tides in the Strait of Gibraltar, Journal of Geophysical Research, **95**, C5, 7313-7335

Davies P A, Dakin J M and Falconer R A, 1995, Eddy Formation Behind a Coastal Headland, Journal of Coastal Research, **11**, 1, 154-167

Dyer K R, 1973, Estuaries: A Physical Introduction, John Wiley & Sons, London

Ferrier G and Anderson J M, 1997, The application of remotely-sensed data in the study of frontal systems in the Tay Estuary, Scotland, UK, International Journal of Remote Sensing, **18**, 9, 2035-2065

Gill A E, 1982, Atmosphere-Ocean Dynamics, Academic Press, London

Godin G, 1972, The Analysis of Tides, Liverpool University Press, Liverpool

Griffiths R W, 1986, Gravity Currents in Rotating Systems, Annual Review of Fluid Mechanics, **18**, 59-89

Griffiths R W and Linden P F, 1981, The stability of buoyancy-driven coastal currents, Dynamics of Atmospheres and Oceans, **5**, 281-306

Jonsson I G, 1975, The wave friction factor revisited, Technical University of Denmark, Institute of Hydrographics and Hydraulic Engineering Progress Report 37

Lamb H, 1932, Hydrodynamics, 6th ed., Cambridge University Press, London

Longuet-Higgins M S, 1970, Longshore currents generated by obliquely incident sea waves, Journal of Geophysical Research, **75**, 6778-6801

McDowell D M and O'Connor B A, 1977, Hydraulic Behaviour of Estuaries, The MacMillan Press, London

Muir-Wood A M and Fleming C A, 1981, Coastal Hydraulics, 2nd edition, The MacMillan Press, London

Ryther J H, 1969, Photosynthesis and fish production in the sea, Science, **166**, 72-76

Sears F W, Zemansky M W and Young H D, 1982, University Physics, 6th edition, Addison-Wesley, Reading, Massachusetts, USA

Signell R P and Geyer W R, 1991, Transient Eddy Formation Around Headlands, Journal of Geophysical Research, **96**, C2, 2561-2575

Taylor J R, 1993, Turbulence and Mixing in the Boundary-Layer Generated by Shoaling Internal Waves, Dynamics of Atmospheres and Oceans, **19**, 1-4, 233-258

Thorpe S A, 1994, Observations of Parametric-Instability and Breaking Waves in an Oscillating Tilted Tube, Journal of Fluid Mechanics, **261**, 33-45

Tritton D J, 1988, Physical Fluid Dynamics, Oxford University Press, Oxford

Verron J, Davies P A and Dakin J M, 1991, Quasigeostrophic flow past a cape in a homogeneous fluid, Fluid Dynamics Research, 7, 1, 1-21

Wu T Y and Lin D M, 1994, Oceanic Internal Waves - Their Run-Up on a Sloping Seabed, Physica D, 77, 1-3, 97-107

Mathematical modelling of coastal processes

Ping Dong

University of Dundee

1. Introduction

The coastal water environment including both the sea and estuaries has long been recognised as among the most valuable natural resources. Over the centuries, it has supported many vital industries such as transportation, trade, fisheries and, more recently, power generation and recreation. With the rapid population growth in the coast region, increasing threat of flooding due to long-term sea level rises and the need of environmental conservation, it is recognised that coastal management can no longer be confined within the narrow engineering domain and conducted in a piecemeal manner. It has to be not only more rational but also more regionally based. The central issue within modern coastal management strategy is to resolve the conflict between development and preservation of coastal zones in order to ensure their long-term sustainability.

Although the argument for development is a convincing one, to achieve sustainability is not an easy task. Apart from socio-economic issues, the natural coastal environment also poses some of the most difficult scientific and engineering problems. The coastal zone is highly dynamic, affected constantly by wind, waves and currents. The flows and the movement of water-borne substances exhibit both systematic trends and large irregularities in space and time. Interactions between the sea and the land result in various bed forms, some of which tend to evolve at a slow rate such as undersea sand banks while others can respond rapidly to water movements such as ripple formation and beach profile changes. It is this complexity and interdependency of the physical processes that makes the use of numerical models necessary. These models are developed with the understanding of the varying temporal and spatial scales involved in coastal processes and, once developed, will help to crystallise this understanding by making quantitative predictions.

For example, in order to develop long-term coastal management plans it is necessary to make predictions about the future changes of the coastline that may occur naturally or as the result of human interventions. In some natural coasts where suitable long-term-data are available, observational evidence alone may be sufficient for making useful predictions by standard data extrapolation techniques. A recent review of this area is given by Reeves and Fleming (1997). However, in most cases where either the measured data are scarce or the coastal evolution is known to be influenced by long-term changes in the environment, including that caused by man-made coastal structures, more process-based analysis techniques must be used. The techniques used for such predictions would involve numerical models of some sort, supplemented by observational evidence (historical or recent data) and occasionally physical models. The models usually deal with one or more of the fundamental processes such as waves, water levels, currents and sediment transport. In other problems, the main processes may be winds, temperature, concentration of chemicals and even flora and fauna.

Thus we see that numerical modelling of coastal processes plays an essential part in coastal engineering design and management from process analysis to scheme design. Due to the multiplicity and complexity of the physical processes involved, it is not yet possible for a single model to deal with all these processes. Therefore, a large number of numerical models have been developed, some of which are designed for solving specific problems while the others have relatively wider applicability. For nonspecialists this situation could be very confusing. This chapter is intended to help to bridge the gap between the modeller/researcher and potential model user/manager. By focusing on a number of key coastal processes it explains the mathematical basis on which these numerical models are based and the necessary simplifications commonly adopted. The usefulness as well as the limitations of these models will be assessed from the perspective of both users and researchers through a number of carefully selected examples. The discussion will be restricted to the mathematical modelling of hydrodynamic and associated morphological processes. The emphasis is on the process-based modelling approaches. Due to the page limitation we can only give an overview of the typical numerical models used in both coastal research and engineering. Both established modelling systems and more recent developments are discussed. Mathematical equations have been deliberately avoided and the readers who are interested in them are encouraged to consult the text books and technical papers given in the references.

2. Numerical models and physical reality

Numerical models used in studying coastal hydrodynamics and morphology can range in complexity from simple, highly parametrised models to detailed process-based models which require solving the dynamic equations in three dimensions. However, as it can be appreciated, no matter how simple or complicated the numerical models may be, they can only represent the physical reality to a certain degree. It is not possible to develop a model that includes every detail in a single coastal process, let alone all coastal processes. Therefore, it is useful to know more about how a model is constructed and what predictions actually mean.

As a general rule with few exceptions, coastal model development typically involves a combination of rigorous idealised models for specific simple processes, physically motivated approximations and parametrisations for more complicated processes, and empirically based "constants" for processes not yet fully understood. As a consequence, the accuracy is usually achieved at the expense of the versatility the model can attain. This conflict between the accuracy in model predictions and complexity that the models intend to resolve has been elegantly demonstrated by Hunt (1991) in a schematic diagram as shown in Figure 1. Although this diagram was originally devised in the context of fluid mechanics, it is believed to be equally applicable to coastal process modelling. The different curves in the diagram correspond to how specific the flows are and how much the approach depends on the working assumptions. Each line corresponds to a particular technique and the fuzzy edge defines how far the technique can be applied to more complex flows. By applying these methods across this line their limitations can be established and improved methods can be developed to give resulting lines which are further on the right of the existing lines.

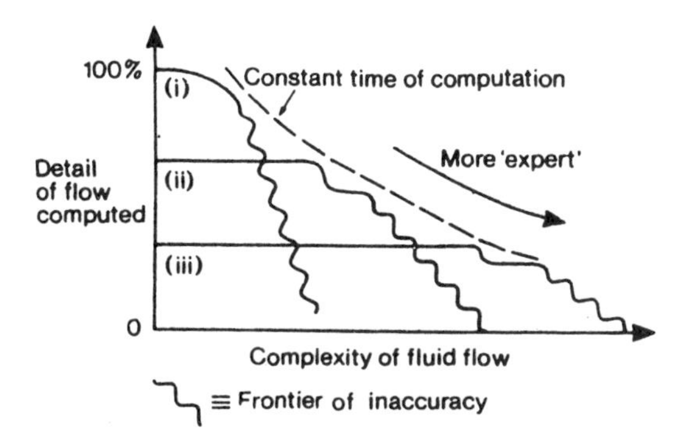

Figure 1. Schematic diagram showing computer modelling and the physics they represent (Hunt 1991)

In the context of coastal process modelling, one of the key considerations in model development is that the model resolutions should be appropriate to the temporal and spatial scales of the processes they are intended to represent. Similarly, in practical applications, coastal scientists and engineers must select the best available models for solving a specific problem as more sophisticated models do not necessarily give more accurate results. To use the models appropriately one needs to know on what principles these models are based, what are their ranges of applicability, what data are required as input to the models and what outcomes would constitute 'correct' results.

Finally, it should be realised that progress in the development and testing of numerical models can frequently far outpace the improvement of the knowledge of physical processes, which is particularly true for sediment transport, turbulence and bed friction. This can be seen by a quick examination of published materials in the open literature in which there are many papers describing new methods of model building and new schemes for numerical solutions but far fewer describing systematic comparison of computed coastal processes with good quality field data. Therefore, it is necessary to guard against the tendency of using numerical models as a blunt instrument merely for getting quick answers to problems without understanding the limitations and potential errors inherent in the model predictions.

3. Surface waves

Most coastal processes are directly or indirectly the result of wave action. An understanding of wave phenomena including wind generation and propagation is therefore fundamental to the study of shoreline processes.

3.1 Wave generation

In order to determine wave conditions at a nearshore location, it is necessary to know the wave characteristics in deep water where the effects of the sea bed on wave propagation are small in comparison with those of the wind. Since wind waves are choppy and irregular, consisting of waves of different sizes, origins and frequencies that superpose on each other, offshore wave conditions derived by numerical models are usually in the form of a 'directional spectrum' which specifies the distribution of wave energy in both frequency and direction. Over a suitably small region, the statistical properties of the waves can be regarded as spatially homogeneous. It is then sufficient to consider wave properties at a single point within this region, concentrating on their variation in time with wind forcing. However, over a large region or when land boundaries exert significant influence on the wave growth, it would be necessary to determine waves at a number of points.

Over the past 40 years, various wind-wave prediction models have been developed and routinely used in coastal engineering practice. The key components in such models are empirically-determined directional wave spectra which provide the link between wind input (speed, direction, duration and fetch) and wave energy. The most widely used spectrum for the North Sea is that put forward during the Joint North Sea Wave Project, commonly known as the JONSWAP spectrum. Among the models which use the JONSWAP expression to predict wave conditions from wind data are the JONSEY and hindwave models developed by Hydraulics Research Wallingford. These kinds of models, characterised by empirical parametrisation of spectral shape, are commonly referred to as the first- and second-generation wind-wave models. In practical applications it may be necessary to calibrate and fine-tune the models using simultaneous measurements of wind and wave data for a short period (e.g. 1 year). Once the model is properly set-up and calibrated, the wave time series can be generated for any given wind records. If the input wind data are of sufficient length to contain long-term statistical information, the statistical properties of the waves can be derived and used to predict more extreme events by extrapolating the data to the levels required.

Although the first- and second-generation models are still popular because of their computational efficiency, major shortcomings of these two groups of models were demonstrated during a comparison study among nine such models by the SWAMP group (SWAMP Group 1985). Two notable examples of shortcomings are a limitation to deep water and a small temporal gradient in the wind field. This study motivated the development of third-generation models that remove any restrictions on the spectral shape (Hasselmann *et al.* 1988; Cavaleri *et al.* 1989). These models are based on the so called radiative-transfer equation which governs the evolution of the wave energy spectrum. All the external sources and sinks of wave energy are specified as the summation of a number of separate influences such as the energy input from the wind, energy dissipation and nonlinear wave-wave interaction. The equation is solved on a predetermined computational grid system and its output are the time series of wave spectral at each grid point. The wind-wave model operated by the UK Meteorological Office is such a model. It is run routinely to give predictions of hourly wave time series and monthly wave statistics on a 30km model grid system covering the entire waters around the British Isles. Although the third-generation models have proved to be successful during numerous hindcasting studies, many uncertainties still remain to be resolved. Processes such as nonlinear interactions and energy dissipation due to breaking waves still require empirical corrections to fit field data sets. The considerable computational effort require to run such models makes it impractical for most standard engineering applications.

3.2 Wave propagation

As distant storm waves leave the generation zone and enter into nearshore waters, they will undergo various changes through shoaling, refraction, diffraction, reflection and ultimately breaking. Nonlinearity and wave-current interaction become more important. Due to the complexity of wave propagation, modern research on regular and irregular wave propagation is being conducted in parallel. Generally models are initially developed for regular waves to study the behaviour of such waves in detail. Once sufficiently refined, the models are then extended to irregular waves.

For monochromatic waves, early models were based on the ray theory which accounts for wave shoaling and refraction due to changes in bathymetry or steady current fields. Since models of this type involve the solution of only first-order equations they can be applied efficiently to a large area of coastal water. A typical example of the results of such calculations is shown in Figure 2.

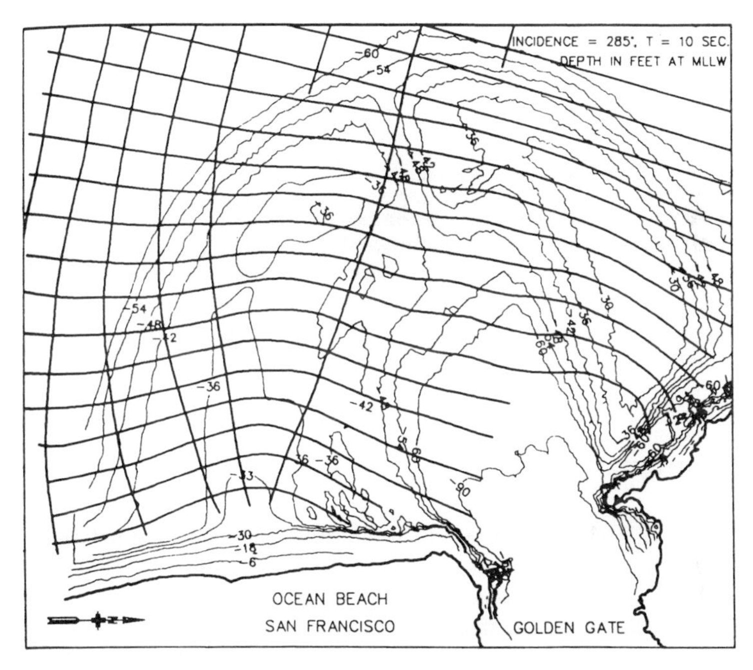

Figure 2. Wave refraction over San Francisco Bay (From Battalio and Trivedi 1996)

However when the bathymetry is complicated the ray theory can break down giving unrealistic predictions in regions where ray crossing occurs. This is due to the neglect of diffraction in the theory. This problem has been overcome by the development of models based on the mild-slope equation (Berkhoff 1973) which allow both wave refraction, diffraction and reflection. Such models are widely used in port and harbour design. Since the mild-slope equation is in elliptic form, its numerical solution is time-consuming. In order to overcome this problem various parabolic equations were derived from the original equation by retaining the effects of refraction and diffraction but neglecting reflection. As a result, efficient forward marching methods can be used instead of the more computationally intensive methods involving the solution of the whole wave field simultaneously. The early parabolic models are restricted to small wave angles of incidence while the more recent development (Dalrymple and Kirby 1988, Dalrymple *et al.* 1989 and Li 1997a) has largely overcome this problem allowing the wave angles up to ± 70 degrees.

Despite the success and popularity of linear wave models their suitability in shallow water has been constantly questioned. This is because the nonlinear effects that are neglected in linear wave theory can be significant in shallow water with wave profiles exhibiting pronounced vertical or horizontal asymmetries. Weakly nonlinear propagating waves techniques have been developed to include nonlinear terms in the parabolic equations (Kirby and Dalrymple 1993). Nonlinear versions of mild-slope equations have also been reported by Nadaoka *et al.* (1994), Isobe (1994) and Tsay *et al.* (1996). A more advanced modelling system suitable for general wave transformation prediction in fairly shallow water is based on the two-dimensional Boussinesq equations. These models are generally capable of reproducing the combined effects of most wave phenomena of interest such as shoaling, refraction, diffraction, reflection and complex sea bed bathymetries. The more recent versions can also deal with irregular waves (Freilich and Guza 1984) and have improved dispersion characteristics (Nwogu 1994, Chan and Liu 1995). This last property is important since the new Boussinesq equations do not have the same depth restrictions as that of the original equations; they can be applied to water of much greater depth. Despite their popularity, Boussinesq equations can only deal with weak nonlinearity and they cannot be used to model waves that are near breaking.

Recently a full three-dimensional model for regular waves was developed (Li, 1997b) which solves the Laplace equation using multigrade techniques. This model does not require the bed slope to be slowly varying as in the case of the mild-slope equation or Boussinesq equation models. Initial tests have demonstrated its accuracy and potential appeal to engineering applications. However, considerable research is required to subject it to the tests of simulating much stronger nonlinear waves, to include the effects of energy dissipation and to extend the model to irregular waves.

4. Coastal circulation

Coastal currents or circulation usually exist as a mixture of different types of aperiodic and periodic water movements. Typical examples are tidal currents, wind-induced circulation, density currents and wave-induced currents near the coastlines. These currents are responsible for the transport of heat and matter of all kinds such as liquid pollutants and sediment particles. Due to the strong influence of land and sea bed boundaries on the water movements, variability of speed and direction of the coastal currents can be significant in both space and time.

4.1 Large-scale currents

The expression large-scale currents refers to those currents which vary very slowly compared with surface waves. Typical examples are tidal currents and wind-induced circulation. Traditionally, numerical models for simulating large-scale water motions are based on the shallow water equations in depth-averaged form. These equations were derived by integrating the Navier-Stokes equations over the water depth assuming the pressure to be hydrostatic. The equations can be solved using various numerical techniques (finite element, finite different, boundary element etc.) to predict the surface elevation and two components of the horizontal velocity. Such models have been routinely used to calculate tidal currents and wind-induced currents at any point in time and in space. A vast amount of literature is available on this subject (see Abbott 1979 for a review). Although the accuracy of predictions at a particular coast still depends on the appropriate specification of forcing terms together with suitable boundary and initial conditions, the depth-averaged models are generally considered adequate if the flow conditions change fairly slowly in space and time (i.e. there are no sharp variations in bed levels or in wind speeds/directions).

However, for those situations where flow conditions change rapidly due to either the wind forcing or bottom topography, it is much more appropriate and accurate to use a 3-dimensional model which resolves the variations in the vertical dimension as well as in horizontal dimensions. Computations based on the 3-dimensional shallow water equations are more recent developments and noticeable progress has been made in developing numerical techniques for solving these equations. Figure 3 shows a typical example of computations by such a model.

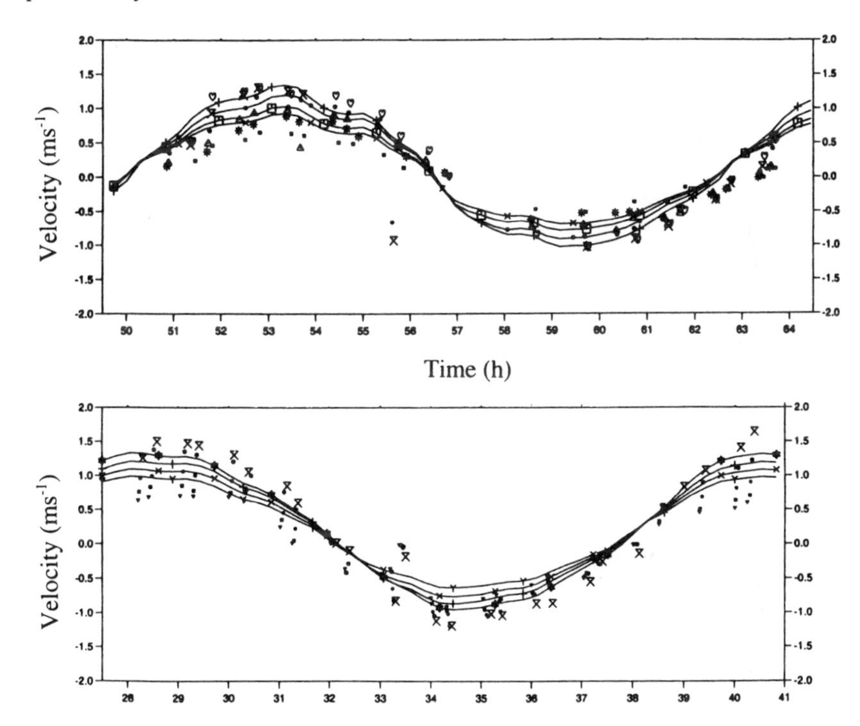

Figure 3. Comparison of predicted (solid lines) and data measured for flow velocities at the surface and down to depths of 8m for (a) Sunk Channel and (b) Middle Shoal in the River Humber (See Lin and Falconer, 1997, for details)

Clearly the predictions are quite good qualitatively and reasonable quantitatively. The main difficulties in making further progress must lie in the better representation of turbulence mixing in the water column and detail representation of bed frictions. For further information on the subject see Stansby and Lloyd 1995 or Lin and Falconer 1997.

4.2 Wave-induced currents

There are three wave-induced current systems in the nearshore zone which dominate the water movements in addition to the to- and fro- motions produced by waves directly. These are

- a cell circulation system of rip currents and associated longshore currents,
- longshore currents produced by oblique wave approach to the shore and

- vertical circulation of water commonly referred as 'undertows' caused by mean water level gradients.

All three current systems owe their origin to the wave breaking which converts part of the mechanical energy carried by the organised wave motions to mean currents.

Over the past thirty years, significant advances have been made in the development of a rational basis for studying wave-induced currents because of their importance in transporting sediments on the beaches. The standard modelling approach is based on a set of time-averaged and depth-integrated equations of motion for waves and currents. Using the radiation stress concept (Longuet-Higgins 1964) and linear wave theory, together with empirical treatments of breaking dissipation, these equations are capable of reproducing most of the important phenomena in the nearshore region, such as wave set-up and set-down, longshore currents and rip currents. In the past ten years, following the work of Svendsen (1984), the theoretical formulation for undertow induced by normal incident waves has been accomplished. Recent research has been focused on extending the radiation stress formulation to three-dimensional situations (Koutitas 1986, Li and Maddrell 1994) or quasi-3-dimensional situation (Southgate and Nairn 1993). Due to the difficulty of specifying the spatial distribution of breaking waves and turbulence, a satisfactory three-dimensional model for arbitrary bottom bathymetries is yet to be developed. Its application to practical engineering problems is still long way off.

5. Transport processes

Coastal and estuarine flows have great capacity for transporting mass, momentum, heat and water-borne substance of all kinds. Tidal currents, wind-driven currents and surface waves all play important parts, although the temporal and spatial scales on which they operate can be quite different.

5.1 Transport of passive substance

Models for predicting the transport of passive substances such as pollutants and temperature are generally referred to as water quality models. These models attempt to simulate changes in the concentration of substances in space and in time. If the substances are sufficiently inert for their concentration to be regarded as unchanging except by dilution, they are referred to as conservative substances. However, many substances in water are subject to change in concentration due to physical, chemical and biological processes. Therefore, a general water quality modelling system will consist of equations governing fluid flows together with scalar transport equations which govern the concentration of organisms, temperature, salinity and biological/chemical reactions. Among these, temperature and salinity distribution are the easiest to deal with, while biological and chemical processes are the least predictable.

Over the past twenty years, a large number of water quality models have been developed for predicting water quality in lakes, estuaries and coastal oceans. Most of these models are 2-dimensional models but more recent ones are 3-dimensional models. Traditionally all the processes are treated as deterministic but statistical approaches are gradually gaining popularity. The key problem faced in model development is the existence of a large number of quality criteria to be considered and in most cases the level of each criterion is the result of complex interactions. Since these interactions can involve a wide range of temporal and spatial scales it is not easy to determine a priori the appropriate scales that models need to resolve. The situation is further exacerbated by the difficulties of any experimental approach to forecasting water quality. All these problems mean that most water quality models are more capable of predicting the relative changes of various variables rather than predicting absolute values.

5.2 Transport of heavy particles

Sediment transport plays a vital role in many aspects of marine engineering. The movement of sediments can influence the viability and economy of harbour construction, the quality of water entering the cooling water intakes of power stations, the flushing capacity of storm water sewers and the stability of coastlines and offshore installations such as pipelines due to erosion and scours. Other aspects relating to coastal sediment dynamics include coastal morphology and the transport of heavy metals and toxic waste via absorption of contaminants by sediment particles. Therefore, it is essential to have a good understanding of sediment transport mechanisms and the capability to predict these processes as accurately as practicable.

Broadly speaking, there are three types of sediment transport methods, apart from purely empirical ones, that form the basis of present time sediment transport modelling.

- The first is based on the estimation of shear stresses at the bed assuming that bed load transport and sediment concentration respond instantaneously to, and are a direct function of, these stresses. Once the shear stresses are determined at each grid point using wave or current hydrodynamic models, the sediment transport rate at that point can be evaluated.
- The second method is based on energy flux or stream power concept. The total work done on a unit area of stream bed in unit time has two components; one is work done in keeping material in suspension and in overcoming resistance to bed load movement. The other is work done in overcoming bed resistance not associated with bed movement. The sediment transport rate is formulated as a function of the former fraction of the work.
- The third method is to model sediment transport in a way similar to that for any other water-borne substance. Sediment concentration is determined at each grid point with specific treatment of the exchange of particles between the mobile bed and water column. The volumetric transport rate is calculated as the product of the local flow velocity and sediment concentration. Clearly this method is only applicable to suspended sediment transport.

Within each of the above three methods many formulations with varying degrees of sophistication have been developed which are applicable to situations such as current only, wave only and combined wave and currents. The physical basis and experimental validations behind these formulations can be found in a number of excellent monographs on the subject (Fredsoe and Deigaard 1992, Nielsen 1992). For practical applications the book by Soulsby (1997) is most useful; it also contains some new formulae developed recently at Hydraulics Research Wallingford, UK.

The most frequently used coastal sediment transport models are primarily two dimensional in plan. These models involve solving the depth-integrated or depth-averaged two dimensional sediment transport equations to describe the suspended sediment transport. In most models (eg. Galappatti & Vreugdenhil, 1985) the bed load was determined empirically using local hydrodynamic and sediment parameters. In order to take into account the nonlocal effect of sediment exchange at the near bed region both bed load particles and suspended particles should be treated as transport processes. A model of this kind was proposed by Li and Chesher (1997) which solves two coupled transport equations, one for suspended particles and the other for bed load particles. Both theoretical and experimental evidence has shown that two-dimensional models are fairly accurate if the flow conditions change slowly in space. In situations where flow conditions change rapidly, the fundamental assumption that the vertical scale is much smaller than the horizontal scale will be violated. It is more appropriate to use a 3-dimensional model in which the local suspended sediment distribution is dependent on both horizontal and vertical mass transport rates. A number of models are now available in the literature which contain similar physics but vary greatly in the numerical techniques used.

The most recent ones include that by Olsen and Skoglund (1994) and Lin and Falconer (1997), both of which used finite-difference schemes. Judging from the limited published results so far the predictions by these models seem to agree rather well with the laboratory data but only qualitatively with the field measurements as shown in Figure 4.

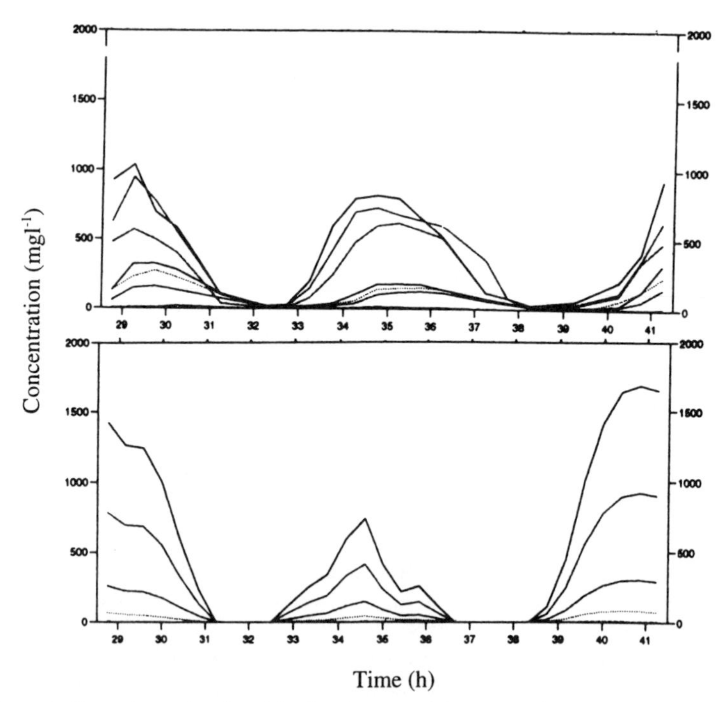

Figure 4. Comparison of predicted (top) and measured (bottom) sediment concentration for River Humber over a period of 13 hours. (From Lin and Falconer 1997). The upper curve refers to the bed (depth approximately 5m) and the lowest refers to the surface;

6. Morphological changes

Coastal morphology is constantly shaped by natural processes such as wind, waves and currents. Understanding these changes and their long-term impact on the environment are of vital importance for meeting the conflict demands on coastal resources from development and conservation.

Before attempting to make predictions of future shoreline changes it must be realised that the present shorelines of the world, however, are not the result of present day processes alone. Nearly all coasts were profoundly affected by the rise in sea level caused by the melting of the Pleistocene glaciers, between 15,000 and 20,000 years ago. The rising sea flooded large parts of the low coastal areas, and shorelines moved inland over landscapes formed by continental processes. The configuration of a given coastline may, therefore, be largely the result of processes other than marine, and may originally have been shaped by stream erosion or deposition, glaciation, volcanism or Earth movement. Therefore, so-called long-term mathematical modelling should be understood in the context of engineering rather than geology. It is largely based on the present day processes.

Morphological evolution models can be classified into various categories. Broadly speaking the models could be either parametric or physical process-based. Parametric models consist of a set of empirical relationships that characterise the observational trends of a particular morphological process, be it beach profile, shoreline or bathymetry. Most of such models are site-specific and have only limited predictive capabilities. Contrary to the parametric models the process-based models are based on general physical laws, at least partially, and therefore potentially have much wider applicability. The most commonly used morphological evolution models are (i) Coastal Profile Models which predict the movement of a single cross-shore profile, (ii) Beach Planshape Models which predict the evolution of a single shoreline, (iii) Coastal Area Models which calculate the changes of bed contours due to depth averaged wave-induced flows and (iv) 3-D Models.

All the models discussed so far have only limited applicability because they involve certain idealisations in the movement of flows and sediments. Coastal profile models cannot be applied to situations with large longshore gradients, while the coastal area models would fail in situations where the vertical profile of flow or sediment concentration is not uniform. A more complete treatment would be needed which describes some or all processes in three dimensions, especially when local problems such as scours around structures and sediment in-fill of dredged channels are concerned. In certain cases the more sophisticated models do not necessarily give better results as shown in Figure 5.

7. Conclusions

From this brief review it is clear that numerical models have become indispensable tools in coastal process analysis and management. These models can be applied to solve a wide range of problems which are either too complicated or too time consuming to be dealt with using methods other than numerical models. Furthermore the advance in computer technology, including hardware as well as and software, has made most of these models both readily available and affordable. On the other hand, great efforts are still required to improve the capability of the existing numerical models. The specific tasks will range from better understanding of nonlinear interactions and random processes through the improvement of numerical schemes for solving governing equations to the collection of good quality field data.

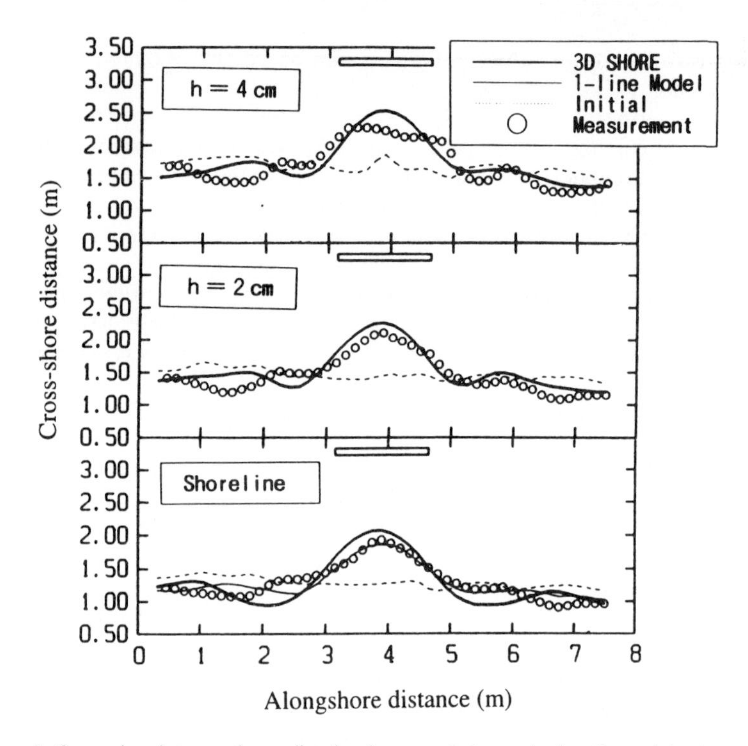

Figure 5. Comparison between the predicted and measured changes in shoreline and depth contours (after Shimizu *et al.* 1996)

The principal conclusions arising from the above review are that for model developers, greater attentions are now required to be paid to the understanding of fundamental physical processes rather than merely adjusting model parameters to fit the data while for model users it is important to appreciate the applicability of numerical models as well as their limitations. With future improvements of modelling techniques and increased use of models for dealing with real world problems, it is certain that the values of numerical models will be even more widely appreciated and the models will be used more as an aid to understanding rather than as a blunt instrument.

References

Abbott M B, 1979, Computational Hydraulics: elements of the theory of free-surface flows, Pitman, London

Battalio P E and Trivedi D, 1996, Sediment transport processes at ocean beach San Francisco, California, Proceedings of 25th Coastal Engeering Conference, **3**, pp 2691-2704

Berkhoff J C W, 1973, Computation of combined refraction-diffraction, Proceedings of the 13th Coastal Engeering Conference, Vancouver, 1972, ASCE, **1**, pp 471-490

Cavaleri L, Bertotti L, and Lionello P, 1989, Shallow water application of third generation WAM wave model, Journal of Geophysical Research, **94**, C6, 8111-8124

Chan Y and Liu P L-F, 1995, Modified Boussinesq equations and associated parabolic models for water wave propagation, Journal of Fluid Mechanichs, **288**, 351-381

Dalrymple R A and Kirby J T, 1988, Models for very wide-angle water waves and wave diffraction, Journal of Fluid Mechanics, **192**, 33-50

Dalrymple R A, Suh K D, Kirby J T and Chae J W, 1989, Models for very wide-angle water waves and wave diffraction, Part 2. Irregular bathymetry, Journal of Fluid Mechanics, **201**, 299-322

Fredsoe J and Deigaard R, 1992, Mechanics of coastal sediment transport, **3**, World Scientific Publishing, Singapore

Freilich M H and Guza R T, 1984, Nonlinear effects on shoaling surface gravity waves, Philosophical Transactions of the Royal Society of London, Series A, **311**, pp 1-41

Galappatti R and Vreugdenhil C B, 1985, A depth-integrated model for suspended sediment transport, Journal of Hydraulic Research, **23**, 745-762

Hasselmann S, Hasselmann K, Koman G K, Jonssen P, Ewing J A and Gardone V, 1988, The WAM model - A third generation ocean wave prediction model, Journal of Physical Oceanography, **15**, 1777-1810

Hunt J C R, 1991, Industrial and environmental fluid mechanics, Annual Review of Fluid Mechanics, **23**, 1-41

Isobe M, 1994, Time-dependent mild-slope equation for random waves, Proceedings of the 24th International Conference on Coastal Engeering, ASCE, pp 285-299

Kirby J T and Dalrymple R A, 1983, A parabolic equation for combined refraction-diffraction of Stokes waves by mildly varying topography, Journal of Fluid Mechanics, **136**, 153-466

Koutitas C, Kitou N and Katopodi I, 1985, A model for circulation and sediment transport by winds and waves around groyne system, International Conference on Numerical and Hydraulic Modelling of Ports and Harbours, 23-25 April 1985, organised by BHRA, Birmingham, U.K., pp 175-180

Li B, 1997a, Parabolic model for water waves, Journal of Waterway, Port, Coastal, and Ocean Engineering, **123**, 4, ASCE, 192-199

Li B, 1997b, A three dimensional multigrid model for fully nonlinear water waves, Coastal Engineering, **30**, 235-258

Li F and Chesher T J, 1997, Non-equilibrium sediment transport and seabed deformation model, submitted to 26th International Coastal Engineering Conference

Li B and Fleming C A, 1997, A three dimensional multigrid model for fully nonlinear water waves, Coastal Engineering, **30**, 235-258

Li B and Maddrell R, 1994. A three-dimensional model for wave-induced currents. Proceedings of 24th International Conference on Coastal Engineering, ASCE, Japan, 2297-2310.

Lin B and Falconer R A, 1997, Numerical modelling of three-dimensional suspended sediment for estuarine and coastal waters, Journal of Hydraulic Research, **34**, 4, 435-456

Longuet-Higgins M S and Stewart R W, 1964, Radiation stresses in water waves: a physical discussion with applications, Deep-sea Research, **11**, 529-562

Nadaoka K, Beji S and Nakagawa Y, 1994, A fully-dispersive nonlinear wave model and its numerical solution, Proceedings of the 24th International Conference on Coastal Engineering, ASCE, pp 427-441

Nielsen P, 1992, Coastal bottom boundary layers and sediment transport, Advanced Series On Ocean Engineering, **4**, World Scientific Publishing, Singapore

Nwogu O, 1994, Alternative form of Boussinesq equations for nearshore wave propagation, Journal of Waterway, Port, Coastal and Ocean Engineering, **1198**, 6, ASCE, 618-638

Olsen N R B and Skoglund M, 1994, A three-dimensional numerical model of water and sediment flow in a sand trap, Journal of Hydraulic Research, **32**, 833-844

Reeves D E and Fleming C A, 1997, A statistical-dynamical method for predicting long term coastal evolution, Coastal Engineering, **30**, 259-280

Shimizu T, Kumagai T and Watanabe A, 1996, Improved 3-D beach evolution model coupled with the shoreline model (3D-Shore), Proceedings of the 27th Coastal Engineering Conference, **3**, pp 2843-2865

Soulsby R L, 1997, Dynamics of Marine sands, A Manual for Practical Applications, Report SR 466, Hydraulic Research Wallingford

Southgate H N and Nairn R B, 1993, Deterministic profile modelling of nearshore processes, Part 1, Waves and currents, Coastal Engineering, **19**, 27-56

Stansby P K and Lloyd P M, 1995, A semi-implicit lagrangian scheme for 3D shallow water flow with a two-layer turbulence model, International Journal for Numerical Methods in Fluids, **20**, 115-133

Svendsen I A, 1984, Mass flux and undertow in a surf zone, Coastal Engineering, **8**, 347-365

SWAMP Group, 1985, Ocean wave modelling, Plenum Press, New York, N.Y.

Tsay T-K, Liu P L-F and Wu N-J, 1996, A nonlinear model for wave propagation, Proceedings of the 25th International Conference of Coastal Engineering, ASCE, pp 589-601

Remote sensing of coastal environment and resources

Vic Klemas

University of Delaware

1. Introduction

The coastal zone, as defined in Chapter 1, consists of an area of land and sea which is of not insignificant area and which is of great environmental and human importance. It plays a vital role in nutrient assimilation, geochemical cycling, water storage, and sediment stabilisation and sustains the majority of neighbouring marine finfish and shellfish resources. Unfortunately, these coastal areas are under severe threat: wetlands are being destroyed by dredge and fill operations and impoundments, estuarine and coastal waters are being contaminated by pollution and subjected to eutrophication due to the increasing concentration of people, commerce and industry either in or adjacent to this zone and fisheries are being exhausted, see Table 1. At worst, these assaults may lead to a collapse of coastal ecosystems and their natural resources, and it is urgent that coastal states develop standardised and rapid procedures for the better monitoring of these trends so that effective management can be introduced.

- Eutrophication
- Habitat modification
- Hydrologic and hydrodynamic disruption
- Exploitation of resources
- Toxic effects
- Introduction of non-indigenous species
- Global climate change and variability
- Shoreline erosion and hazardous storms
- Pathogens and toxins affecting human health

Table 1: Major coastal environmental issues

How a monitoring strategy for the coastal zone is developed depends on the investigator's objectives. A general programme encompasses (a) the monitoring of physical, chemical and biological properties of the water column, (b) the study of marine processes, such as tides, circulation patterns, etc. and (c) and the assessment of living marine resources and their habitat, including the surrounding wetlands and the uplands that can have an impact on water quality.

Specific indicators of change and disturbance for Coastal Zone Management (CZM) include (a) the physical characteristics of water such as salinity, temperature and amount of dissolved oxygen, (b) quality and quantity of freshwater discharge to coastal waters, (c) rapid changes in biomass and primary production, (d) eutrophication, (e) high concentration of toxic materials, (f) increased suspended sediment concentrations, (g) bleaching of coral reefs, (h) rapid changes in plant and animal community composition, (i) changes in areal extent and type of the plant community, (j) changes in wetland structure and its hydrologic conditions and (k) coastal erosion. Vertebrates, invertebrates and plant communities, when analysed in conjunction with selected abiotic parameters, serve as ecological indicators of change. The sources of the disturbances are often distant from the affected coastal areas, but they can cause direct effects such as loss of plants and animals or indirect losses through changes in habitat. Resource managers will have to identify the most serious concerns arising from local problems and select the appropriate indicators for assessment.

2. The Need for Remote Sensing in Coastal Research

Traditionally, coastal and estuarine surveys have used ship and field data and made very little use of remotely-sensed data. Ship and field data provide relatively accurate point samples for a variety of parameters. Bio-optical profiling systems allow rapid sampling of optical, physical and biological properties of the waters. Pollutant levels in seafood supplies and in coastal environments can be monitored by periodic sampling of indicator organisms such as mussels and oysters. Plankton and benthos sampling requires simple equipment, but supportive knowledge of the different plant and animal species and monitoring of vertebrate communities can evaluate habitat quality. Wetland hydrology is difficult to assess accurately because of the high costs and long time frame required for characterisation and quantification. Coastal erosion, bleaching of coral reefs, wetland biomass and primary productivity, plant community composition and wetland structure, areal extent and type are some of the indicators that can be assessed more efficiently by remote sensing than by using conventional methods.

Moored ships can provide relatively accurate point data of a wide variety of desired variables and obtain vertical measurement information, including samples, from a range of depths in the water column. Buoys are utilised to provide both long-time series data at selected locations and information as a function of depth in the water column. Shipboard collected data complement the limited depth sampling capabilities of remote sensors, yet remote sensing technologies provide the synoptic and large-scale views that place ship field data into a broader context. In most cases, a combination of remote sensing and the more conventional shipboard data collection techniques is recommended to assess the coastal zone.

Many coastal and estuarine phenomena vary too rapidly in space and time for observation by conventional ship and field techniques. Compared with the open ocean, lagoons and estuaries are very small and undergo rapid changes due to tidal effects. Consequently, observation requirements for coastal features differ significantly from those for open ocean investigations since both spatial and temporal resolution requirements become more demanding as one moves closer to the coast. This is illustrated in Figure 1, which compares the sensor spatial and temporal resolution requirements that are needed to meet various coastal and open ocean applications. For instance, the meanders of coastal currents and ocean-dumped waste plumes on the continental shelf need to be updated about every four hours with 50m resolution, while studies of the movements of tidal-induced estuarine fronts and pollution plumes require still more precision with one-half hourly observations at about 10m resolution. On the other hand, it suffices to map coastal wetlands and land use about once every three years with a resolution of 10-20m.

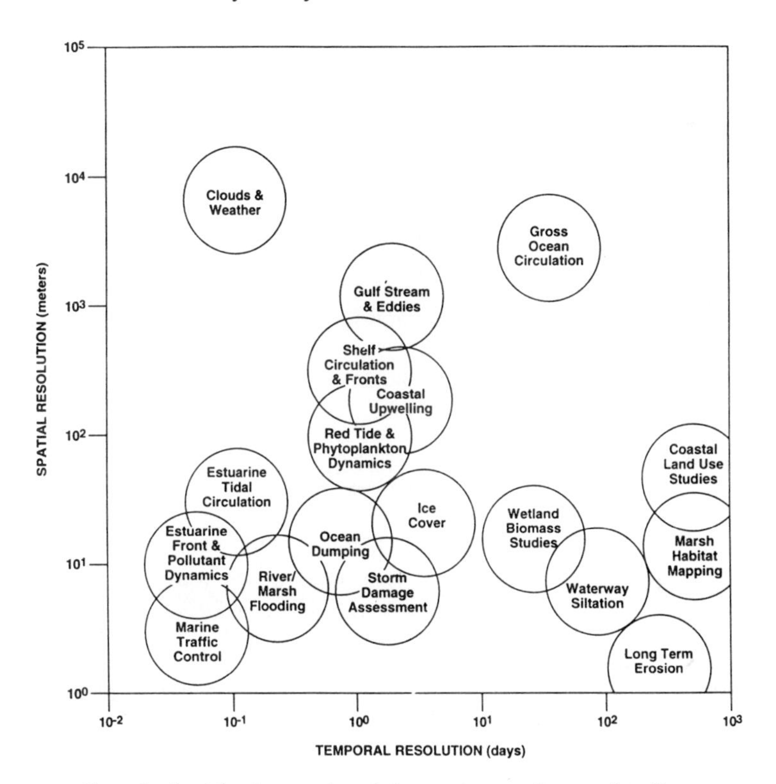

Figure 1. Spatial and temporal resolution requirements for coastal studies.

3. Remote Sensing Techniques

Coastal applications of remote sensing require a wide assortment of sensors, including film cameras for beach erosion and vegetation mapping, multispectral scanners for wetlands stress and estuarine water quality studies, thermal infrared scanners for mapping sea surface temperatures and currents, synthetic aperture radar for wave studies and microwave radiometers for salinity measurements. The most popular instruments used in ocean remote sensing are identified in Table 2 and Table 3 summarises the ability of each sensor to detect key coastal and environmental features from aircraft or satellites. For instance, coastal vegetation and land cover can easily be mapped with film cameras and multispectral scanners. Sea surface temperature is routinely mapped with thermal infrared scanners or radiometers. Ocean wave spectra can be derived effectively from Synthetic Aperture Radar (SAR) imagery. On the other hand, according to Table 3, chlorophyll concentrations are more difficult to sense remotely since they require unique algorithms and complex atmospheric

Altimeter -	a pencil beam microwave radar that measures the distance between the spacecraft and the Earth. Measurements yield the topography and roughness of the sea surface from which the surface current and average wave height can be estimated.
Colour Scanner -	a radiometer that measures the intensity of radiation reflected from within the sea in the visible and near-infrared bands in a broad swath beneath the spacecraft. Measurements yield ocean colour, from which chlorophyll pigment concentration, and diffuse attenuation coefficient, and other bio-optical properties can be estimated.
Infrared Radiometer -	a radiometer that measures the intensity of radiation emitted from the sea in the infrared band in a broad swath beneath the spacecraft. Measurements yield estimates of sea surface temperature.
Microwave Radiometer -	a radiometer that measures the intensity of radiation emitted from the sea surface in the microwave band in a broad swath beneath the spacecraft. Measurements yield microwave brightness temperatures, from which wind speed, water vapour, rain rate, sea surface temperature, and ice cover can be estimated.
Scatterometer -	microwave radar that measures the roughness of the sea surface in a broad swath on either side of the spacecraft with a spatial resolution of 50km. Measurements yield the amplitude of short surface waves that are approximately in equilibrium with the local wind and from which the surface wind velocity can be estimated.
Synthetic Aperture Radar -	a microwave imaging radar that electronically synthesises the equivalent of an antenna large enough to achieve a spatial resolution of 25m. Measurements yield information on features (swell, internal waves, rain, current boundaries, and so on) that modulate the amplitude of the short surface waves; they also yield information on the position and character of sea ice from which, with successive views, the velocity of sea ice floes can be estimated.

Table 2: Spaceborne ocean-sensing techniques

SENSOR	Plat-form	Veg Land Use	Biomass & Veg. Stress	Coast-line Erosion	Bottom Feature SAV*	Depth Profiles	Susp. Sed. Pattern	Susp. Sed. Conc.	Chloro phyll Conc.	Oil Slicks	Surf. Temp.	Water Salinity	Current Circ. Patterns	Wave Spectra	Surf. Winds
Film	A	5	2	5	4	3	4	3	3	4	0	0	3	3	2
Cameras	S	4	2	4	3	2	4	3	2	3	0	0	3	2	1
Multispectral	A	5	4	5	4	3	5	4	4	5	0	1	3	3	2
Scanners	S	4	3	4	3	3	4	4	3	3	0	1	3	2	1
Thermal IR	A	3	3	2	0	0	2	1	0	4	5	2	4	0	2
Scanners	S	2	2	1	0	0	2	1	0	3	5	0	4	0	1
Laser	A	0	0	3	3	4	1	0	0	1	0	0	0	5	2
Profilers	S	0	0	1	1	2	0	0	0	0	0	0	0	2	0
Laser	A	1	0	1	0	1	2	3	4	4	1	2	1	0	0
Flourosensors	S	0	0	0	0	0	1	1	2	2	0	0	0	0	0
Microwave	A	1	0	1	0	0	1	1	1	4	4	4	3	3	4
Radiometers	S	0	0	0	0	0	0	0	0	2	3	3	2	3	4
Imaging	A	4	2	4	0	1	2	0	0	4	2	2	3	4	3
Radar	S	3	1	3	0	1	1	0	0	3	1	1	2	3	2
Altimeter	A	0	0	0	0	0	0	0	0	4	0	0	4	5	4
(Radar)	S	0	0	0	0	0	0	0	0	3	0	0	4	4	4
Scatterometer	A	1	0	0	0	0	0	0	0	4	1	1	2	2	5
(Radar)	S	1	0	0	0	0	0	0	0	1	1	1	2	2	5
CODAR (Radar)	G	0	0	0	0	0	0	0	0	0	0	1	4	4	3
RADS Acoustic)	G	0	0	3	4	3	4	4	0	1	1	1	4	4	1
UW Camera	G	0	0	3	3	3	3	2	2	3	0	0	1	0	0

Table 3: Performance of remote sensors for estuarine and coastal studies The first column denotes the platform with abbreviations: A - aircraft (medium or low altitude), S - spacecraft (satellite) and G - ground (boat or field). The numbers indicate the staus of the sensor: 5 - operational; 4 - functional, not yet operational; 3 - Demonstrated potential, field tests required; 2 - Potential utility, research needed; 1 - Limited utility; 0 - Not applicable. (* SAV denotes Submerged Aquatic Vegetation)

1. **Water Colour**	**Multispectral Imaging Camera**
Suspended Sediment	Xybion Solid State Videocamera
Chlorophyll	(6 Spectral Bands)
Dissolved Organics	
Fronts and Plumes	**Spectroradiometer**
Phytoplankton Blooms	Spectron Spectrometer
Turbidity Maxima	ODAS (3 Band Radiometer)
Etc.	
2. **Water Surface Temperature**	**Thermal IR Radiometer**
	Barnes PRT-5
3. **Water Salinity**	**Microwave Radiometer**
	U. Mass. Design (Cal Swift)
4. **Surface Features**	**Airborne Radar**
Oil Slicks	SAR
Organic Slicks	
Fronts	
Currents, Waves	

Table 4: Small aircraft instrument package

corrections. Only microwave radiometry can measure water salinity and this can only be done from low or medium altitude aircraft.

Aircraft can provide frequent overflights at good spatial resolution. New sensor packages small enough to fit on single-engined or two-engined aircraft are being developed and can operate at a tenth of the cost of traditional four-engined aircraft (Table 4). Deployed in conjunction with satellite sensors such as the Advanced Very High Resolution Radiometer (AVHRR), these airborne sensors should be able to observe tidal, seasonal and annual variations and spatial distributions of phytoplankton blooms, sediment plumes, estuarine fronts, circulation patterns and other estuarine phenomena.

Once identified and collected, information on the various environmental indicators used in monitoring the coastal zone must be systematically organised and presented in an accessible and appropriate format for analysis by scientists and decision makers. The inclusion of temporally dependent data and the desire to provide for long-term monitoring also requires that the integration of these data be easily maintained and updated. One of the powerful tools currently used in resource management for the storage, retrieval, and analysis of environmental data referenced by geographical location is the Geographical Information System (GIS). It is dependent upon the establishment of an accurate standardised georeference coordinate system, such as is provided by a Global Positioning System (GPS).

The contents of a GIS typically include a wide range of data which may be referenced spatially and temporally. For coastal zone management, data layers should not be limited to observable environmental parameters only, but should include information on the impact of human activities and programmes on natural resources. Data may represent rapidly changing phenomena or may be derived from disparate sources involving some uncertainty, and their comparability and reliability may therefore be limited (Figure 2). The importance of data integration capabilities and spatial analysis tools in a GIS cannot be overstated. Few applications are possible solely with data retrieval and display tools; most demand the capacity to link diverse data together using spatial keys. Some key GIS analysis techniques are given in the next section.

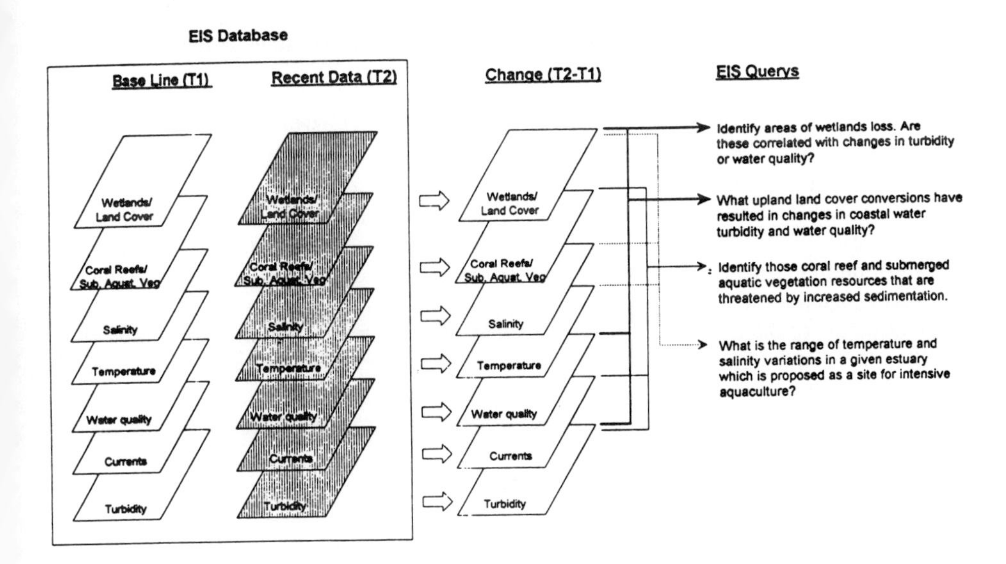

Figure 2 An example of an EIS system for coastal zone management

4. GIS analysis techniques and examples

Overlay Analysis

1. CAD-CAM Packages (Computer Aided Design/Computer Aided Mapping)

2. Boolean Algebra: AND, OR, NOT, etc.

Modelling

1. Cartographic (Careful planning and detailed flow charts)
2. Simulation Approach (Evaluate areas good wildlife habitat and improve it)
3. Predictive Modelling (Predict high probability location of archaeological sites)

Buffering

(Wetlands/Uplands buffers, hazardous waste site buffers, etc.)

Network Analysis

Linear path to represent flow. (Hydrology, transportation, communication, etc.)

Error Analysis

1. User Forms (Age, scale, coverage, relevance, etc.)
2. Measurement/Data Errors (Instrument error, field error, natural variation, etc.)
3. Processing Errors (Precision, interpolation, conversion, digitisation, etc.)

In practice, the uses of GIS range from automated cartography to environmental modelling, policy formulation and management (Figure 3). They serve to underpin routine operations, such as making inventories and monitoring change, and also to support policy reviews and the overall long-term planning process. A comprehensive coastal GIS can aid policy makers and regulators in issues such as the over-exploitation of fisheries resources, the degradation of coastal and marine ecosystems, declining water quality and the potential loss of endangered marine species and coastal wildlife (Table 5). While the effective management and resolution of these problems is primarily dependent on policy decisions and enforcement, a GIS system can be instrumental in the decision making process and formulation of suitable regulations and zoning laws.

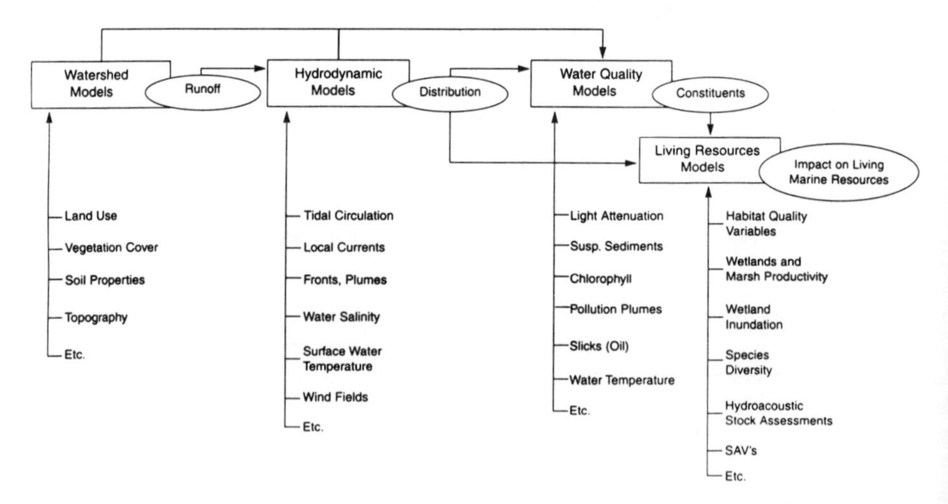

Figure 3 Examples of data used in watershed, hydrodynamic, water quality and living resource models.

Florida (Florida Marine Research Institute and Federal Agencies)
• Marine Mammal Research (Manatee Habitat)
• Boat and Diver Use Patterns in the Florida Keys (Reef Carrying Capacity)
• Fisheries Management (Shrimp Closure and Management)
• Oil Spill Planning and Response (Clean-up, Damage Assessment, etc.)
• Site Selection for Testing of Explosives (Navy Mine Cleaning)
Alaska (State and Federal Agencies)
• State Coastline for Oil Spill Planning and Response
• Alyeska Pipeline Spill Contingency Plan
• Fisheries Research (Fish Habitat and Evaluation)
• Sea Otter Research (Oil Spill Injury to Sea Otters)
Louisiana (State Agencies and Universities)
• Wetland Habitat Changes (Flora and Fauna)
• Coastal Change (Shoreline and Bathymetry)

Table 5: Examples of coastal applications of GIS

A GIS that integrates traditional cadastral data with environmental parameters can be very effective in protecting and ensuring the continued benefits of a rich coastal zone. This goal can be achieved through such measures as (a) control of industrial development to minimise environmental impacts on critical habitats and water quality standards, (b) identification of mangrove and other important coastal habitats for the purpose of protection and restoration, (c) targeting of enforcement and education efforts in areas with high numbers of illegal fishing incidents and known over-exploitation and (d) monitoring of coastal land and sea uses that are economically vital, but often mutually disadvantageous, to the inhabitants of coastal areas.

5. Current and Future Research Applications

More recently remote sensing techniques (Table 6) are becoming particularly attractive, because a wide range of improved sensors and analysis tools are becoming available to coastal and other data users. As shown in the table, marine biology is going to receive a major boost with the launch of SeaWiFS and of further ADEOS satellites, each of which has an ocean colour scanner on board for fisheries related studies of productivity and water quality. These scanners will help us to map concentrations of chlorophyll, suspended sediments, dissolved organics etc. ERS-1 and Radarsat are providing us with new radar data which help physical oceanographers study ocean dynamics, including ocean currents and waves. The high resolution of these radar systems makes them valuable for studies of coastal

SYSTEMS	ATTRIBUTES	APPLICATIONS	PROBLEMS
1. SeaWiFS, ADEOS/OCTS (Multispectral Scanners)	Ocean Colour (Chlorophyll, seston, dissolved organics, pollutants, etc.)	Ocean Productivity, Fisheries Management, Pollution Control, Marine Biology	Calibration, Algorithms, Atmospheric Correction, Cloud Cover
2. ERS-1, Radarsat, (Radar Imagers, Scatterometers, Altimeters etc.)	Ocean Dynamics (Waves, currents, fronts, etc.)	Global Climate, Pollutant Movement, Coastal Processes, Physical Oceanography	Difficult to Interpret, Inadequate Algorithms
3. High Resolution Satellite Systems	Spatial: 1-5m Spectral: 200+ bands	Water Pollutants, Submerged Aquatic Vegetation, Land cover, urban change	High Data Rates
4. Small A/C Sensors (MSS Video Cameras, Thermal IR, Microwave Radiometers)	High resolution, Multitemporal Coastal-Estuarine Processes (Colour, temperature, salinity plumes, fronts, etc.)	Tidal Effects, Plankton Blooms, Pollutant Dynamics, Wetlands Change	Small Swath-width, Small Sensor Payloads, Near-Shore Range
5. EOS (NASA Mission to Planet Earth)	Long-term Systematic Earth Observations	Global Change, Global Climate	Data Management, Data Distribution
6. GIS Databases, EOS-DIS	Multilayer Data Sets (standardised, spatially interrelated, graphically displayed, easily accessed)	Trend Analysis, Resource Management, Environmental Modelling	Standardisation Digitisation, Reformatting, Registration
7. Models, Algorithms	Improved Understanding of Physical/Biological Processes and Relationships	Environmental Prediction, Management Decisions	Inadequate Data Sets to Test/Calibrate Models
8. New Analysis Techniques	Sub-pixel Analysis, Fuzzy Logic, Neural Networks	Small Feature Detection, Improved Classification, Water Auditing Algorithms	More Complex, High Cost

Table 6: Recent remote sensing systems

processes and watershed topography/hydrology. Satellites with high spatial (1-5m) and spectral (hyperspectral) resolutions across the visible and infrared spectra will improve our ability to measure water pollutants and vegetative stress (Table 7). Small sensor packages on single-engined aircraft will decrease operating costs and make remote sensing platforms accessible to a larger number of coastal scientists and managers. Improved models and algorithms used with Geographical Information Systems will improve our ability to measure, model and manage valuable coastal resources. Thus during the next five years we can expect major advances in the application of remote sensing and GIS techniques to monitoring the coastal environment and managing marine resources.

	Owner/Operator	Launch Date	Sensor
RADARSAT	RADARSAT (CDN)	4Q/95	SAR (8m)
EarlyBird	EarthWatch (US)	1Q/97	Pan/MSS (3m/16m)
ADEOS	Japan	1996	Pan/MSS (8m/16m)
Almaz-2	Russia	1996	Radar (5m)
Lewis	TRW/NASA (US)	1996	Hyper-Spec (5m/30m)
Clark	CTA/NASA (US)	1996	Pan/MSS (3m/15m)
QuickBird	EarthWatch (US)	4Q/97	Pan/MSS (1m/4m)
CRSS	Space Imaging (US)	4Q/97	Pan/MSS (1m/4m)
OrbView	Orbimage (US)	1997	Panchromatic (1m)
Resource-21	R-21/Boeing (US)	1999	Pan/MSS (3m/10m)
			As of 8/1/96

Table 7: New satellite systems

Bibliography

Cracknell A P and Hayes L W B, 1991, Introduction to Remote Sensing, Taylor & Francis, London, 293 pp.

Ikeda, M and Dobson F W (editors), 1995, Oceanographic Applications of Remote Sensing (CRC Press Inc., Boca Raton, FL), 492 pp.

Jensen, J R, 1996, Introductory Digital Image Processing: A Remote Sensing Perspective, 2nd edition (Prentice Hall Series in Geographic Information Science, Upper Saddle River, NJ), 316 pp.

Lillesand, T M and Kiefer R W, 1994, Remote Sensing and Image Interpretation, 3rd edition (John Wiley & Sons, Inc., New York), 750 pp.

Richason Jr B F (editor), 1978, Introduction to Remote Sensing of the Environment (Kendall/Hunt Publishing Co., Dubuque, IA), 496 pp.

Airborne remote sensing of fish habitat, channel morphology and riparian conditions

A Roberts, K Bach, C Coburn and M Haefele

Simon Fraser University

1 Introduction

This chapter outlines a pilot study to specify and evaluate aerial photographic interpretation, photogrammetric measurement and remote sensing analysis procedures that would significantly contribute to assessments of fish habitat, channel morphology and riparian conditions (Anon. 1996, Hogan *et al.* 1996 and Johnston and Slaney 1996). Although the work that is described here is for inland rivers, the techniques are also applicable in wetland and coastal environments where the sizes of the phenomena or objects to be studied are so small that it makes little sense to try to use satellite data.

The study area is the Sowaqua Creek drainage area and its associated corridor and flood plain, with additional data being utilized from the Harrison, Horsefly, Nicola and Stein Rivers and Lake Garibaldi in British Columbia, Canada (see Figure 1).

The study involved a number of drainage areas since it was not possible to acquire comprehensive new remote sensing imagery within the time frame of the work described here. Data from the Harrison and Horsefly rivers included multispectral digital imagery as well as digitally scanned colour and colour infrared aerial photography. Much of this imagery was also acquired with coincident ground observation data in order to provide accuracy evaluations. The Nicola River was used to provide examples of specific field conditions that can be interpreted from aerial photography. The Sowaqua and Stein river data included airborne multispectral digital imagery as well as colour and colour infrared aerial photography; these data were used to examine a number of riparian forest classification parameters in areas involving mature and old growth forests as well as channel morphology assessments in Sowaqua Creek.

An evaluation for riparian assessment utility was undertaken to estimate and outline the potential, procedures and limitations for using: (1) aerial photographic interpretation, (2) photogrammetry and (3) digital image processing of multispectral airborne imagery to estimate and assess the parameters listed in the table overleaf.

For all evaluative procedures estimation and measurement accuracy procedures are outlined and, where possible, examples using data from the selected study areas are presented for illustration of each procedure and the related performance evaluations.

---

[1] The Plates referred to in this chapter may be found after page 174.

River characteristics	Disturbance	Substrate
reach classification	large woody debris	substrate classification
reach subdivision	erosion parameters	substrate area estimate
channel type	upslope impact potential	
channel disturbance indicators	upslope impact potential	**Bathymetry**
potential barriers		shallow water bathymetry
barriers	**Vegetation**	mean water depth
stream order classification	cover type and percent	mean bankfull channel depth
pool type	aquatic vegetation	maximum pool depth
long profiles of channels	riparian vegetation type	pool outlet depth
cross sectional channel profiles	riparian structural stage	residual pool depth
elevation determinations	over stream canopy closure	
floodplain topography		**Fish**
channel gradient	**Sediment**	over winter dewatering
mean wetted width	suspended sediment concentrations	fish population estimates
	sediment sources	habitat units
		habitat characterization
		off-channel fish habitat

Parameters assessed by airbourne measurements

New aerial photography, interpreted by itself, cannot provide complete information about, for example, most river bank erosion changes since the accuracy of the river bank as outlined on the TRIM (Terrain Resources Information Management) digital maps will not be sufficient to permit such evaluations. TRIM is a complete DEM (digital elevation model) and digital topographic map of British Columbia at a scale of 1:20 000 with x, y and z accuracy of $\pm$10m; it is owned and maintained by the Government of British Columbia. An historical analysis of existing aerial photography from archival sources will permit an evaluation of the following: (1) historic trends in habitat quantity and quality, (2) historic trends in water quality and quantity and (3) historic evaluation of riparian habitats. Additional considerations include requirements for ground observation data when new aerial photography and/or multispectral electro-optical remote sensing imagery (e.g. video, multispectral video, digital multispectral) is to be flown for comparison with earlier photography.

2 Procedures

This section addresses the utility of photogrammetry, aerial photographic interpretation and digital image processing procedures in assessing and measuring specific riparian parameters. In order to show the variety of techniques and their usefulness, examples using different procedures and several study areas were used.

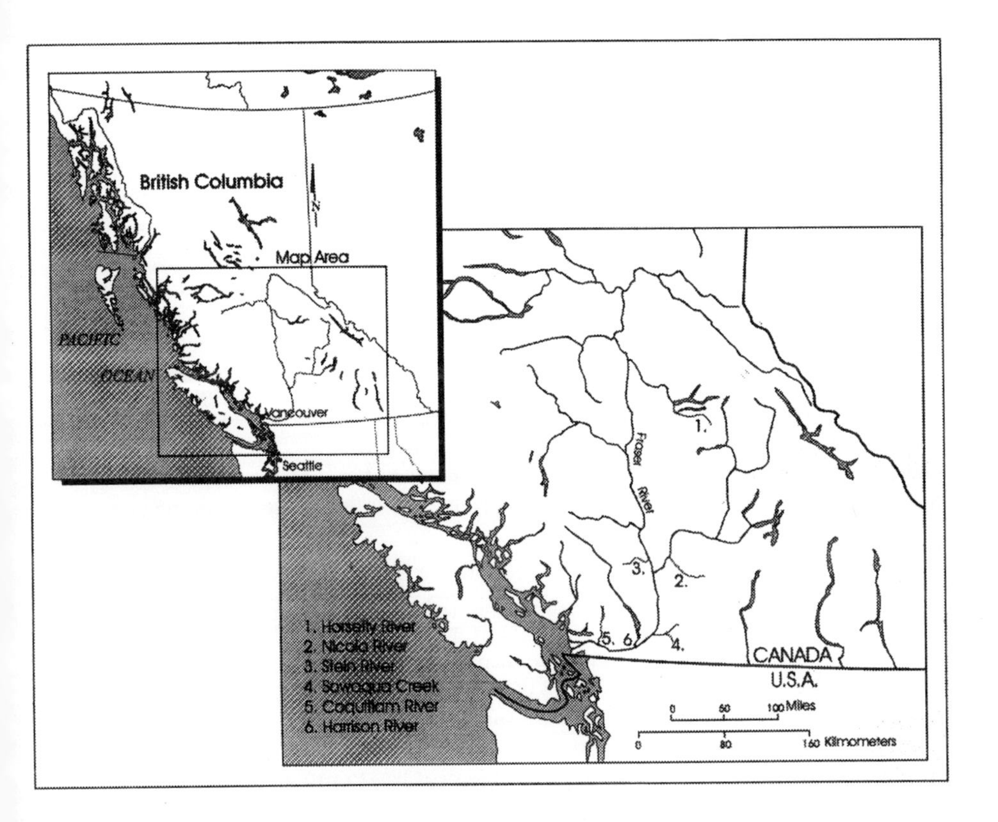

Figure 1 Location map of study sites.

2.1 Photogrammetry

Photogrammetry is defined as the science and art of obtaining reliable measurement by means of photography. Photogrammetric procedures range from simple measurements taken directly from aerial photographs to advanced analytical techniques using specialized photogrammetric equipment. These procedures allow accurate measurements of elevations, areas and distances, including removing or compensating for tilt and spatial distortion in aerial photographs.

The physical property of aerial photographs that allows these types of measurements to be made is related to the projection nature of photography and stereoscopic parallax. The techniques of orthophotography for obtaining two-dimensional information, and the theory of determining heights from parallax measurements are well known and are described in a number of textbooks (e.g. Avery and Berlin 1992, Dickinson 1979, Lillesand and Kiefer 1994). The measurements of the photographs can be made with a stereoscope and parallax bar, with a stereocomparator or with a digitising system and a computer software package. Two-dimensional maps can be produced with analogue or digital plotters.

2.2 Aerial photographic interpretation

The process of aerial photographic interpretation is a visual skill that is acquired through training and experience. It is a combination of an art and a science to distinguish and identify features. Photographic interpretation relies upon the skill training and background of the interpreter and the use of deductive processes to determine "what goes on here" to extract information contained in an aerial photograph. The use of the three-dimensional stereoscopic model is very important in such interpretations and is clearly a situation where the whole is greater than the sum of the parts; interpretative information can be much more accurately and comprehensively extracted from a stereo model than from looking at the two individual aerial photographs separately.

Each aerial photographic stereoscopic model contains detailed information that can be extracted at an appropriate level of complexity for the task at hand. A wide variety of types of photographs (black and white panchromatic, black and white infrared, colour, and colour infrared) are available; each of these different photographic records contain different data. There are two major sources of information contained in all photographic records, spatial information and radiometric information.

Spatial information

Spatial information is present on a photograph in the arrangement of tones which provide spatial clues. Tones that occur within a defined distance and in various combinations define shape, size, pattern and texture. Although these items appear to describe very different things they are all related by the fact that they are spatial attributes. These features are always used in combination to extract information. For example, a forest is composed of trees and each tree has a particular size and shape. In groups the trees have a pattern and a texture which can make up a further set of sizes and shapes. This attribute is known as self similarity, or the fractal property, of a feature.

Radiometric information

The different tones represented on a photographic record represent different intensities of reflected light. The lighter the object appears the greater the amount of reflected radiation. For example, if a black and white photograph is taken of a red house and several trees with a red filter, the house would appear as a light tone (a large amount of reflected red electromagnetic energy recorded) and the trees as a dark tone (not much reflected red recorded). This change in photographic tone with different targets helps the interpreter identify and map features of interest.

Both black and white and colour aerial photography can also be done using infrared film that is sensitive to the near infrared portion of the electromagnetic spectrum. A common misunderstanding is that infrared films record the heat differences in objects; this is not correct. Such films record differences in reflected infrared electromagnetic energy in the near infrared spectral region (700 - 900nm) immediately beyond the visible spectrum (400 - 700nm). This type of photography is very useful for detecting subtle reflectance differences in vegetation (e.g. healthy deciduous trees in full leaf reflect a great deal of near infrared, healthy conifers less and stressed vegetation typically has a lower infrared reflectance). Figure 2 shows the spectral response curves of several different types of vegetation and other surface targets. In the visible spectrum there is less difference between the vegetation features (they are green) and the other surface targets. There is a greater distinction between the vegetation in the near infrared portion of the spectrum in comparison with the other surface targets. In this near infrared region it is evident that each different type of vegetation has a distinctly different amount of reflection. This is a property that can be exploited to provide considerable assistance in interpreting riparian vegetation patterns.

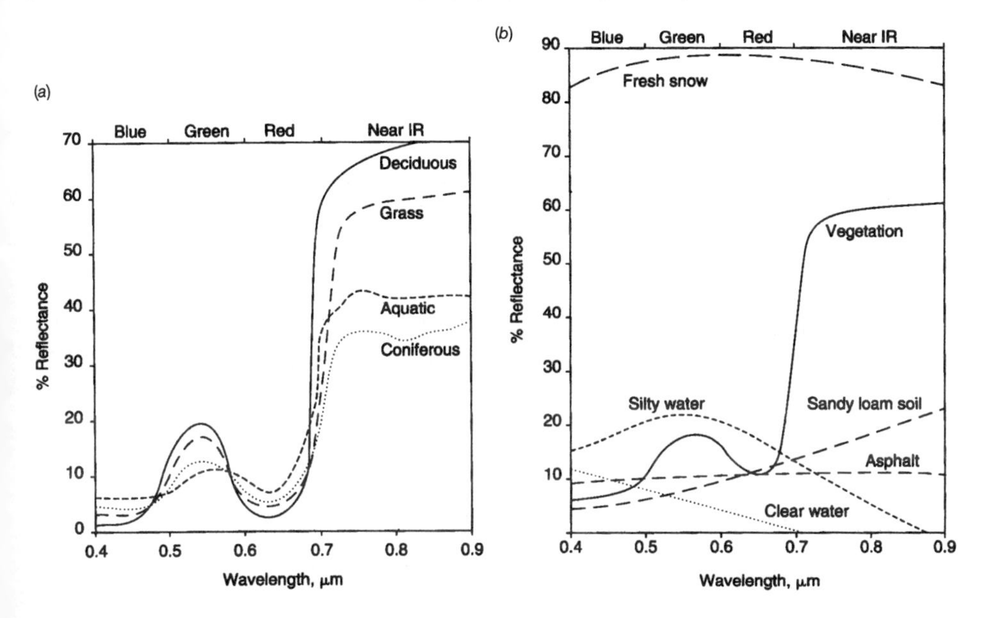

Figure 2 Spectral response curves for (a) vegetation, (b) six different surface features.

Colour photography

All colour photography uses multiple emulsion films and colour dyes to represent spectrally different features in the images. Normal colour film has three film emulsion layers that are generally sensitive in the blue, green and red spectral regions. Yellow, magenta and cyan dyes are respectively attached to these film layers to provide a representation of true scene colours in the colour diapositive or on prints from negative film. Colour infrared film is a false colour film because the same three dyes are attached respectively to green, red and near infrared emulsion layers on the same film base. This results in green targets appearing blue, red targets appearing green and targets with high near infrared reflectance appearing red. Blue light is filtered such that there is normally no blue exposure on these films. Since healthy vegetation reflects most in the near infrared it will appear bright red on colour infrared film. Such imaging involves a spectral component (colour) to assist with the interpretation of features on the photographs.

Digital image processing

Spectral information provides excellent interpretative assistance through the use of colours. When viewing various images the interpreters can readily see differences in vegetation, sediment load, water depth, etc. This is very useful because features like pools and changes in riparian vegetation can readily be identified. Digital image processing permits the interpreter to go beyond the intuitive classification of such features and to quantify specific target features (e.g. suspended sediment concentrations, specific water depths) using personal computers and appropriate image processing software. Training and experience are important contributors to an interpreter's success when using such procedures. Furthermore some aspects of this work are highly experimental and it requires an experienced and well trained team to evaluate the utility of a digital processing procedure in these cases.

3 Parameters of river characteristics

Reach classification - stream reaches are classified as a homogeneous section of the stream channel characterised by discharge, gradient, channel morphology, channel confinement and stream bed and bank materials. Reaches also have a specific repetitive pattern of features. This task has previously been defined as a predominantly photointerpretive task. With the use of photogrammetric tools, accurate lengths and gradients can be measured and used in place of, or as a supplement to, field observations. Photogrammetry can also provide reliable measurements given adequate ground control. The time required to survey an area with suitable photogrammetric procedures and an adequate sampling effort is considerably less than would be required for the same survey using only ground survey techniques.

Plate 2 shows an unrectified aerial photographic mosaic of a section of Sowaqua Creek. This type of image is useful in interpreting features along the stream as it gives a broader overview without sacrificing spatial detail.

Reach subdivision involves the partitioning of larger reaches into smaller and more homogeneous sections. This procedure is based on specific attributes that are important to a specific study. For example, different riparian structural stages, or changes in substrate, might be used to further divide a reach. Many of the criteria used to perform this classification task can be derived from aerial photographs. In this case, reconnaissance photography (normal colour, colour infrared, digital, or black and white panchromatic), flown with optimal illumination conditions (Sun angle and time of year), can provide excellent data for this type of distinction. Plate 3 is an example of the same area shown in Plate 2 except that this image is a multispectral digital reconnaissance colour infrared composite image rather than a scanned colour aerial photograph.

Because of the specific spectral properties of vegetation, data captured in the infrared region of the spectrum yields valuable information for the discrimination of different vegetation types. This difference is most evident in the contrast between deciduous and coniferous vegetation. For example the old logging road (next to the current logging road) is distinctly seen on the colour infrared image (Plate 3)) while it is less noticeable on the normal colour aerial photograph (Plate 2).

Stream **channel type** describes the channel typology and morphology. Channel typology is determined based on the shape of the channel (step-pool, cascade-pool and riffle-pool) with modifications based on substrate and debris. In areas where the channel is clearly visible, and at times when the suspended sediment concentrations are low, aerial photographic interpretation is an ideal tool for this type of classification. Specific substrate size determinations should be conducted at selected accessible locations in order to provide suitable ground truth for this type of classification. For further information on substrate classification see the Section 7 below.

Channel type and disturbance indicators are used to identify specific channel morphologies that are suitable habitat for spawning and to evaluate disturbance levels. Of specific importance to this determination is the effect of logging activities on habitat condition. Areas that were formerly prime areas for spawning and have since been degraded through logging activity are of particular interest. Aerial photographic interpretation is an appropriate tool for this type of analysis. In areas with previous historic coverage, the effect of logging on spawning habitat can be quantitatively evaluated through a study of change over time. In areas where canopy or shadow limit the use of aerial photographs, field surveys should be substituted. Figure 3 shows a recent cut-block by Sowaqua Creek. The effects of logging activities in the area can be clearly seen. There are several slope failures around the road causing an increased sediment load in the creek.

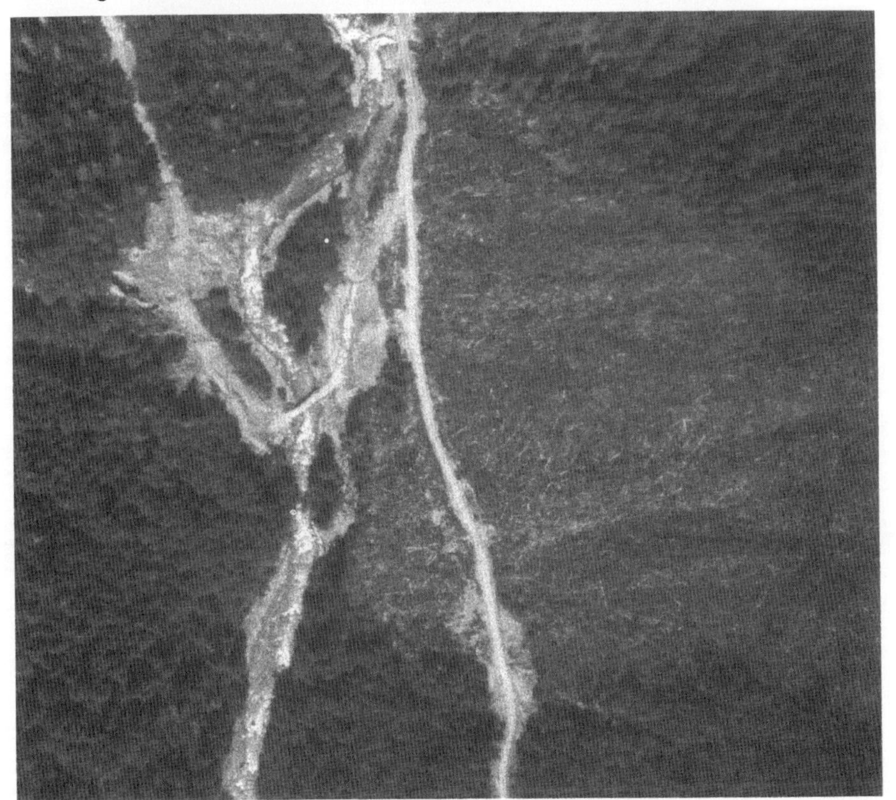

Figure 3 Low altitude reconnaissance photography acquired in June 1997 over a cut block beside Sowaqua Creek. This image shows disturbance caused by logging roads and the resulting slope instability.

Potential barriers - features that might prevent, or constrict, juvenile or adult salmonids are mapped as potential barriers. These features are then field checked to assess the nature of the barrier. The following areas are mapped as potential barriers: (1) culverts and disused bridges (Figure 3), (2) landslides and bank slumping (Figure 3), (3) log jams (Figure 4), (4) beaver dams, (5) falls greater than 2m, (6) cascades or chutes (white water) and (7) gradients greater than 20%. Many of these features can be identified on aerial photographs. Parameters that require measurement (falls and gradient barriers) should be measured using photogrammetric procedures for greater accuracy. If the area in question is in an area where tilt or relief displacement are evident, or likely, then measurements should be made using analytical photogrammetry as both tilt and relief displacement can introduce an unacceptable level of error into these measurements.

Figure 4 A log jam in the Stein Valley - the jam is clearly visible in the channel.

The evaluation of **barriers** is similar to the evaluation of potential barriers. Barriers are features that can be identified as restricting access to salmonid habitat. In this case, aerial photographic interpretation is a useful tool in the determination of barriers, providing the area of interest is not obstructed from an overhead view. Close attention must be paid to time of year and time of day considerations for flight planning in order to reduce the effect of shadows over the stream (since this obstructs viewing of features).

Stream order classification is a scale-dependent measure. In most circumstances stream order is determined within a single drainage basin and is therefore derived from aerial photographs at a smaller scale (i.e. 1:50000) than is required for stream habitat evaluation. This is a straight-forward photographic interpretation procedure, since specific information about each stream is not part of the classification system. Most of these systems (for example the Strahler stream ordering system) are based solely on the numbering of streams. In the Strahler system all streams that have no up-stream tributaries (smallest detectable channel that would intermittently carry water) are labelled as first-order streams. The stream order is not incremented until two streams of the same order meet. For example, the stream that results from the joining of two first-order streams is a second-order stream, two second-order streams make a third-order stream and so on. If a first-order stream meets a second-order stream the stream remains a second-order stream.

Pool types can be mapped from aerial photographs as the type of pool (scour, dammed or unknown) is a qualitative assessment based on the surrounding attributes of a specific pool. For example, a scour pool is formed by scouring of bed material around or adjacent to an obstruction. If the obstruction is visible on the aerial photograph then the pool can be accurately typed. With the use of large-scale aerial photography many subaqueous features can be interpreted in clear water. Pools appear as darker tones when compared to lighter (shallower) riffles. In many cases subtle variations in bathymetry, for example salmon spawning redds, are visible in optimal imaging conditions (see Figure 5).

Figure 5 This is a stretch along the Horsefly river. There are brighter patterns visible in the water. The spawning salmon move the gravel into the horseshoe-shaped nests. The water on the crests of these nests is shallower and they appear brighter than the surroundings.

Long profiles of channels - the calculation of profiles along channels is possible using remotely-sensed data and digital techniques. Channel profiles require information on the depth and bedform of a specific area for which the profile is desired. Using photogrammetric techniques the slope and length of the channel can be computed. Using these data in conjunction with shallow water bathymetry can yield comprehensive channel measurements. Shallow water bathymetry techniques, using reconnaissance multispectral imagery and digital image processing, require experienced personnel and are still at an experimental stage (see later section on shallow water bathymetry).

Channel gradient is the rise of the channel over a specific distance. The measurable variables (elevation and distance) require field measurement or the use of an analytical stereoplotter and suitable ground control. The use of a stereoplotter, over a parallax bar, for these measurements is required as the distortions present in tilted aerial photographs can lead to unacceptable levels of accuracy (rivers appearing to run up-hill are not uncommon when a gradient is measured from unrectified photography). The primary reason for this difficulty is that stream gradients are often very subtle and a small amount of tilt displacement can give an erroneous sense of slope. Standard photogrammetric techniques require that some positional and elevation control exist for the desired area. In areas where these data are not available, field survey methods must be employed to obtain the necessary ground control.

Cross-sectional channel profiles can be constructed using a combination of photogrammetry and shallow water bathymetry. With the aid of an analytical stereoplotter all features above water level (including bars) can be accurately surveyed. This information can then be combined with shallow water bathymetry to yield a cross-sectional channel profile. Shallow water bathymetry techniques are experimental in many contexts and are covered below.

The **mean wetted width** is a measurement of the horizontal distance perpendicular to the channel axis from the water's edge on one side of the stream to the water's edge on the opposite side of the stream. Provided the edge of the stream is in full view on the aerial photography, this is a straight-forward measurement for photographic interpretation and photogrammetry. If the stream is relatively small on the aerial photograph, this distance can be measured directly from the photograph using a suitable precision device as distortions will be negligible. If, however, channel size is relatively large, then an analytical stereoplotter should be used in order to remove the tilt, radial displacement and scale distortions.

Mean bankfull channel width, which is similar to the measurement of mean wetted width, describes the distance between the banks of the river at the tops of the river banks. This measurement is more difficult to obtain using photogrammetric methods as the tops of the banks are often obscured by vegetation. If vegetation is not a factor then this measurement can be obtained in the same manner described for mean wetted width.

Elevation determination is a primary task of photogrammetry. All topographic maps compiled in Canada and the United States rely on photogrammetry for their accurate depictions of elevation and area. Elevations can be measured in forest environments as the trees tend to mirror the underlying topography and an experienced photogrammetrist can measure the average height of the trees and subtract this value when measuring elevation at the tops of the trees to get a measurement of the topography.

Floodplain topography can be characterised using aerial photographic interpretation. Geomorphologists specifically trained in landform evaluation traditionally have used aerial photographic interpretation as a primary tool. When combined with ground truth the landforms present in the flood plain can be accurately mapped. Figure 6 is a stereogram of the confluence of Sowaqua Creek and the Coquahalla River. When viewed with the aide of a pocket stereoscope the area of the stereogram can be seen stereoscopically. A stereoscopic view allows superior identification, more accurate delineation of landforms and produces more accurate maps.

4 Disturbance parameters

Large woody debris - the identification and classification of large woody debris is a straightforward photo interpretative task. The detection of large woody debris is scale dependent, if the scale of the aerial photographs is too small then accurate identification of large woody debris is not possible.

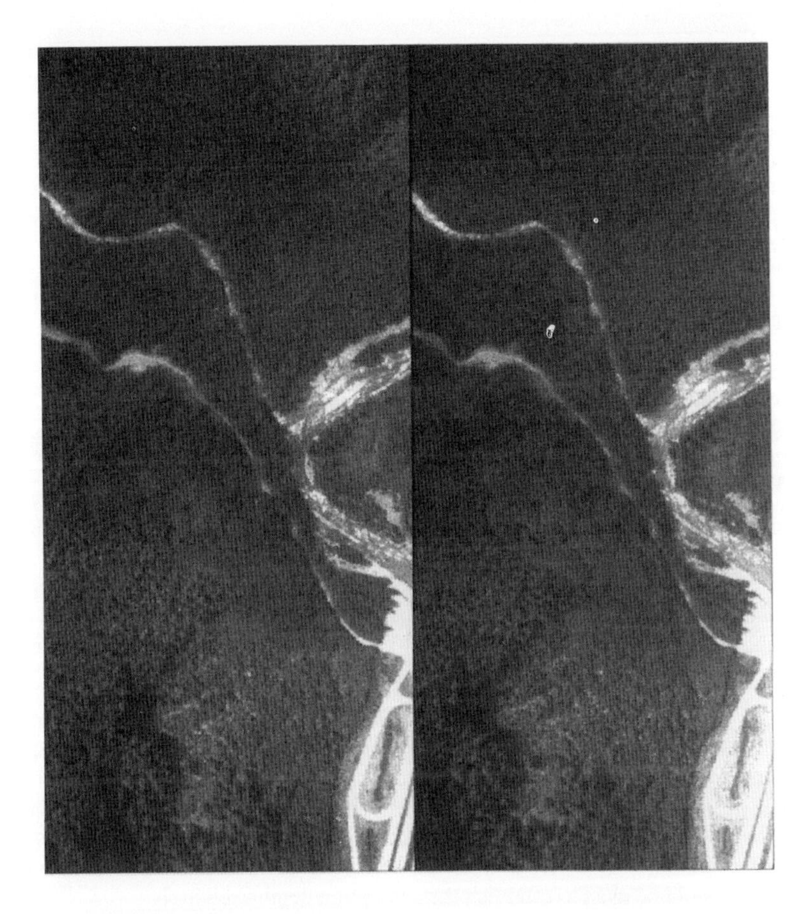

Figure 6 A digitally created stereogram of the confluence of Sowaqua Creek and the Coquihalla River.

On large-scale aerial photography (1:5000 - 1:20000) large woody debris present within the bankfull channel width can easily be seen on aerial photography provided the logs have not become so sun-bleached that they fade into the background. The number of trees and logs can be estimated and the size of the pile measured using simple photogrammetric procedures. Plate 4 shows an area of large woody debris along Sowaqua Creek. This 35mm reconnaissance aerial photography colour image was flown in June 1997 and scanned at 1200dpi for digital display and analysis; on the original colour diapositive the individual logs are much more clearly visible. Scanning of aerial photography for digital analysis alters the contrast characteristics (usually increases contrast) and reduces the spatial resolution of the original photography.

Erosion parameters - an experienced geomorphologist/photogrammetrist can assess erosion parameters using photogrammetric measurements and photo-interpretative techniques. One of the most important criteria for erosion susceptibility is slope angle. Slopes can easily be measured from aerial photographs. In areas where relief displacement or tilt are factors, slopes should be calculated with the aid of an analytical stereoplotter. In many cases geomorphologists can identify and classify the slope material by looking at the vegetation patterns and other site associations (drainage, landforms or erosion features). Most photographic interpretation of this type should be ground checked to validate the observation whenever it is deemed necessary (see also the section on sediment sources below).

Upslope impact potential can also be determined using aerial photographic interpretation. In all instances involving important or critical interpretations, they should be done by an experienced photo-interpreter (see also the section on sediment sources below).

5 Bathymetry parameters

Bathymetry generally involves the measurement of water depth in marine or riverine environments. Water depth is an important factor for many parameters in riparian evaluations and fish habitat assessment.

Shallow Water Bathymetry

Aerial photographic interpretation can be used to assess water depths intuitively. Generally, the deeper the water is, the darker it looks on an aerial photograph. If the water is shallow, the bottom of the river can be seen on the photograph. In deeper water, the texture of the bottom will not be visible; deeper water will look smoother and darker. This relation between image brightness and texture and water depth can be used to find the locations of pools, and/or identify deeper and shallower reaches along a river.

The use of colour infrared film, and related digital imagery, has potential for qualitative water depth assessment. Water is a strong absorber of near infrared electromagnetic energy. Starting at a certain depths, depending on water and substrate conditions (30-90cm depth), the incident near infrared energy will be mostly absorbed and any water deeper than the absorption depth will appear black on the imagery. This could be used to distinguish easily between water shallower and deeper than the near infrared absorption depth. Ground truth measurements should be used to establish the absorption depth accurately.

Plate 5 shows a comparison between the reflectance characteristics of water for normal colour film and colour infrared film. The shallower areas in the river delta can be distinguished from the deeper areas as they appear brighter. The underwater topography of the delta can be clearly seen.

The relation between water depth and image brightness can also be evaluated quantitatively when digitally scanned colour aerial photography or multispectral digital electro-optical data are used. Certain brightness values (called digital numbers or DNs) on the digital images can be directly associated with specified water depths. Ground truth observations (water depth measurements along selected calibration sites) are essential to establish accurately the quantitative calibration relation between DN values and water depth. Simple regression techniques can be used to establish the calibration curve that becomes the basis for further water depth estimations on the same imagery. (This approach has been used widely over shallow areas at sea.)

Figure 7 shows an example of such a curve from a study site along the Harrison River. Image brightness values were selected from the imagery for the locations of the water depth measurements and a regression curve was established. It is evident that there is a strong relation between the two variables (i.e. the DNs are lower the deeper the water becomes).

This is used to produce a calibration curve that can then be used for water depth prediction in other areas on imagery from the same riverine environment, flight conditions and exposure values. For example, a DN of 25 would predict a water depth of 6cm. A map of the water depth area for an image can be generated by applying the regression curve to the digital numbers of the image. A map of the topography of the river bed can be developed from such a procedure to show estimated bottom contour lines. These types of procedure are still highly experimental and require precise ground observation data and trained experienced personnel.

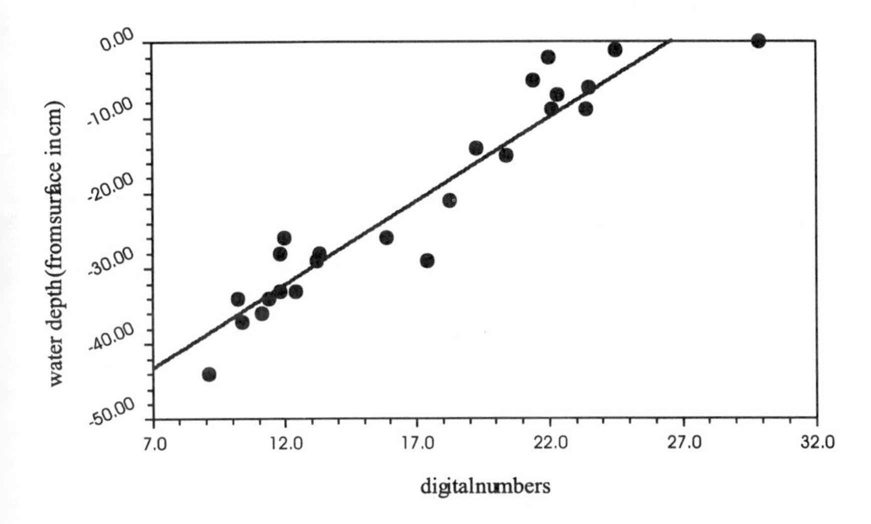

Figure 7 Harrison River regression curve of digital numbers vs water depth. Water depth can be predicted from the digital numbers.

Procedures have been developed for acquiring the necessary ground truth observations for coordination with pixel DN values from the multispectral digital images. It is crucial to identify accurately the DNs that represent the locations of the ground truth water depth measurements to be used in the calibration. If the correct location is not accurately determined, the correlation between the two variables (DNs and depth) will be degraded. In most circumstances, it is very difficult to identify an exact location to within a few cm. As a consequence, the DN values used are averages of several pixels in the target area. This procedure has performed quite well and is the approach currently being used. Further research is necessary to improve calibration as well as identification of target locations and to assess the accuracy of the procedure for predicting water depth.

Apart from the usual limitations (weather and atmospheric conditions), this quantitative procedure is challenged by several factors. Surface roughness will cause sunlight to be reflected directly into the camera while imaging and cause specular reflection (see Section 6 on sediment below) that will increase the average DN values as well as alter the calibration curve and subsequent depth estimations. Also, large numbers of spawning fish in the water at sub pixel resolution can be a problem because the water will appear shallower than it actually is (see Section 9 on fish below). Changes in substrate (see Section 7 on substrate) and turbidity (see Section 6 on sediment) will change overall brightness levels and influence the depth estimates. Most of these problems can be solved by using appropriate calibration procedures. Selecting the right time of the day for image acquisition, avoiding times of heavy spawning activity in the river and not imaging during periods of high sediment concentrations will reduce water depth estimate errors.

Mean water depth - in fish habitat assessment, the water depth within a habitat unit is usually determined by averaging three depth measurements along a transect to portray average conditions within that habitat unit. This is the most efficient procedure, although using the average of only three measurements reduces accuracy (Johnston and Slaney, 1996).

The above procedure for shallow water bathymetry will improve the measuring of mean water depth significantly since a map of water depth, covering the entire habitat unit, can be generated once the calibration has been established.

Mean bankfull channel depth -is usually determined by first measuring the vertical distance from a horizontal line at the height of the bankfull width to the water surface and then measuring the mean water depth. The two measurements are then added to yield the mean bankfull channel depth. The first parameter can be measured from aerial photography using photogrammetry. Shallow water bathymetry procedures can give the mean water depth.

Maximum pool depth - is measured or estimated at its deepest point. The procedure for obtaining shallow water bathymetry estimates can be used to measure this parameter.

Figure 5 shows a stretch along the Horsefly River. The sinusoidal patterns in the water are the redds from the spawning sockeye salmon. In a similar fashion pools will be visible on multispectral digital imagery and their depths can be estimated (see also Figure 8).

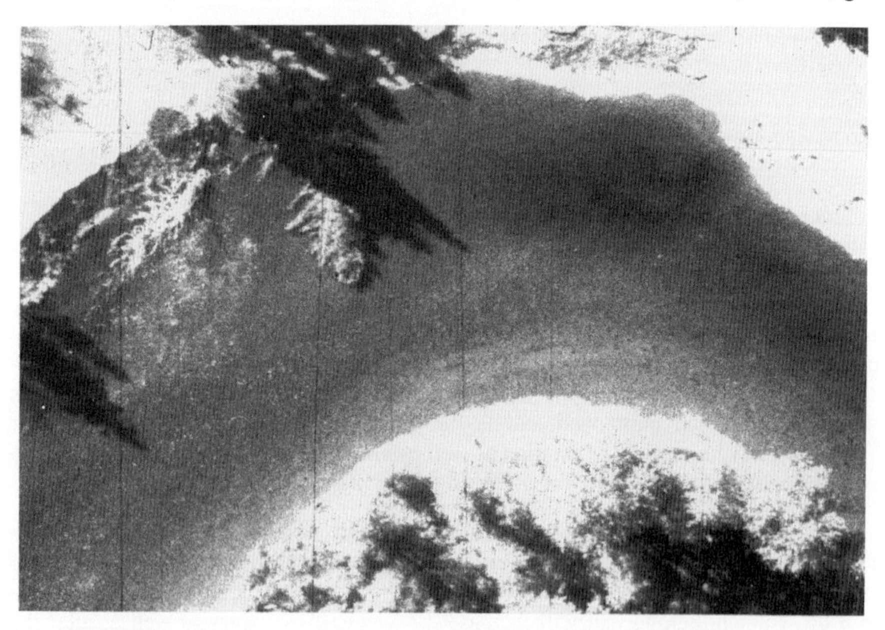

Figure 8 A portion of a spawning ground, where the fish are visible as dots. The image has been enhanced to emphasize water features. The darker areas in the water are pools.

Pool outlet depth - as in shallow water bathymetry and maximum pool depth above.

Residual pool depth - is maximum pool depth minus the outlet pool depth (see above).

6 Sediment parameters

High sediment loads adversely affect fish spawning and rearing habitats (Groot and Margolis 1991). Sediment introduced into streams through human activity, such as agriculture or logging, is therefore of considerable concern for fisheries management (Department of Fisheries and Oceans 1986). The sediment load of a river is commonly measured as its suspended sediment concentration, SSC. Another readily available measure is turbidity which is influenced not only by SSC but also by dissolved substances. At high SSC, turbidity has been shown to be predominantly determined by sediment (Goodin *et al.* 1993).

There is considerable interest in remote sensing of water quality in the literature (e.g. Gitelson *et al.* 1994, Van Stokkom 1993). Generally researchers focus on marine or lacustrine environments (Gitelson *et al.* 1993) with some examples from large, deep streams (Ferrari *et al.* 1996). Our current research focuses on the detection and monitoring of SSC in the highly dynamic riparian environments of salmon spawning and rearing habitats.

The simplest way to employ airborne remote sensing for SSC monitoring is through a visual interpretation of suitable aerial images. Figure 9 shows the confluence of the Little Horsefly River with the Horsefly River. Despite the relatively high flying height and, therefore, low spatial resolution, one can clearly see that the Horsefly River carries significantly more sediment than the Little Horsefly.

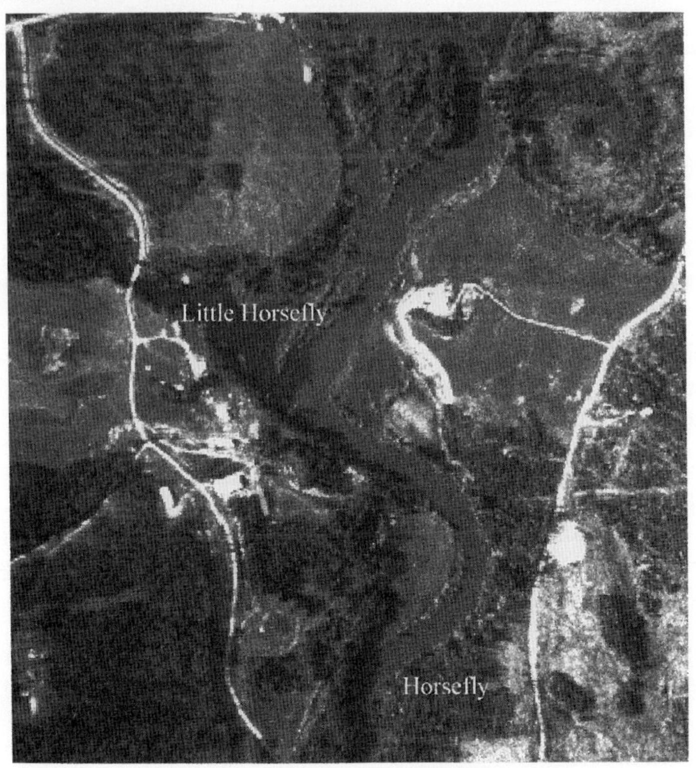

Figure 9 The confluence of the Horsefly River with the Little Horsefly River showing the higher sediment load of the former. Visual image interpretation allows relative concentration estimates.

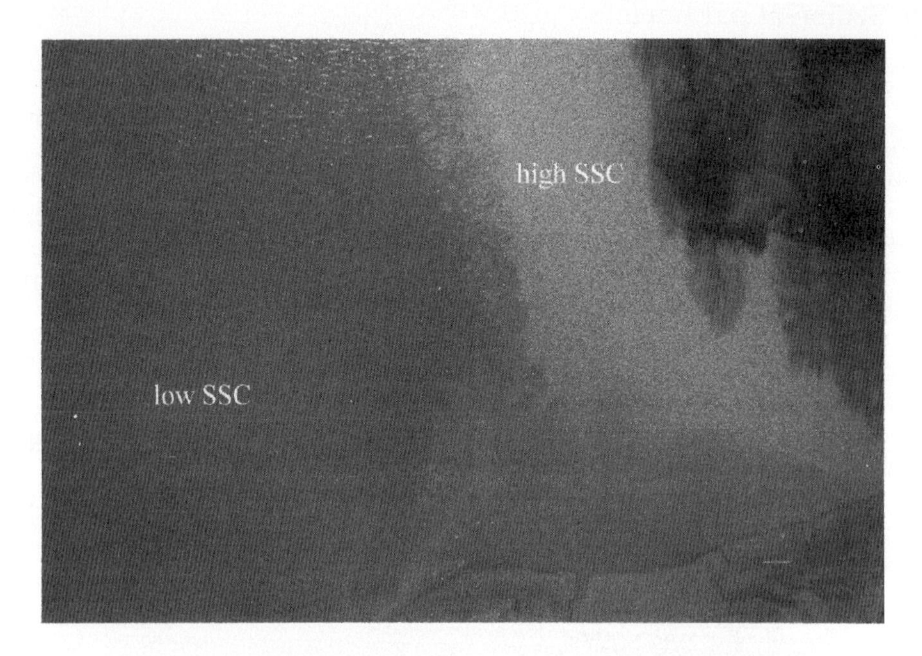

Figure 10 The confluence of the Coquitlam River with the Fraser River. The Coquitlam River introduces large amounts of sediment into the Fraser river. A mixing zone is visible and subtle changes in sediment concentration can be detected.

Figure 10 shows the confluence of the Coquitlam River with the Fraser River. At this lower flying height, subtle changes in SSC are visible. At the very least, one can distinguish between low, medium and high concentrations.

Visual interpretation only allows relative estimates: for example, the Horsefly carries more sediment than the Little Horsefly. After its confluence with the Coquitlam the Fraser carries more sediment close to its right bank. No intuitive estimates of quantity are practical and visual interpretation does not allow for the monitoring or mapping of actual SSC concentrations in terms of mgl^{-1}. It is, however, of considerable practical value because it can significantly reduce the amount of field work when looking for sediment sources and variations in SSC. For a watershed like the Horsefly, one can identify the tributaries that introduce sediment into the river, as well as dynamic changes in SSC. A subsequent detailed field study to determine actual SSC values can then be limited to a few locations of interest. Similarly photo-interpretation helps in monitoring streams that have not been identified as problem watersheds. Then after inspection, and only if the visual inspection reveals changes in the sediment regime, one can examine more closely a particular stream and location. The British Columbia Forest Service, Caribou Forest Region, currently uses visual inspection from the air to monitor remote streams before logging activities start.

In addition, qualitative monitoring can be applied in retrospect utilizing historic aerial photography and imagery to establish trends in sediment loads for specific watersheds. Since for most streams there are no historic sediment data available, a visual comparison between different images over time is the only way to examine past SSC variations. From a resource management point of view, streams historically carrying high sediment loads can then be separated from those with recent SSC increases.

In addition to relative estimates of SSC using visual inspection, airborne remote sensing can provide quantitative estimates of SSC and turbidity. Our previous research reveals that quantitative estimates are difficult at low concentrations (< $25mgl^{-1}$), especially in shallow streams where depth and bottom substrate play an important role (e.g. Roberts *et al.* 1995). Under these conditions, the estimated values are generally more variable than the measured SSC and require adjustment for bottom reflection and depth variations. If the SSC is sufficiently high (obscures bottom reflection), the estimated SSC values become more stable. These parameters are dealt with in separate sections (see Section 5 on bathymetry and Section 7 on substrate).

Multispectral digital and photographic imagery of the Horsefly River watershed during the 1997 freshet were flown for SSC analyses. The digitally converted 35mm colour reconnaissance aerial photography has been analysed and used to illustrate these values. The ground truth consisted of sediment samples integrated over the whole water column (integrated SSC), sediment samples from near the water surface (surface SSC) and turbidity measurements. Due to limited access and the strong river current, all ground truth samples and measurements were taken from bridges. These ground truth data were compared with digital numbers (DNs) extracted from the scanned aerial photographs.

We initially compared raw DNs with the ground truth data in an attempt to correlate the two. The DNs show little reaction at low concentrations but respond well to higher ones until the image saturates. One can see quite a range of DN values at each turbidity level, stemming from the extraction of several values for each sampling location (Figure 11a).

In a second step (Figure 11b), all the DN values for each sampling location were averaged. This resulted in an improved correlation. Figure 12 shows the results for the SSC estimates with lower correlation values, especially from the integrated samples. The expected function does not fit well here and the best fit function was a reciprocal logarithm for surface SSC and a reciprocal model for the integrated SSC. The sample size for integrated SCC was much smaller than for turbidity and sampling occurred only at times of low concentration. This would have a considerable impact and partially explains the curve differences. Similarly the values for surface SSC were concentrated at low SSC levels and this would alter the shape of the curve.

Some of the DN values included in this analysis probably contained an error component due to imaging conditions and should be excluded. These errors can be due to a number of factors, such as bottom reflection, shadow, sunglint, atmospheric conditions, uneven illumination and vignetting (light fall-off across the image due to camera lens parameters). In this study, obvious problem areas have been excluded from many of the analyses but, because of changing weather conditions and low SSC throughout the study area, significant errors were introduced and resulted in reduced estimate accuracies. Figure 13 shows a slight improvement in the correlation between turbidity and DN values after applying a simple radiometric calibration technique, which accounts for unequal illumination of different images (Pellika 1996).

A.

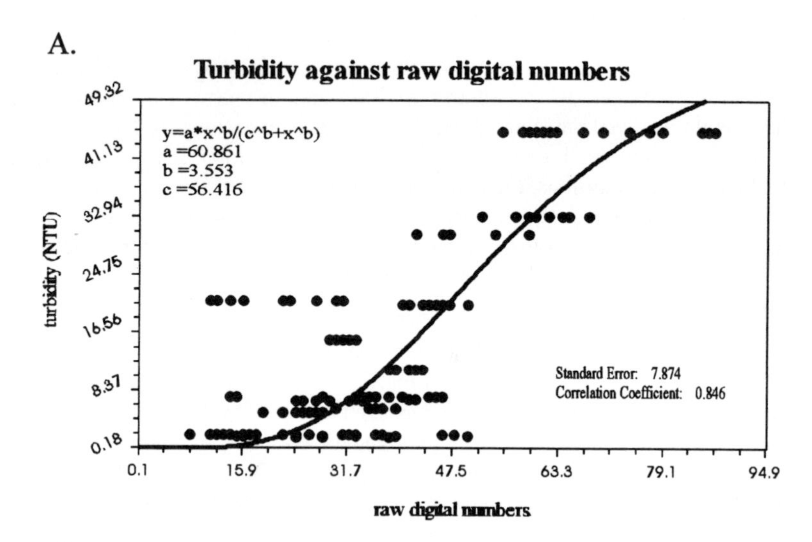

B.

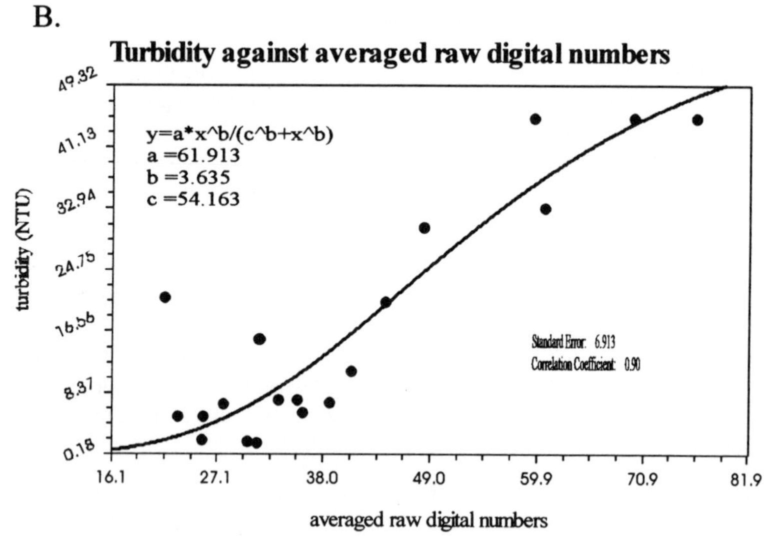

Figure 11 Turbidity correlation with digital numbers. Averaging the digital numbers extracted from different locations in the vicinity of the turbidity sampling location improves the correlation. The curve shown represents the expected response of digital numbers to changes in turbidity.

A.

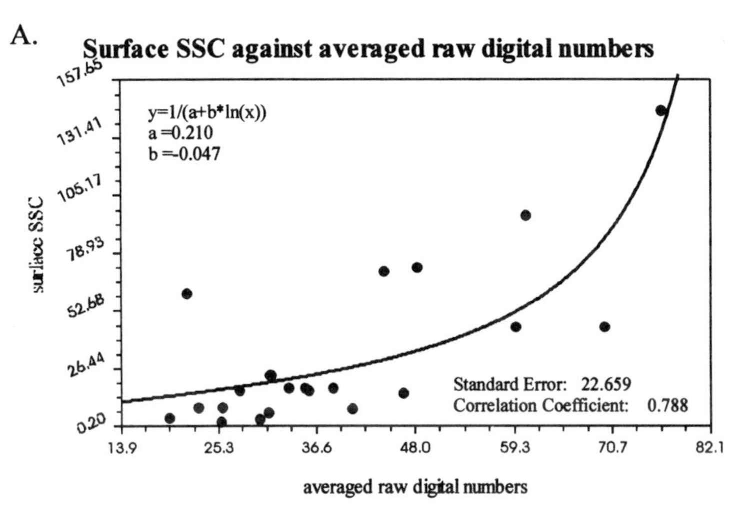

B.

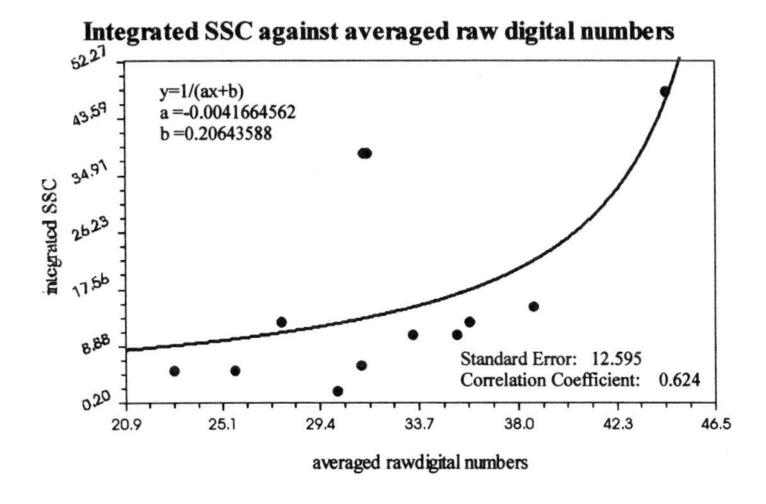

Figure 12 Correlation between suspended sediment concentration and digital numbers is weaker than between turbidity and digital numbers. The expected curve does not fit in this case. A possible reason is the overall low level of concentrations.

A.

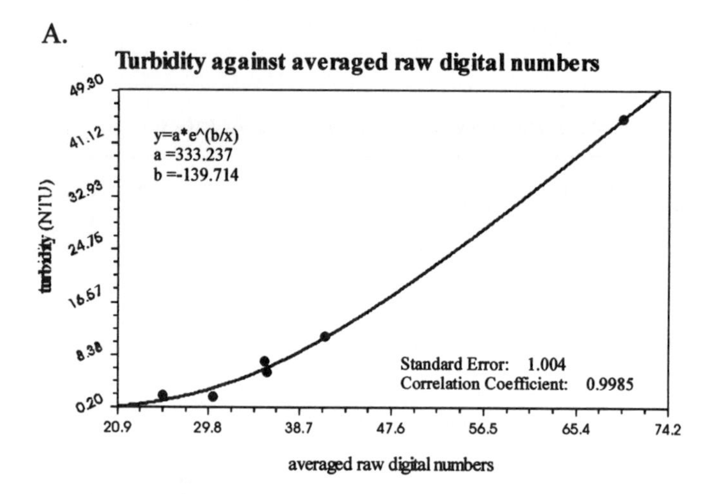

B.

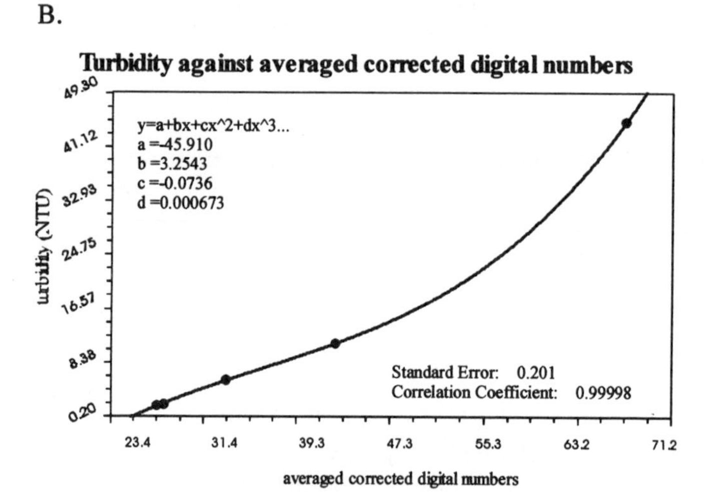

Figure 13 A simple radiometric calibration technique, addressing different lighting conditions between images, improves the correlation between turbidity and digital numbers only to a small extent. To achieve better results uneven illumination within a given image must be addressed.

In conclusion, based on raw DN values, turbidity can be predicted in most situations with reasonable accuracy and better than other parameters. Surface SSC has adequate potential for estimation calibration, while integrated SSC does not perform to the same levels of accuracy. Turbidity is an optical measure, whereas SSC is a physical one. Since remote sensing also is an optical technique, the better correlation with turbidity is not surprising. Unless the water is quite clear, remote sensing can only detect features at or near the surface, therefore this suggests that surface SSC should produce a more accurate calibration than integrated SSC values. However, most salmon streams, like the Horsefly, are quite clear for most of the year. In fact, the integrated SSC samples for this study were taken under relatively clear water conditions. Bottom reflection would have affected the values and would have contributed towards the reduction in integrated SSC performance.

Operational quantitative remote sensing of suspended sediments and turbidity in shallow streams has the potential of becoming a readily available procedure using reconnaissance multispectral imagery. For most studies it would be initially an experimental procedure and would require experienced personnel and a detailed research programme. Important considerations will include:

- the application and testing of further radiometric calibration techniques, such as white region normalization, to reduce ambient light flux problems,
- establishing and testing criteria for the selection of appropriate pixel read-out locations for improved correspondence with ground truth,
- analysis of multispectral electro-optical digital imagery and comparisons with the performance of digitally converted reconnaissance aerial photography. Since CCD electro-optical cameras are more responsive to subtle changes than photographic film, an improvement in correspondence with surface observations would be expected; however, these higher contrast characteristics make accurate image exposures more difficult and such CCD-based systems are more sensitive to light flux problems,
- development and testing of correction procedures for varying water depth and bottom substrate conditions (see shallow water bathymetry and bottom substrate).

Sources

For the identification and mapping of sediment sources, two approaches exist. Conventionally, aerial photographic interpretation as well as field surveys are used to identify and map all potential sources, such as roads, clear cuts or land slides within the watershed. A large proportion of these potential sources, however, may not introduce any sediment into the river under most circumstances. Whether a potential source is in fact a real sediment source needs to be verified in the field. Digital image analysis of multispectral remote sensing data may be used to monitor SSC. When a change in SSC within the river or a sediment introducing tributary is detected one can search for the sediment source within a comparatively small area. This approach is liable to miss small sediment source but allows for faster and more efficient identification of significant sources.

In summary, qualitative monitoring of SSCs and sources using airborne multispectral reconnaissance remote sensing is operational. In addition, such remote sensing data have the potential to provide detailed and comprehensive quantitative estimates of SSC. However, for any particular study, an experimental approach involving calibration and further research is essential for reliable performance and to explore and develop this potential fully. We expect such reconnaissance remote sensing to play a significant role in future quantitative sediment monitoring projects in coastal riparian environments. Although airborne remote sensing of SSC has a larger margin for error than in the case with ground data, the synoptic capability and access to remote areas, provides an important contribution.

7 Substrate parameters

Classification

As discussed in the sections on bathymetry and suspended sediment, the type of bottom substrate affects the accuracy and precision of quantitative estimates of suspended sediment concentrations and of water depth. Several attempts at classifying substrate types have been made by other researchers (e.g. Lyzenga 1978 and Peddle *et al.* 1996), some with considerable success. A comparison of the images in Figure 8 and 18 shows that the distinction between spawning gravel (Figure 8), sandy bottom (Figure 14a) and coarse gravel or boulders (Figure 14b) is possible without field observations. Under the right conditions, a more precise classification of bottom substrate is possible; however, at this time, no research has been undertaken towards making it operational.

In addition to the size and type of the bottom material, algae cover and the presence of aquatic vegetation will have a considerable impact. Changing water depth, overhanging vegetation, sunglint, and other light flux factors have an impact on the classification of bottom substrate in the same way that they can hamper other procedures.

Area Estimates

Since remote sensing provides a synoptic view, successful substrate classifications can easily be translated into area estimates using photogrammetric methods. Sunglint, overhanging vegetation and shadow can prevent a successful substrate classification and it is not possible to expect to accurately map substrate along the entire length of a river using only remote sensing data. However, remote sensing allows convenient substrate mapping of substantial portions of most rivers. Considering limited access, especially in remote locations, remote sensing has considerable advantages over field surveys.

8 Vegetation parameters

Cover type and percent are components in all vegetation classifications, traditionally conducted with the aid of aerial photographic interpretation and field verification. The spectral response of varying vegetation shows the greatest separation in the near infrared region (Figure 2)and stressed plants often appear as a different colour from healthy plants. This property is routinely exploited for accurate mapping of vegetation.

Until recently, colour infrared film for small format reconnaissance camera systems was difficult to obtain and expensive to process. Eastman Kodak has just released a new colour infrared film. This colour infrared film can be processed using the E-6 process (rather than E-5A) which considerably reduces the expense and processing turn-around time, as it permits local processing in most photo-finishing laboratories. Plate 6 shows a normal colour and a false colour infrared image acquired over the Coquitlam River. This example shows the ability of the colour infrared film to distinguish between water and plants.

A. Coarse Gravel (Sowaqua Creek, BC)

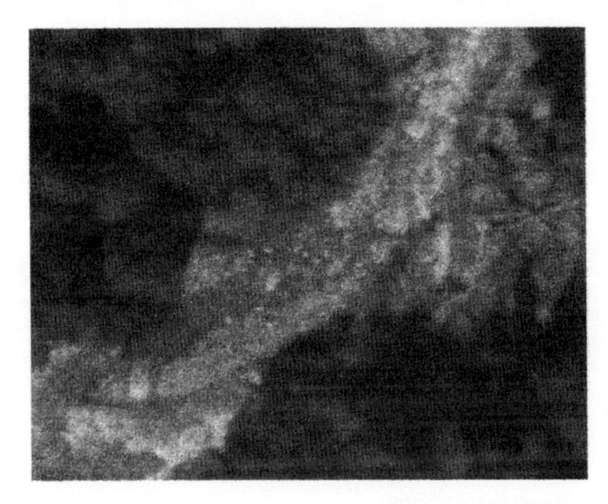

Substrate mainly made up of coarse gravel and boulders. Compare figure 37 which shows finer gravel (spawning gravel).

approx. 30 m

N

Flying height:	1000 ft
Focal length:	45 mm
date:	June 3, 1997
type:	normal colour

B. Sand (Horsefly River, BC)

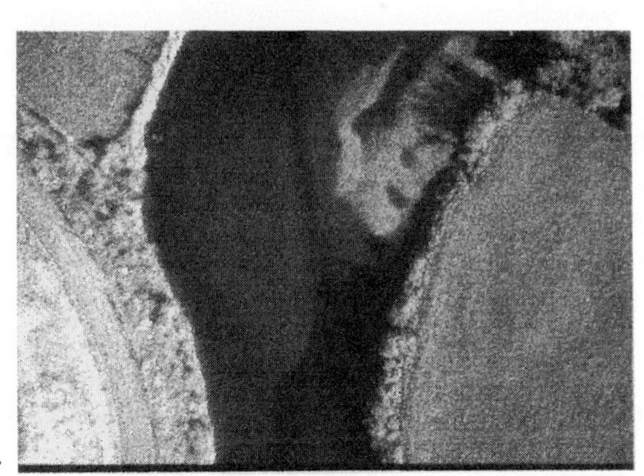

Substrate composed mostly of sand. In addition to the visibly finer sediment the smooth water surface indicates slow moving water in which fine sediment can settle out.

50 m

N

Flying height:	1000 ft
Focal length:	45 mm
date:	June 3, 1997
type:	normal colour

Figure 14 Examples of substrate: (a) Sowaqua Creek, where the substrate consists mainly of coarse gravel and boulders and (b) Horsefly River, where the substrate is composed of sand. Note in the Horsefly River case in addition to the visibly finer sediment, the smooth water surface which indicates slow moving water in which fine sediment can settle out.

Riparian vegetation - the mapping of riparian vegetation types and structural stages is a standard application of aerial photographic interpretation. It is usually done in several stages: preliminary mapping and interpretation, field reconnaissance and subsequent finalization of the interpretations. An example can be found in Morantz and Haefele (1996) where, in addition to type and stage, levels of health are reported as well. Figure 15 represents a portion of their classification system which includes vegetation, land use, bank erosion, encroachments and other parameters.

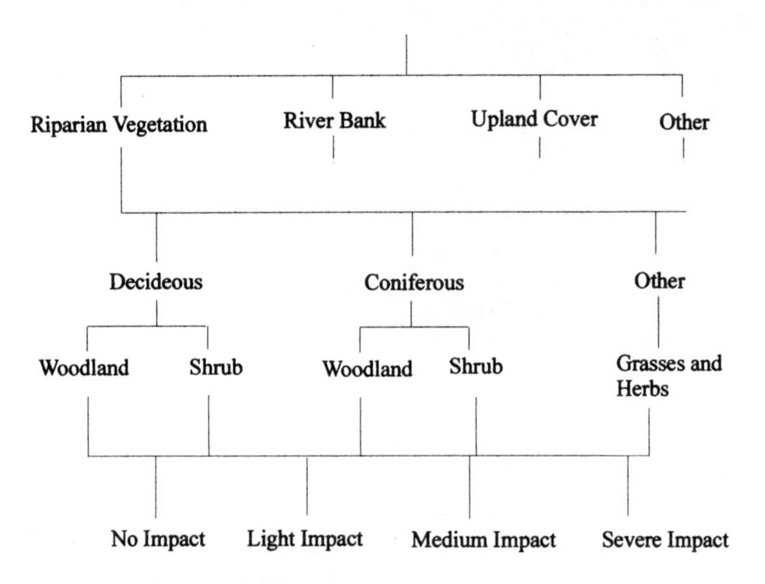

Figure 15 A portion of the Nicola River classification system.

Aquatic vegetation - images taken within the visible spectrum can generally be used to detect objects in clear water up to several metres in depth. So it is possible to identify locations with aquatic vegetation. The colour image in Plate 7 shows a portion in the Horsefly River with aquatic vegetation clearly visible in the channel.

With adequate ground truth, it is possible to identify the species of the vegetation by using digital image classification procedures. Furthermore, colour infrared film has the ability to detect more subtle vegetation differences than normal colour film. It can be used to detect aquatic and submerged vegetation that is close to the surface. The colour infrared image in Plate 8 is an example from the Horsefly River flown in the Spring of 1997. The channel has flooded over the banks and covered the adjacent floodplain. Vegetation that is not submerged appears in shades of red. Vegetation that is submerged in very shallow water (on top of the levee) appears in bluish tones. The deeper water looks dark.

9 Fish parameters

Over-winter dewatering

The extent of over-winter dewatering is a critical measure for the availability of fish habitat. It can be determined through the mean water depth at the lowest water level compared to the mean water depth at other times during the year. The lowest water level, at which mean water depth is conveniently measurable, usually occurs shortly before freeze up. Obtaining several mean water depth measurements throughout the year, however, is redundant unless the bed form changes significantly. Bedform changes are most likely in the spring freshet or during major flooding events. Conditions for measuring mean water depth are best in late summer when water levels are relatively low, SSC is low and weather conditions are fair for reconnaissance flights and collecting field measurements.

Remote-sensing-based methods to determine the bathymetry of a river not only yield mean water depth but allow mapping of the river's bedform, provided it is not too deep (see Section 5 on bathymetry). Once established the bedform can be assumed constant at least until the next major flooding event, which would not be likely to occur until the next spring. Therefore, monitoring of the water level will suffice for determining mean water depth on subsequent dates by taking into account previous bedform data. In remote areas, multiple imaging flights to determine the location of the waterline may replace the monitoring of actual water levels to provide an estimate.

Mapping the waterline can be achieved through aerial photographic interpretation as well as through digital image processing. The generally high contrast between dark water and brighter land lends itself to detection using digital classification procedures. This land-water contrast is highest in the near infrared portion of the spectrum and colour infrared film or near infrared channels of multispectral imaging systems are well suited for this application. Figure 16 presents a portion of the Horsefly River on three different dates: spring, early autumn and early winter. Generally this kind of image analysis can be carried out without any field work. Note that overhanging vegetation as well as shadow from vegetation can introduce significant error. Since there is no need to know the water level and the water line at all locations, the analysis may be restricted to areas with favourable conditions. Also note that the extreme contrast between dark water and fresh snow (Figure 16c) makes proper image exposure difficult and may prohibit shallow water bathymetry estimates.

Off channel habitat

Off-channel habitats are back-waters, lakes and ponds that may or may not be connected with the river's main channel. These habitats play an important role as rearing or over-wintering habitat. In addition, flooded areas during the spring freshet or after major rain storms in summer provide important feeding grounds for smolts. Aerial photographic interpretation and digital image processing provide convenient means of mapping such habitats.

Water bodies away from the river are easily detected on all kinds of imagery, with near infrared imagery providing the best contrast. Depending on the image scale, one may not, in some instances, be able to identify connections between off-channel habitats and the main channel. In such cases field visits are necessary to determine the level of utility of the off-channel habitat.

Horsefly at Airport Site (Black Creek)

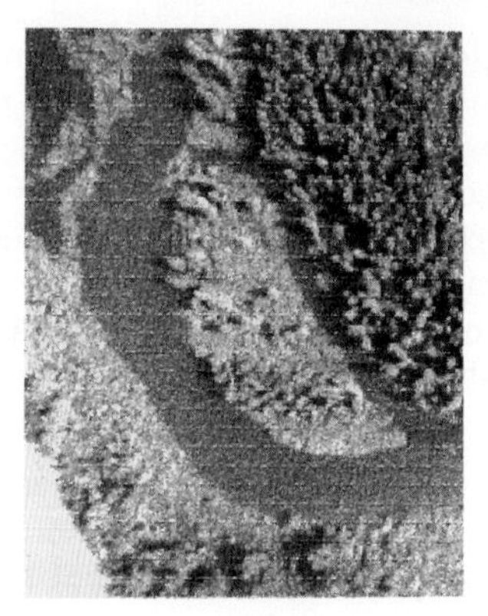

A.

N

approx. 100 m

Flying height:	1000 ft
Focal length:	24 mm
date:	June 3, 1997
type:	normal colour

B.

Flying height:	1000 ft
Focal length:	24 mm
date:	September, 1993
type:	normal colour

N

approx. 100 m

C.

approx. 100 m

N

Flying height:	1000 ft
Focal length:	24 mm
date:	Novemberm 4, 1994
type:	normal colour

Three images of the same location at different times of the year. Note the profound difference in wetted witdth between spring and early fall, while the difference between fall and winter is only minimal

Figure 16

Fish population estimates

Currently the Department of Fisheries and Oceans estimates the number of spawning salmon using manual counts on the ground, supported by occasional helicopter reconnaissance flights. Ground-based manual counting is probably the most accurate and precise method available. However, it is labour intensive and time consuming, and therefore costly. Similar to the monitoring of sediments, airborne multispectral remote sensing promises to reduce the amount of field work necessary to conduct fish population estimates.

Aerial photography can now already provide fish counts if the following conditions are met:

- very low flying height (high resolution),
- shallow, clear water,
- no obstructions (e.g. overhanging trees),
- high sun angle (to avoid sunglint and facilitate water penetration).

The conditions mentioned above severely limit the use of photo-interpretation for population estimates. Furthermore, the use of aerial photography may increase the turnaround time compared to field methods due to processing and interpretation constraints. An automated estimate of the number of spawners from multispectral imagery or from scanned photographs has, to our knowledge, never been attempted. We are, however, currently implementing three different experimental methods to estimate the density of spawners in riparian habitats. They range from the analysis of local variance within an image to the use of spatial statistics, to a simple model to simulate an image under fish free conditions, which is then compared to an actual image with fish present. These methods will be tested for the first time on the 1997 Horsefly River sockeye run.

10 Conclusion

For most riparian features there is some level of operational remote sensing to assist with and improve resource management in such environments. In a general sense most of these operational procedures involve photogrammetry and aerial photographic interpretation although, with experienced remote sensing trained personnel, digital analysis of multispectral remote sensing imagery can also be utilized at the operational level. This is obviously constrained by the type and scale of imagery and in general digitally scanned colour and colour infrared aerial photography may be the most reliable and robust product for digital analysis for operational resource management purposes.

An important distinction must be made between operational and experimental procedures for the remote sensing of fish habitats and riparian features. Resource management requires reliable procedures that are robust and can be undertaken without highly specialized training under normal field operating conditions. A number of new remote sensing and image processing procedures are being developed and tested in operational environments and it is expected that these, mainly quantitative, evaluations, using airborne electro optical imagery and image processing procedures, will be refined for future routine use in resource management.

References

Anon., 1996, WRP Riparian Assessment Field Guide, First Approximation, Riparian Assessment and Prescription Procedures, Watershed Restoration Program, Ministry of Environment, Lands and Parks and Ministry of Forests, Victoria, BC, Canada

Avery T E and Berlin G L, 1992, Fundamentals of Remote Sensing and Air Photo Interpretation, 5th Edition, Maxwell MacMillan Canada, Toronto

Department of Fisheries and Oceans, 1986, Policy for the management of fish habitat, Department of Fisheries and Oceans, Ottawa, ON

Dickinson C G, 1979, Maps and Air Photographs, Edward Arnold, London

Ferrari G M, Hoepffner N and Mingazzini M, 1996, Optical properties of the water in a deltaic environment: prospective tool to analyse satellite data in turbid waters, Remote Sensing of Environment, **58**, 96-80

Gitelson A, Garbuzov G, Szilagyi F, Mittenzwey K-H, Karnielli A and Kaiser A, 1993, Quantitative remote sensing methods for real-time monitoring of inland waters quality, International Journal of Remote Sensing, **14**, 1269-1295

Gitelson A, Szilagyi F and Mittenzwey K-H, 1994, Improving quantitative remote sensing for monitoring of inland water quality, Water Research, **27**, 1185-1194

Goodin D G, Han L, Fraser R N, Rundquist D C and Stebbins W A, 1993, Analysis of suspended solids in water using remotely sensed high resolution derivative spectra, Photogrammetric Engineering and Remote Sensing, **59**, 505-510

Groot C and Margolis L (ed.), 1991, Pacific salmon life histories, UBC Press, Vancouver

Hogan D L, Bird S A and Wilford D J, 1996, Channel Conditions and Precriptions Assessment (Interim Methods), Watershed Restoration Technical Circular No. 7, Watershed Restoration Program, Ministry of Environment, Lands and Parks and Ministry of Forests, Victoria, BC, Canada

Johnston N T and Slaney P A, 1996; Fish Habitat Assessment Procedures, Watershed Restoration Technical Circular No. 8, revised April 1996, Watershed Restoration Program, Ministry of Environment, Lands and Parks and Ministry of Forests, Victoria, BC, Canada

Lillesand T M and Kiefer R W, 1994, Remote Sensing and Image Interpretation, 3rd Edition, John Wiley & Sons, Toronto

Lyzenga D R, 1978, Passive remote sensing techniques for mapping water depth and bottom features, Applied Optics, **17**, 379-383

Morantz D and M Haefele, 1996, Nicola River drainage riparian condition survey and mapping report, Department of Fisheries and Oceans, Fraser River Action Plan, unpublished

Peddle D R, LeDrew E F and Holden H M, 1996, Remote estimation of coral reef abundance and depth from SPOT image spectral mixture analysis and ocean profile spectra, Fiji, Proc 26th Int. Symp Remote Sensing of Environment, ISRSE (International Symposium on Remote Sensing of Environment) and CRSS (Canadian Remote Sensing Society), Vancouver, BC, pp 555-558

Pellika P, 1996, Multitemporal relative spectral calibration of airborne video data, Proc. 2nd Airborne Remote Sensing Conference and Exhibition, Environmental Research Institute of Michigan (ERIM), Ann Arbor, MI, **3**, pp 17-26

Roberts A, Kirman C and Lesack L, 1995, Suspended sediment concentration estimation from multi-spectral video imagery, International Journal of Remote Sensing, **16**, 2439-2455

Van Stokkom H T C, Stokman G N M and Hovenier J W, 1993, Quantitative use of passive optical remote sensing over coastal and inland water bodies, International Journal of Remote Sensing, **14**, 541-563

Acquisition of satellite based remote sensing data for coastal applications

Paul S Crawford

University of Dundee

1. Choice of technique

The study of coastal regions and the dynamics of the systems which affect their state can be greatly enhanced by remote sensing techniques. By taking the 'ground truth' data from *in situ* measurements and developing models which extract similar information from remotely-sensed data, aircraft or satellite based instruments can provide a powerful and cost effective tool for the analysis and monitoring of large areas. The practicalities of this approach often depend on the cost and availability of suitable data; these are improving with technology.

There are advantages and disadvantages to all techniques and the best approach for any given situation will normally involve the fusion of various data sources into one coherent model. Briefly, the three techniques considered here are:

- *in situ* (ground truth) measurements,
- airborne instruments and
- spaceborne instruments.

1.1 Ground truth data

The *in-situ* measurements used to provide the ground truth data are typically floating buoys (either with data recording and/or radio links for telemetry) or ship-based equipment. Buoys can provide good temporal coverage for local measurements, but are not always at the required location. Ships are expensive but can carry heavy complex equipment and follow a track over a wide area of interest.

These *in situ* techniques can provide a direct measurement of the parameter(s) required and offer unique possibilities, such as underwater measurements of sea bed profile, *etc.*, which cannot be directly obtained from other sources.

1.2 Airborne sensors

The airborne and spaceborne techniques are both real remote sensing techniques in that the instrument measures certain parameters which are remotely observable (for example, reflected sunlight) and these are used to infer certain other geophysical or biological information. The difference between the aircraft and satellite techniques relates to the coverage and view point provided.

The aircraft, like the ship, is an expensive tool to operate. It offers the advantages of flexible instrumentation together with good coverage of the area of interest. By flying close to the ground (relative to a satellite orbit) the instrument can view very high spatial detail and is less affected by the Earth's atmosphere, cloud cover, etc. Due to the high flying cost and need to refuel, the temporal coverage is often quite limited. Similarly, remote regions beyond normal airports become difficult and expensive to observe.

1.3 Spaceborne sensors

Satellite based instrumentation offers global coverage and the potential for observing very large areas with good temporal coverage. The actual coverage obtainable depends on a number of factors:

- satellite orbit,
- instrument coverage area,
- spacecraft operating constraints,
- weather and solar illumination (for passive measurements) and
- availability of reception and data handling facilities.

2. Choice of satellite

There are two principal orbits used for remote sensing applications: the geostationary orbit and the polar orbit.

In the case of a geostationary satellite, the orbital plane is the equatorial disk and the height is chosen to match the orbit period to the sidereal rotation rate of the Earth (23h 56m 4.091s). This provides a nearly constant view of the full Earth disk, around 160° of longitude, and so the spacecraft can provide excellent temporal and spatial coverage. The disadvantage is the considerable distance (35,784km or 5.61 Earth radii) which limits the spatial, radiometric and/or spectral resolution achievable. Typically, the geostationary satellites carry passive visible and infra-red imagers with 1–5km resolution at the sub-satellite point and a repeat cycle of around 30 minutes for each full Earth scan.

The polar orbiting satellite is much closer to the Earth, typically 840km (or 0.13 Earth radii) and this improves the achievable resolution. Due to the relatively high resolution required for a major number of coastal applications, this orbit is the most important category. These orbits are inclined by approximately 9° to the polar axis. This exploits the Earth's oblatness to precess the orbital plane by 360°/year so the Sun-Earth-satellite angle remains fairly constant; this is known as a sun-synchronous operation. A polar orbiting satellite orbits the Earth around 14 times per day. The number of these within reception range of a given ground station depends upon the latitude; in the case of Dundee we can receive around 10 orbits per day although some are of very limited contact time.

The coverage area provided by the instruments flown on the polar orbiting satellites vary from a wide swath for instruments with medium resolution to a narrow swath for high resolution instruments. In the case of the Synthetic Aperture Radar (SAR), the coverage area is always offset from the sub satellite point, but this can occur with some passive instruments as well.

As an example relevant to Dundee, the NOAA AVHRR instrument's swath covers almost 3,000km with around 1.1km sub-satellite resolution and an area approaching $18 \times 10^6 km^2$ is observed on each pass. Any given area around the UK is usually covered once or twice every 6 hours by the two operational NOAA satellites.

For coastal applications, high spatial resolution is of great interest and there are a number of very high resolution optical satellites planned for launch soon. These will offer optical data with up to 1m resolution, a feature which formally was in the domain of military spy satellites only. The disadvantage of these systems is the high cost and the very limited coverage area in any one data set, or in the temporal coverage for studying moving features.

2.1 Availability of satellite data

Even though there is a satellite which is capable of observing the area of interest, it is not always possible to do so. The reasons for this include:

- The spacecraft has a limited power capacity and the chosen instrument will only be turned on if the user's request fits the operational schedule.
- The orbit only takes the instruments' field of view over the user's area on a limited number of occasions.
- The instrument is a passive imager and requires cloud free sky (and often solar illumination).
- When the user's area of interest has been imaged, there are no suitable ground stations to receive the data.

A final constraint in most applications is the cost of the data and the speed of delivery to the end user. This can be an overriding factor in the choice of data, particularly for an real time monitoring application which will use a large number of data sets in the project life time.

If we consider a few of the common Earth observation satellites, they fall in to two broad categories:

1. Low data rates (less than a few Mbit sec^{-1}) operating in the now crowded VHF (~137MHz) and L-band (~1.7GHz), such as the NOAA weather satellites. These can be received with relatively cheap equipment and the data is often free.
2. High data rates operating in the X-band (~8.2GHz) for active radar (SAR, wave profile and altimeter such as ERS, RadarSat, etc.) and high resolution passive sensors (Landsat, SPOT, etc.). Most of these are operated for some form of profit and the reception and data handling is still difficult. As a result, the data is often quite expensive (around £1–2,000 per scene).

2.2 The data handling system

On board the spacecraft, there are a number of data sources. These include both the actual Earth observation instrument(s) and the on-board electronics which provide all of the essential support systems, for example attitude control systems and the power supply monitoring from the solar panels and batteries.

The data from these sources is typically combined by some on board computer into a stream of parallel digital words. For transmission, these must be converted into a serial stream of bits which can be sent using a radio link to the ground station. This requires frame formatting where synchronisation words are inserted to allow the receiver to identify the position of the bits in the serial data stream. Additional operations might include:

- Data compression to reduce the volume of data sent (not yet common on board spacecraft).

- Encryption to prevent unauthorised access.

- Forward Error Control (FEC) coding to reduce the required signal power and to help combat interference.

2.3 On-board data collection and formatting

The older generation of satellites are noted for each system employing a different (and in some cases unique) standard for the data encoding, etc. This makes a multi-mission ground station expensive due to the need for multiple hardware or complex programmable systems. The most recent satellites, such as METEOSAT 2nd generation (see EUMETSAT 1996) or NASA EOS, have opted for a relatively common standard based on the CCSDS recommendations (CCSDS 1989).

One of the notable features of this recommendation is the use of a *concatenated code* for the FEC system. This is a very powerful technique which uses two error correcting codes to obtain the best features of both types.

Normally the parallel data words are grouped in blocks first and encoded using a Reed-Solomon block code. This appends a set of *check words* to the end of each block allowing a number of errors to be detected and corrected. The recommended system uses 223 data bytes (8-bits) and 32 check bytes to form a total of 255 and can correct up to 16 words in error.

The next stage is the *frame formatting* where 4 groups of 255 coded words are interleaved and a 32-bit synchronisation word is added to the start of the block. The interleaving process is used so that a burst of errors is divided evenly between the 4 blocks. Since each block can correct up to 16 errors, the system is now tolerant of 64 bytes in error (512-bits in series). See Figure 1 for an example block diagram of such a system.

The use of encryption to prevent unauthorised data access is becoming common, even for meteorological data which is of low commercial value. There are two general types of data encryption: the block cipher and the stream cipher.

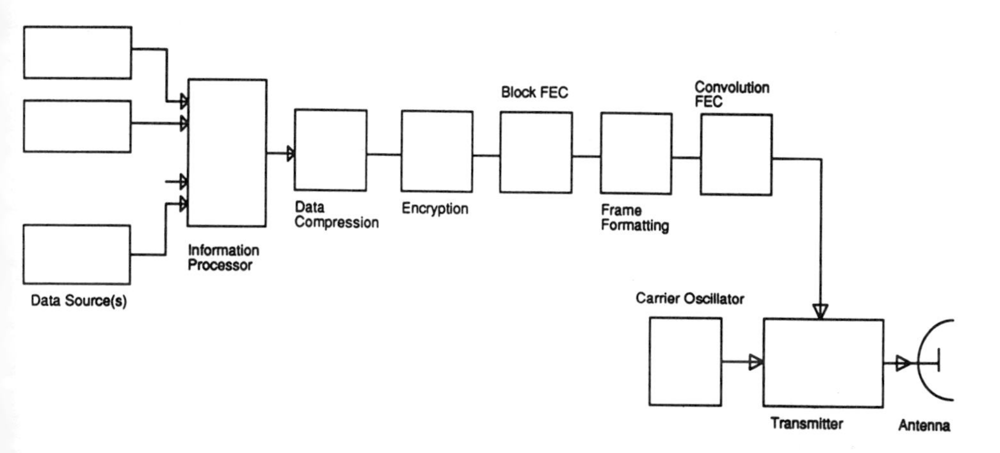

Figure 1. Typical spacecraft on-board processing.

The block cipher uses a function which maps one block of data in to another block of the same size; this is a one-to-one mapping and is reversible. The exact form of the mapping is determined by a secret key. As a general rule, the strength of a cipher should depend on the security of the key, and not on the secrecy of the algorithm. For example, the DES algorithm (NIST 1993) has 64 bits of data and each of the 2^{64} DES input values maps to one of the 2^{64} unique output values. The key is 56 bits long (8 check sum values to make 64) and approximately 2^{54} attempts would be required in a 'brute force' attack.

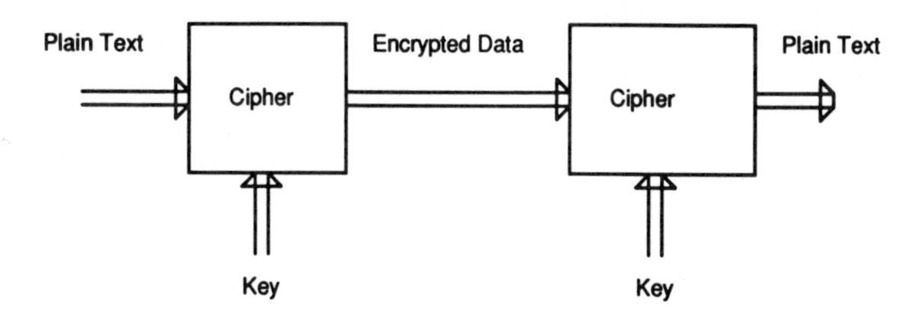

Figure 2. Example block cipher system.

The problem with a block cipher for satellite data is that of error propagation. A single bit in error at the input to the deciphering process and the whole block is lost. An alternative to this is the stream cipher. Here the data is exclusive-OR'd with a pseudo-random sequence at the sending end and the same process at the receiving end reverses the encryption. This is much more attractive for unreliable links and for ease of use, a conventional block cipher such as the DES can be used to generate the pseudo-random sequence.

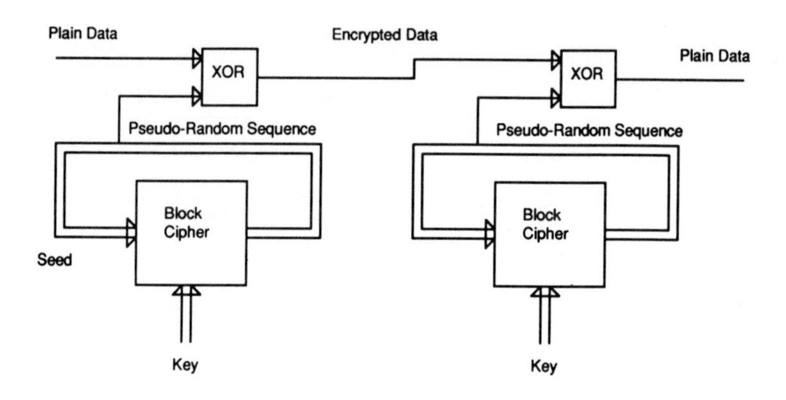

Figure 3. Example stream cipher system.

2.4 Reception system

The first task of a reception system is the tracking system. This must point the reception antenna at the satellite with sufficient accuracy to avoid signal loss. As the data rate increases, the required signal strength also rises and the power limit on the spacecraft shifts the responsibility to a larger ground station antenna. This, together with the higher frequency of the X-band, decreases the antenna beam width which increases the pointing accuracy required compared to low rate satellites. In addition, the greater reflector size increases the wind loading and mass of the moving system, further increasing the cost of the mechanical servo system. An approximate guess is the cost of a reception system increases almost exponentially with reflector size. To achieve cost effective operation, high performance signal processing and very low noise amplifiers can be used to minimise the size of reflector required.

The next task is the reception signal processing system. The task of the reception system is basically to reverse the processing stages on board the satellite to retrieve the original sensor data in a useful form. The higher data rate of advanced satellites pushes the cost of the signal processing electronics up and limits the range of techniques that are practical. As the performance of ICs increase this is less of a problem but the small number of ground stations reduces the economies of scale so common in other areas of electronics. The use of Digital Signal Processing is becoming more common as the demand from other areas of the electronics industry spurs on IC manufacturers. Areas such as computer graphics and mobile phones have driven DSP processor and radio systems up in performance:price ratio, however the data rates from the advances satellites are much higher than all but high definition TV processing. Most specialised radio electronics today is designed to make maximum use of commercial electronics, although the production lifetime of such devices is limited.

A data storage and processing is the third task of the reception system. Today the cost and performance of computers has increased at such a rate there is little need for direct tape based storage anymore. Most systems now ingest directly into the computer and then on to a large disk array. There can still be some surprises, though; a full pass of ERS SAR data could require 11GB of storage, which is beyond the 2GB partition size of a number of current 32-bit operating systems.

Eventually the data must be accessed by the end user, the most common forms of data transfer are magnetic tape (EXABYTE, DLT, etc.) for bulk transfer for slow delivery and network transfer using FTP for near real-time applications.

3. Conclusions

The future of satellite data applications is sure to grow as the environmental impact and cost of human activity must be monitored and controlled. The advances in electronics make ground station availability higher, although pressure on the radio spectrum from commercial applications such as mobile phones is creating serious problems for the future use of this valuable service.

The Internet and World Wide Web, those much-hyped systems today, will bring better access for all to the data as ground stations such as the NERC funded Dundee Satellite Station make their data available in near real-time with this technology. The only problem faced by this service is the declining speed of the Internet and the increasing size of the satellite data products.

References

CCSDS, 1989, Advanced Orbiting Systems, Network and Data Links, CCSDS 700.0-G-2, Green Book, Issue 2, Consultative Committee for Space Data Systems, NASA, Washington, USA

EUMETSAT, 1996, LRIT/HRIT Mission Specific Implementation, Issue 1, EUM/MSG/SPE/057, EUMETSAT, Ground Segment Division, P.O. Box 100555, Am Kavalleriesand 31 64205 DARMSTADT Germany

NIST, 1993, Data Encryption Standard (DES), Federal Information, Processing Standard FIPS PUB 46-2, US Department of Commerce, National Institute of Standards and Technology, December 1993, Washington, USA.

Data fusion

Steve Parkes
University of Dundee

1. Introduction

Remote sensing provides a powerful means for collecting environmental information rapidly and efficiently over wide areas. This source of data can be even more powerful when combined with data from other satellite sensors, airborne sensors and/or other forms of environmental information including GIS (Geographic Information System) data. Multi-sensor data fusion is an evolving technology which encompasses a rich set of techniques for combining different types of data. Neural networks and fuzzy logic methods as well as more conventional techniques may be used to fuse remote sensing image data improving the information content compared to that of a single original image.

This chapter introduces the basic concepts of multi-sensor data fusion, concentrating on the fusion of remote sensing image data. After addressing the need for image alignment an overview of several pixel-level fusion techniques is given and possible sources of data are considered. Some examples will be examined.

1.1 What is data fusion?

A data fusion system transforms data from a set of sensors into a decision about the identity or characteristics of an entity, taking into account the context of the surrounding environment and the relationship of the entity to other entities. Here, an entity is a real-world object, event or phenomenon of interest to the user of the data fusion system. For example a field of wheat is an entity in a land-use application, an aircraft is an entity in a defence application and a cancerous growth is an entity in a medical application. A good definition of multi-sensor data fusion was given by Llinas and Hall (1993):

> *"Data fusion is the integration of information from multiple sources to produce the most specific and comprehensive unified data about an entity."*

To begin to answer the question "What is Data Fusion?" it is worth considering the meaning of each word in this definition:

- Integration – combining, merging, bringing together.
- Information – not just image data but other forms of information.
- Multiple – more than one source of information.
- Source – not only information from sensors but a broad range of sources.
- Most – an optimisation process.

- Specific – a goal in mind, unambiguous.
- Comprehensive – including all available information.
- Unified – formed into a single representation.
- Entity – about a real-world object or event.

1.2 Why use data fusion?

Some specific benefits that can be achieved with appropriate use of data fusion techniques are listed below.

Extended spatial coverage. One sensor may cover an area that another sensor is not able to cover. The combined sensor data covers a larger area than is covered by a single sensor. An example of this is the use of Synthetic Aperture Radar (SAR) to image an area obscured by cloud and hence invisible to an optical instrument.

Extended temporal coverage. One sensor may cover an area at a different time or at a different repetition interval than another sensor. One sensor operating alone may completely miss an important phenomenon or event.

Extended spatial resolution. A single spectral band sensor of high resolution can provide enhanced spatial information for a low resolution multi-spectral sensor.

Increased confidence. A second sensor can confirm the result of another sensor, thereby enhancing confidence in the result.

Reduced ambiguity. Complementary multi-sensor data for an entity can reduce the number of classification hypotheses. The microwave reflectivity of an object detected by a SAR instrument may enable discrimination of two objects which appear similar in an optical image.

Improved detection. Integrating information from several sensors increases the likelihood of detecting an entity or event. This is akin to the integration of sensor data to improve detection probability.

1.3 Application to the coastal zone

Multi-source data fusion has widespread applications both within remote sensing and across many other areas. In the coastal zone wherever remote sensing can be used and where there is more than one source of data available, data fusion techniques may be used to advantage. Ice monitoring and flood monitoring are two examples where data fusion has proved beneficial. In ice monitoring, multi-sensor data fusion can provide increased detection performance and increased confidence through the validation of the result of one sensor by another (Armour *et al*, 1994, Ramseier *et al.* 1993).

In flood monitoring, fusion of SAR and optical imagery is useful because the two forms of data are complementary (Corves 1994, Wang *et al.* 1995). Optical data are good for initial land-use and water body mapping before flooding and SAR data are good for measuring the extent of a flood, particularly because a SAR is an active instrument which can image both through cloud and at night.

2. Image data fusion

2.1 Image alignment

It seems obvious that if you combine two or more sets of data about an entity then you will have more information than is available in a single set of data. However, considerable care must be taken in combining the data to avoid significant errors in the fused data. For example if two images to be fused are misaligned then blurring or mis-classification will result.

Image fusion involves taking two or more images (or image bands), aligning them and then either merging and displaying the data ready for operator interpretation or automatic classification of the combined image data. Image alignment requires that each pair of pixels from a common point on the ground appear at the same pixel location in the two images. At first this seems like a relatively easy task – simply identify a few common points in the two images and then warp one image so that it aligns with the other image. During the warping process some form of pixel interpolation will be necessary. This interpolation can also be used to adjust for any differences in the pixel size between the two images.

With some images taken from the same viewing angle this is adequate but for most images this approach will not result in all pixels being correctly aligned. As an example of where this simple image alignment fails, consider a pair of cameras viewing a pair of objects at different distances from the cameras. This is illustrated in Figure 1. The square and the star are at the same depth but the diamond is further away. In the left-hand image they appear in the order square-diamond-star and in the right-hand image they appear in the order square-star-diamond. This is because of the different depth of the diamond which results in an apparent shift in its position relative to the star. This shift in position is known as disparity or parallax. Because of this effect it is impossible to use a simple image transform to align the two images.

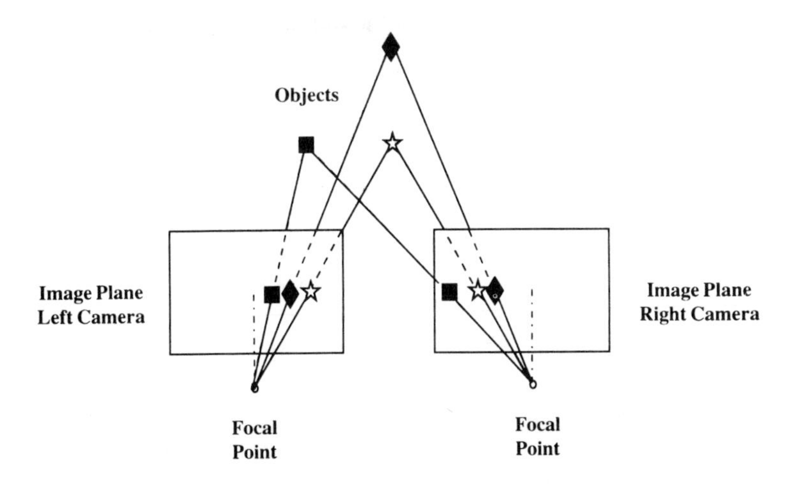

Figure 1 : Images from different viewing angles

The parallax problem arises in Earth observation imagery. If we have a pair of images of undulating or mountainous terrain taken from two different views then disparity will prevent the two images from being aligned correctly. Some means of overcoming the effects of terrain height and the consequential disparity is required.

If the images are taken from the same viewing angle then alignment is simple. With this in mind, one method of aligning images of an area with significant height variation is to use knowledge of the height of each point of the terrain to correct for the effects of height. Each image can be converted using the height information to provide the equivalent view from vertically above the terrain. Such a view is known as an orthoimage. Alternatively, one image can be transformed to simulate the view from the position of the second image sensor. With both images now from the same viewing angle alignment is straightforward.

This process is known as precision-registration or precision-geocoding if the final image projection is onto a map projection. When two or more images are precision-geocoded the pixel position in the image determines the geographic location from which that pixel information was collected. Precision-geocoded images from different sensors may be of different resolutions and this will need to be taken into account when aligning the image data.

The height information used is called a digital elevation model (DEM) or digital terrain model (DTM). A DEM is a set of height data with one value for each point on a map grid. A DEM can be viewed as an image in a map projection where the value of each pixel represents the height of the corresponding terrain point. The complete precision geocoding process is illustrated in Figure 2.

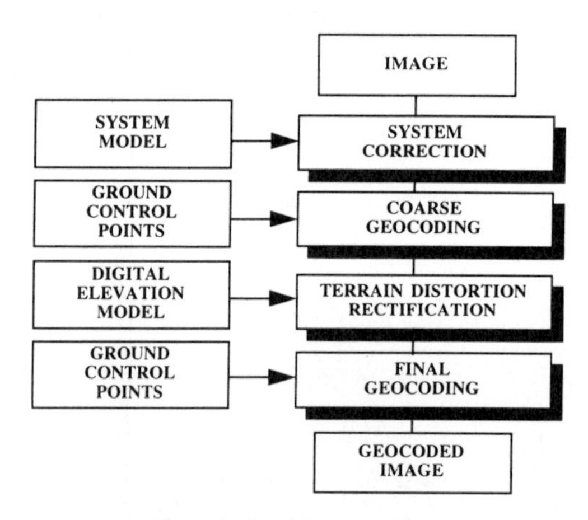

Figure 2 : Precision geocoding

Precision geocoding begins with system corrections to the image using a model of the sensor system. This covers system calibration, radiometric correction and initial geometric correction. A set of ground control points (GCPs) is then used to provide for coarse transformation to a map projection without taking into account terrain relief. A ground control point is a point which is clearly identifiable in the imagery and whose geographical coordinates are known. Using a graphical user interface the image is displayed so that the GCPs can be identified and their positions in the image determined. The collected image positions and the corresponding geographical positions are then used to transform the image to the required map projection.

Once the coarse geocoding has been done a digital elevation model is used to correct for the effects of terrain relief. Additional ground control points may be used to correct for any residual distortion, resulting finally in an image which is accurately aligned with a geographical frame of reference and is ready for fusion.

One book which covers geocoding for several different types of space-borne instrument is by Williams (1995).

Change detection and tracking

For most forms of image data fusion, precision-geocoding and resampling to get the images to the same scale will be sufficient. However, there is one case where this process will still not result in correctly aligned data. That is when the item of interest physically moves or changes in the time between acquiring one image and acquiring the second image. Image fusion may at first seem impossible under these circumstances but there are two ways in which fusion can be applied: for change detection and following entity tracking.

1. **Change detection.** After correctly aligning two images taken at different times they may be subtracted from one another (or ratioed, see later) to highlight the changes that have taken place. A pixel which has not undergone a change will result in a value of zero and appear black in the fused image indicating "no change". A pixel which has changed significantly will appear bright in the fused image.
2. **Entity tracking**. Consider a chemical spill into a river. Using multiple sensors it may be possible to identify the particular chemicals involved. The problem is that the sensors collect information at different times and the chemical spill moves down the river in between sensor data sample times. To overcome this problem the spill must be tracked from one sensor image to the next by predicting the motion of the spill and then adjusting the motion estimate with improved information from the next image. In this way the chemical spill data from multiple sensors can be aligned and fused to aid identification.

2.2 Image fusion

Having aligned the imagery the data are fused by combining the different images (or image bands) into one image. Image fusion either enhances image data for display to a skilled operator who then interprets the image, or classifies the image automatically. Fused image data can improve classification accuracy whether this is done by an operator or automatically.

Several image fusion techniques used for image enhancement will now be introduced, followed by several techniques used for classification.

2.3 Image enhancement techniques

A comprehensive review of data fusion techniques for image enhancement and their applications was made by Pohl and van Genderen (1997). Several of these techniques are described below.

Colour composition fusion

A colour image on a computer display is made up of three colour bands: red, green and blue (RGB). A common image fusion technique is to assign three different sets of image data to the three colour image channels and to display the result on a monitor for interpretation. An example of this is the use of multi-temporal SAR images for classification and change detection. Three SAR images taken of one area at different times are assigned to the R, G and B channels. Areas which have not changed from image to image will appear grey while those areas that have changed will appear coloured.

Optical and microwave data as well as multi-sensor optical and multi-temporal optical and SAR have been fused in this way.

Intensity-Hue-Saturation

The Intensity-Hue-Saturation (IHS) transform operates on an RGB image to form another set of three images: intensity, hue and saturation images. The intensity image captures the spatial information and the hue and saturation images the spectral information (hue provides information on the dominant spectral contribution and saturation provides information on spectral purity). The IHS transform is given by

$$\begin{pmatrix} I \\ v_1 \\ v_2 \end{pmatrix} = \begin{pmatrix} \frac{1}{\sqrt{3}} & \frac{1}{\sqrt{3}} & \frac{1}{\sqrt{3}} \\ \frac{1}{\sqrt{6}} & \frac{1}{\sqrt{6}} & -\frac{2}{\sqrt{6}} \\ \frac{1}{\sqrt{2}} & -\frac{1}{\sqrt{2}} & 0 \end{pmatrix} \begin{pmatrix} R \\ G \\ B \end{pmatrix} \tag{1}$$

$$H = \tan^{-1}\left(\frac{v_2}{v_1}\right)$$

$$S = \sqrt{v_1^2 + v_2^2}$$

The IHS transform has been widely used to enhance the spatial resolution of colour images by combining a colour image with a higher resolution monochrome or panchromatic image. To do this the colour image (three channels - RGB) is transformed to intensity-hue-saturation. The intensity channel is then replaced by a higher resolution panchromatic image. The image replacing the intensity image is normally contrast stretched to match the contrast of the original intensity image. The reverse transform results in a fused image which has the spatial resolution of the high resolution panchromatic image and the spectral content of the colour image.

IHS fusion has been applied with success to the spatial resolution enhancement of SPOT multispectral data (25m resolution) by fusing it with the corresponding SPOT panchromatic data (10m resolution).

Other transforms

There are several other transforms similar to the IHS transform that have been used for data fusion including Hue-Saturation-Value (HSV) and Luminance-Chrominance transforms. In the luminance-chrominance transform the coefficients are based on the sensitivity of the human visual system.

Adding and multiplying

The addition and/or multiplication of images can be used to combine them to enhance contrast and to improve resolution. Many different forms have been used successfully. The basic operations are as follows:-

$$F = w_1 A + w_2 B + c$$
$$F = wAB + c \quad (2)$$

where F is the fused image pixel, A and I are pixels from the two input images, c is a constant and w_1, w_2 and w are weighting factors.

The following fusion technique is used to improve the resolution of SPOT multispectral (XS) imagery using the high resolution panchromatic band (PAN)

$$F_1 = w_1\sqrt{PAN \times XS_1} + C_1$$
$$F_2 = w_2\sqrt{PAN \times XS_2} + C_2 \quad (3)$$
$$F_3 = w_3\sqrt{0.25 \times PAN + 0.75 \times XS_3} + C_3$$

The different weightings used in the calculation of the F_3 band take into account the different correlation between XS_1, XS_2 and PAN and XS_3 and PAN.

After fusion F_1, F_2 and F_3 are assigned to the RGB channels for display and interpretation.

Differences and ratios

Image differences and/or ratios provide another means of image fusion. Large pixel differences between two images are highlighted when one image is subtracted from another. Hence differencing is very useful for change detection. When an image is divided by another the result is one if the two pixels are the same, between one and zero if the divisor is larger than the dividend and greater than one otherwise. With careful setting of a threshold, ratioing can also be used for change detection.

The following difference/ratio method has been found to be useful in the detection of urban change.

$$F = \frac{XS_3 - XS_2}{XS_3 + XS_2} - \frac{TM_4 - TM_3}{TM_4 + TM_3} + C \quad (4)$$

where XS_i is the ith band of the SPOT multispectral imager and TM_i is the i-th band of the Landsat Thematic Mapper.

Ratioing has also been used for spatial enhancement using the following algorithm:

$$F_j = PAN \times \frac{XS_j}{synPAN} \quad (5)$$

where PAN is the SPOT high resolution, panchromatic band and $SynPAN$ is a panchromatic band at the resolution of the SPOT multispectral band produced by reducing the resolution (filtering and resampling) of the high resolution SPOT panchromatic band (resolution changed from 10m to 25m).

Another technique for image sharpening produces the SynPAN band by summing together the SPOT multispectral bands

$$F_j = PAN \times \frac{XS_j}{XS_1 + XS_2 + XS_3} \tag{6}$$

This technique can also be used for the fusion of Landsat Thematic Mapper (TM) bands with SPOT panchromatic by replacing the SPOT multispectral bands (XS_i) with TM bands.

Fusion by principle component analysis

In principal component analysis (PCA) the information in several images or image bands is compressed into just two or three bands. Most of the information is contained in the first principal component image, somewhat less in the second component and much less in the third and subsequent components. Principal component analysis is illustrated in Figure 3.

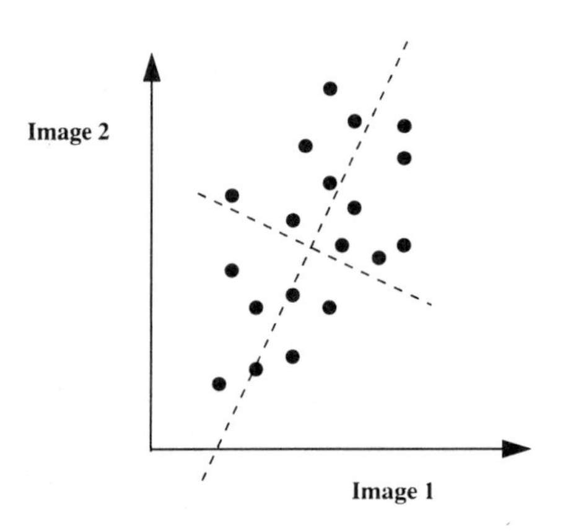

Figure 3 : Principal component analysis

The axes represent the grey-level pixel values of two images. The dots represent particular pairs of pixel values. Principal component analysis calculates a new axis through the data which lies along the line of maximum variance. A second axis is then determined, orthogonal to the first, which is again a line of maximum variance. These two axes can be seen in Figure 3. When combining several images several principal components may be extracted.

Since the first principal component is along the line of maximum variance it will contain most of the information in the data set. The second component will contain less information. Application to a correlated set of images will result in an uncorrelated set of images ordered in decreasing variance.

PCA can be used to fuse images in two ways. The first method is to apply PCA to a set of images and to then view the first three bands as the RGB components of a colour image. In some classification applications the first principal component is discarded and the next three components are assigned to the RGB channels, since the first component contains information mainly related to surface texture and the next three components contain most of the spectral information.

The second method of fusing image data with PCA is for image sharpening. A set of multispectral bands are subjected to PCA. The first principal component which contains the spatial information is discarded and relaced by a higher resolution image which is stretched to have the same mean and variance as the original first component.

2.4 Image classification techniques

The data fusion techniques outlined above all depend upon the human operator to do the image interpretation. Data fusion is used to enhance the image data making it easier for the operator to identify key features in the imagery. To do this, data fusion combines relevant information from several images into a single representation which can be displayed to the operator.

Other image data fusion techniques exist which perform identification or classify features in the imagery into a set of pre-defined classes. An overview of several of these classification techniques will be provided later, but first it is necessary to introduce the basic idea behind classifier operation.

A classifier works on feature vectors. Features are either individual pixels taken from an image or a value derived from a group of pixels. A textural feature is a value derived from an area surrounding a particular pixel, for example, the variance, mean or median of the surrounding area. A set of features corresponding to a single point on the ground is a feature vector. A spectral feature vector is a set of corresponding pixels from several spectral bands. A textural feature vector is a set of texture features derived from the area around a particular pixel. Feature vectors may contain a combination of spectral and textural features.

A feature space is the multi-dimensional space defined by the domain of a feature vector, i.e. the set of possible values that a feature vector may hold. Each feature is one of the dimensions in the feature space. Consider a two-dimensional feature vector. It will define a two-dimensional feature space as illustrated in Figure 4.

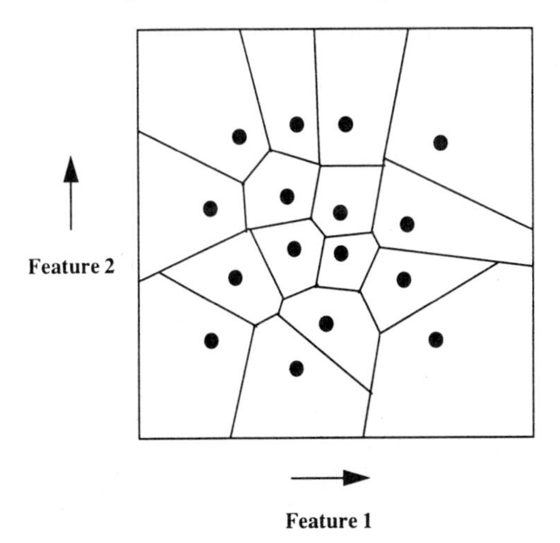

Figure 4 : Classification in a 2D feature space

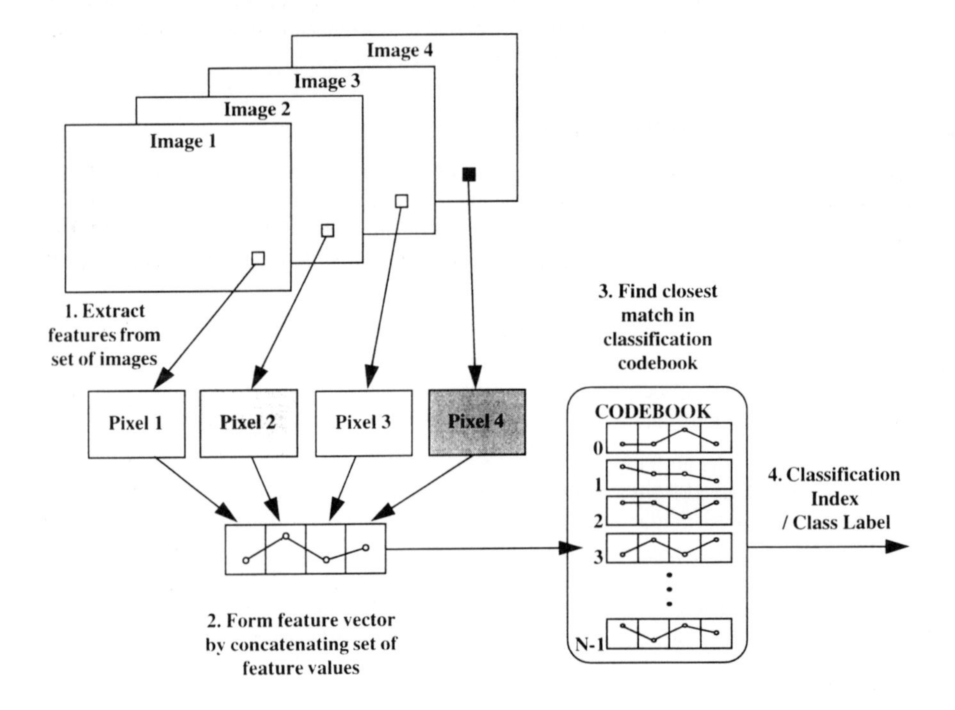

Figure 5 Image fusion with feature vectors

Feature 1 and feature 2 form the two axes. A point in the feature space represents a pair of values of feature 1 and feature 2, or equivalently the value of the feature vector. The feature space of Figure 4 is divided into several regions or classes. A feature vector falling in region-A would be classified as being of type-A. The regions are actually defined by exemplar vectors which are examples of a feature vector belonging to a particular class. The exemplar vectors are illustrated as the dots in the centre of each region. The regions are defined in this case as the boundary midway between each pair of exemplars. An unclassified feature vector is assigned to the closest or nearest-neighbour examplar.

These ideas are readily extended to multi-dimensional feature spaces. Figure 5 illustrates how several images may be fused together using feature vector classification. Corresponding features are extracted from the set of images (in this case the features are simply the corresponding pixel values from each image). The features form a feature vector. The feature vector is compared to a set of exemplar feature vectors held in a classification codebook. The closest match in the codebook determines the appropriate classification for the feature vector. A classification index is used to label each class.

Several forms of classifier will now be described briefly (see for example van der Heijden 1994). There are three common forms: statistical classifier, neural network, fuzzy logic.

Statistical classifiers

A classifier divides a feature vector space into a set of discrimination regions separated by hypersurfaces. Each discrimination region represents a class of the feature vectors and is given a class identifier. A feature vector is classified according to the region in the feature space in which it lies.

A **minimum distance classifier** is a special case of classifier in which the discrimination regions are defined by a set of exemplars or sample patterns. A feature vector is classified by comparing it to each of the exemplars. It is assigned to the class given by the exemplar which it is closest to.

Bayesian classifiers. A single feature vector may represent objects from more than one class. In this case the discriminating functions of the optimal classifier must be assigned probabilistically. The Bayesian classifier organises the discrimination regions according to the *a priori* probabilities of classes and the conditional probability density of features in each class. It minimises the error incurred when an object is classified.

A Dempster-Shafer classifier - overcomes the limitations of the Bayesian approach by allowing uncertainties in the assignment of probability to each class, or to be more precise, to each hypothesis that a feature vector belongs to a particular class. Each proposition is assigned two values - one a measure of *support* for a hypothesis and the other a measure of its *plausibility*.

In a **Φ-classifier** a non-linear function is applied to the feature vector space to transform it to a domain where objects may be discriminated more readily. A special case of the Φ-classifier is the use of a local wavelet transform to extract feature vectors representing local textural information.

In a **learning classifier** the discriminating regions are set by training the classifier with a set of feature vectors accompanied by their correct classification. The classifier learns to discriminate by generalising from the training set. There are two common learning strategies - probability density estimation from the training set and direct loss minimisation where the discrimination regions are adjusted to minimise overall classification error. Learning vector quantisation is a learning form of minimum distance classifier.

Cluster analysis is a form of learning classifier which does not need a training set (unsupervised learning). Feature vectors are grouped together according to their mutual similarity. Each cluster is then assigned a class identifier. Cluster analysis can be used to help identify suitable discriminating features to be included in feature vectors.

Neural networks

Neural networks are based on the behaviour of biological neurons. The perceptron is the basic building block of many neural networks and is equivalent to a linear classifier that classifies input vectors into two classes.

The **multi-layer perceptron** comprises several layers of highly interconnected perceptrons. Typically there are three layers: an input layer, an output layer and an internal or hidden layer. Interconnections between layers represent synaptic connections. A weighting applied to each synapse determines the operation of the network. The neural network learns to classify feature vectors applied to its input layer using a training set and the back propagation algorithm. This algorithm adjusts the synaptic weights to minimise the classification errors.

Once trained the output of the multi-layer perceptron provides a classification for each vector applied to its input.

The **kohonen network** implements unsupervised feed-forward learning in a three layer network. Weights representing clusters of similar input vectors after training are held on the internal layer. During operation an output node will "fire" if the input vector is close to any one of a cluster of vectors held on the internal layer. The kohonen network is useful for assessing features suitable for discrimination.

The **hopfield network** has interconnection weights that are initialised from a set of exemplars during operation. A feature vector to be classified is placed on the output of the network. The network is then allowed to iterate until a stable state is reached. The output of the network will then be the exemplar which most closely resembles the input feature vector.

Fuzzy logic

Fuzzy logic or fuzzy set theory applies mathematical techniques to reasoning with imprecise data. Each member of a fuzzy set has an associated membership value between 0 and 1 which indicates the degree to which a member belongs to the set. Fuzzy logic operates on fuzzy sets with combination rules and inferencing from fuzzy probabilities. Fuzzy logic has been used successfully in the fusion of data from various sensors.

3. Data sources

The selection of data sets to fuse will depend very much on the application. Generally prior knowledge and experience will be used to select images or other data sets which contain information of relevance to the particular application. In some cases statistical techniques may be of value in helping to select images to fuse which are uncorrelated and/or contain most variance.

Data Fusion can use information from a wide range of sources and not only sensor data. The major types of information source for a remote sensing application are considered below:-

Image data

For remote sensing applications of data fusion, sensors which produce image data are natural candidates for data fusion. Visible, infra-red and microwave images can be combined.

Other sensor data

Data obtained from other types of sensor may be appropriate - for example wave spectra data from a wave buoy combined with SAR bathymetry or interferometry in a coastal modelling application.

Environmental data

Meteorological data may be important in analysing the response from a particular sensor. An extreme example of this is cloud cover obscuring an optical image from a satellite. Another example is the effect of moisture on SAR data.

Digital elevation models may be used to correct sensor data for terrain relief variation. This can be of particular importance when merging SAR data with optical data. SAR has a low look angle and suffers foreshortening and shadow effects which can severely limit its suitability for data fusion without geometric correction.

Information from Geographic Information Systems (GIS) can support the identification of entities and is a particularly useful source of non-image data in remote sensing applications because the information is geo-referenced.

Entity signature, behaviour and contextual knowledge

Entity signature, behaviour and context is used for helping to identify entities. Knowledge of the appearance of entities of interest across the different types of data being fused is clearly necessary. The behaviour of an entity and its context can also give important clues to the identity of an entity. An example of an entity signature is the spectral response of a woodland species. An example of contextual knowledge is the fact that a forest fire occurs within or on the edge of a forest. An example of behavioural knowledge is that a forest fire moves in the direction of the wind.

Fusion objectives information

Control information provided to the data fusion process by the operator is also a form of data being used in the data fusion process. This information sets the goals for the fusion process. An example is a list of entity types to be detected and the areas to be searched for those entities.

4. Levels of data fusion

So far image data fusion has been considered at the pixel-level with corresponding pixels being fused together. There are, however, several different levels at which data fusion can be implemented.

- Data Level Fusion
- Feature Level Fusion
- Decision Level Fusion

4.1 Data level fusion

Data level fusion is illustrated in figure 6(a). The data acquired for an object from several sensors are combined together. An identity declaration is then made based on the merged data. To perform data level fusion the data collected for an object from each sensor must be correctly aligned and associated with the appropriate object. For image data, as we have seen, the image from each sensor must be correctly aligned on a pixel-to-pixel basis.

Data level fusion is called pixel level fusion when image data is being merged.

4.2 Feature level fusion

Feature level fusion is illustrated in figure 6(b). The data from each sensor are processed separately to extract a set of features from the data set. These features are represented by a feature vector. The feature vectors belonging to an object are then concatenated to form a composite feature vector which is then classified to provide an identity declaration.

For image data extracted features will include things like edges, areas of similar texture and areas with similar spectral response.

4.3 Identity level fusion

Identity level fusion is illustrated in figure 6(c). Features are extracted and identity declarations are made separately for each sensor. The identity declarations for each object are grouped together and combined to provide an overall identity declaration. It is clearly essential that the data being fused refers to the same object.

In general better accuracy is obtained by fusing information closer to the source. Data level fusion is only feasible for data sources which produce the same type of observation, e.g. sensors which all produce image data.

5. An Earth observation data fusion system

The typical environmental remote sensing data fusion system, illustrated in Figure 7, brings together all the functions discussed previously. Several disparate sources, including active and passive microwave and visible and infra-red optical instruments, provide the raw data for the system.

Generally the raw data must be pre-processed before fusion to adjust for radiometric and geometric distortions inherent in the instrument. Atmospheric correction will need to be applied for optical instruments and speckle reduction filtering may be applied to SAR image data. After pre-processing the images will be aligned with one another taking into account terrain relief effects where necessary. The aligned images are then fused together to provide a composite image map containing relevant information from the different data sources used. This image map is either an enhanced form of image ready for operator interpretation or an automatically classified image ready for visualisation. The data fusion system is supported by a database which handles the image data together with supplementary information like digital elevation models and geographic information.

6. A military data fusion system

6.1 Three levels of fusion

Data fusion technology was developed initially for military applications but it is now used in a wide range of applications including robotics, medicine and, of course, remote sensing. Many of the ideas developed for military use can be applied to Earth observation applications.

Figure 8 shows a functional overview of a military data fusion system (Hall 1992). Inputs to the system are provided by a range of sensors. The sensor data may need to be pre-processed in some way before it is ready for fusion. The data then pass through three distinct levels of fusion which transform them into a form where they can be used to assist management decisions.

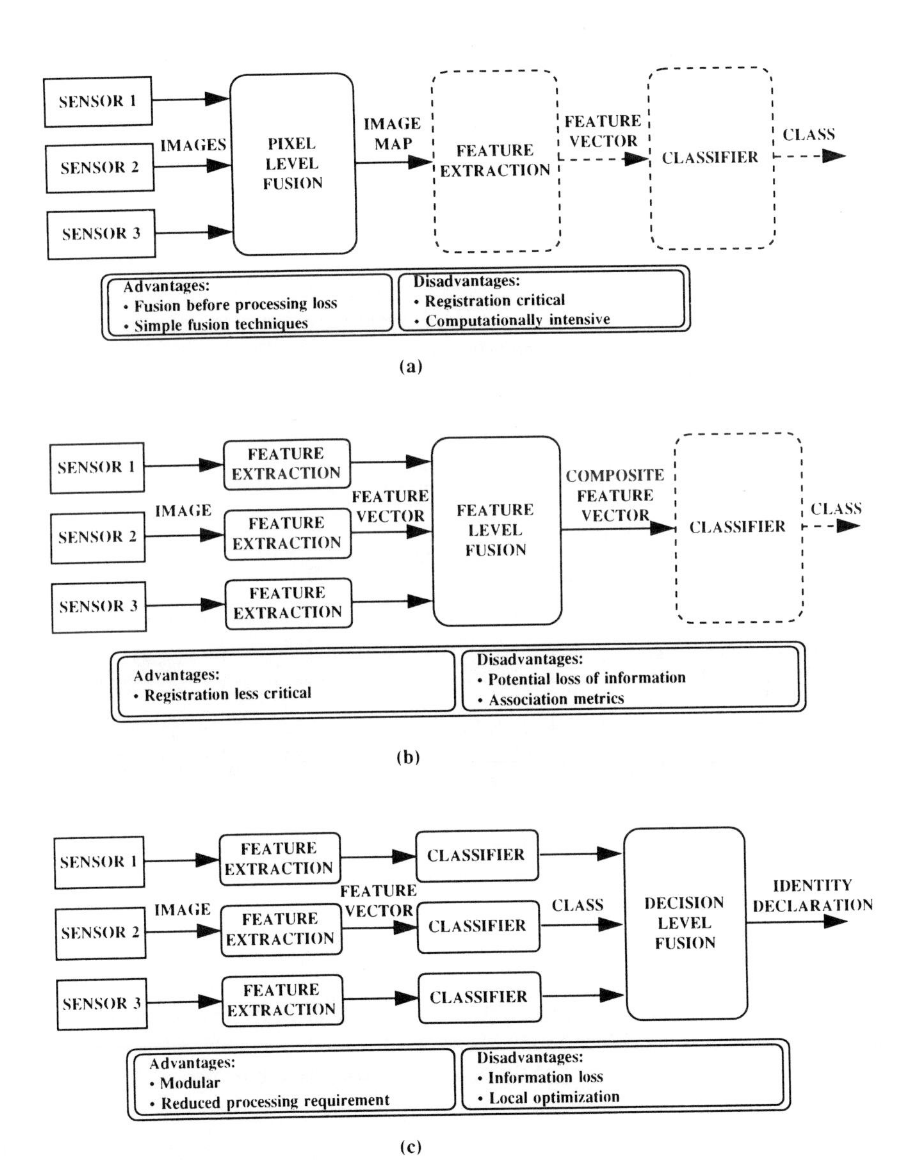

Figure 6 Fusion at (a) pixel level, (b) feature level and (c) identity level

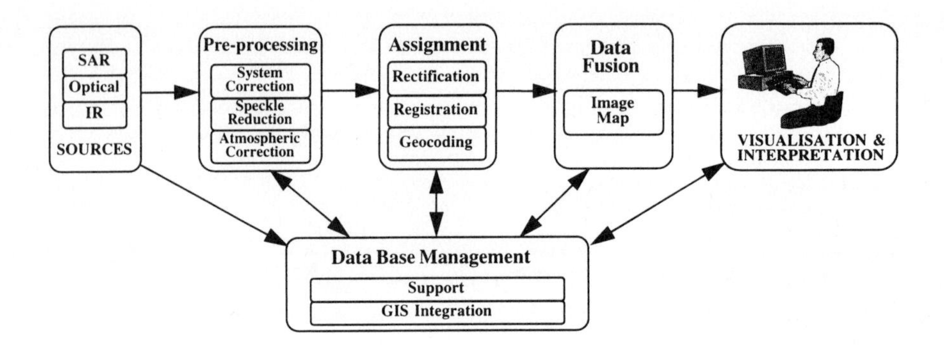

Figure 7: A typical environmental remote sensing data fusion system

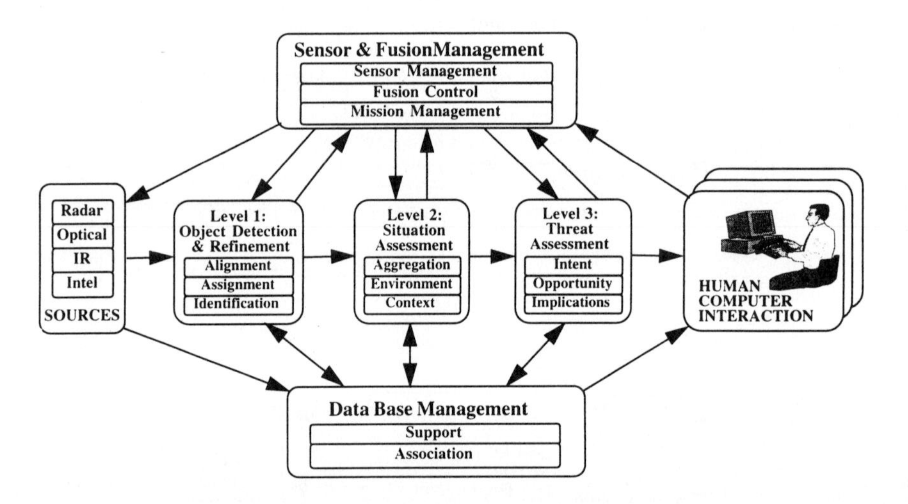

Figure 8 : A typical military data fusion system

A Human Computer Interaction (HCI) interface presents the fused data to the operator in a meaningful manner and supports the assimilation of a potentially overwhelming volume of information. The handling of the data (raw sensor data, processed data, fused data, models, signatures, historical data set and future predictions) is done by a relational database management system.

The three levels of data fusion will now be considered in turn:-

Object identification

Level 1 data fusion aims to transform the source data into a set of object identity declarations. Each object will be accompanied by a set of attributes (e.g. position, size)

which includes its identity. An example for a military application is the identification of an aircraft from Infra-Red (IR) and radar sensor data and estimation of its position and velocity. In an Earth observation application a field of wheat is an object. Its attributes would then be its identity, position and boundary.

Situation assessment

Level 2 data fusion aims to interpret the object information obtained from level 1 into an assessment of the overall situation. It tries to assess the meaning of the level 1 results. To do this the objects identified at level one are related to one another taking into account environmental, contextual and other factors. Objects are aggregated into higher level abstract objects of relevance to the data fusion application. An example for a military application is the aggregation of several objects identified as enemy aircraft into a enemy fighter squadron. The land-use application could, for example, relate fields of a particular crop type susceptible to flood damage which lie in a possible flood area.

Threat assessment

Level 3 data fusion projects the current situation into the future and makes inferences about possible opportunities or threats to support management decisions. Dynamic models of each object or aggregation are used to predict their future state. A series of alternative hypotheses (what ifs) are considered and their outcomes predicted and assessed. An example for the military application is the projection of the current flight path of a fighter squadron towards areas of military vulnerability and an assessment of the risk involved. The land-use example could model various flood levels and consider actions to take. Deliberately opening sluices to flood one area may prevent serious flooding in another area where crops particularly susceptible to flood damage are growing. Another Earth observation application example is the assessment of the likely yield of a crop from an area, given the current situation and various alternative predictions about the weather in the interval before harvest.

6.2 Application to environmental remote sensing

Many of the ideas that have been developed for military systems are being applied to environmental remote sensing. A few of these developing areas are summarised below:-

Entity fusion

Environmental data fusion concentrates on image or pixel-level fusion whereas military data fusion is concerned about discrete, real-world entities. This entity or object-oriented approach can be applied to environmental remote sensing.

In a land-use application a field is a real-world object rather than a pixel covering a part of that field. Pixels from that field must be grouped together to form a set of data gathered about the object. Identification can then proceed based on all the information gathered about the whole field. To be able to process image data in this way it is necessary to determine which pixels belong together, i.e. which pixels are from a single field. This may be done using an association metric which determines how closely related two sensor measurements are. For the current example pixels from a single field should form a continuous region and should have similar spectral and textural characteristics.

Situation assessement / GIS integration

The use of relations between objects and the use of supportive environmental information for situation assessment can also be applied to environmental data fusion. An example of this is in the use of GIS data to support classification. The number of classification hypotheses being applied to an entity may be reduced substantially with the use of supportive information from a GIS.

Threat assessment / decision support

Environmental decision support systems may be developed to support the visualisation and interpretation of the effects of possible management decisions. This type of system would assist in the making of difficult decisions which involve many complex, interacting factors. The development of these systems can benefit from the work done on similar military systems.

7. Conclusion

The basic ideas behind data fusion have been described and several image fusion techniques have been outlined. A comparison of environmental remote sensing and military data fusion systems have suggested several areas where environmental data fusion could benefit from the techniques developed for the military. Data fusion is an exciting area of remote sensing which will play an increasingly important role in the effective management of the environment in the future.

References

Armour B, Ehrismann J, Chen F and Bowman G, 1994, An integrated package for the processing and analysis of SAR imagere, and the fusion of radar and passive microwave data, Proceedings ISPRS Symposium, Ottawa, pp 299-306

Corves C, 1994, Assessment of multi-temporal ERS-1 SAR and Landsat TM data for mapping the Amazon river flood plain, Proceeding of First ERS-1 Pilot Project Workshop, Toledo, Spain / ESA, SP-365, pp 129-132

Hall D L, 1992, Mathematical Techniques in Multisensor Data Fusion, Artech House, Norwood, MA, USA

Llinas J and Hall D L, 1993, Data Fusion and Multi-Sensor Correlation, Course Notes / ATI

Pohl C and van Genderen J L, (To be published) Multisensor Image Fusion in Remote Sensing: Concepts, Methods and , International Journal of Remote Sensing

Ramseier R O, Emmons A, Armour B and Garrity C, 1993, Fusion of ERS-1 SAR and SSM/I Ice Data, Proceedings of the Second ERS-1 Symposium, Hamburg, Germany / ESA, SP-361, pp 361-368

van der Heijden F, 1994, Image Based Measurement Systems : Object Recognition and Parameter Estimation, Wiley, Chichester, UK

Wang Y, Koopmans B N and Pohl C, 1995, The 1995 flood in The Netherlands monitored from space - a multi-sensor approach, International Journal of Remote Sensing, 16, pp 2735-2739

Williams J, 1995, Geographic Information from Space, Wiley, Chichester, UK

Plate 1. *This volume looks at aspects of monitoring of coastal zones of many different types. Here we see typical dune systems in the Netherlands, discussed in the article by Feron (page 315).*

Plate 2 *Uncontrolled photomosaic of a section of the Sowaqua Creek. Photomosaics are useful tools for reach classification (Roberts, page 126).*

Plate 3 *Uncontrolled photomosaic of a section of the Sowaqua Creek using the Multispectral Video (MSV) system (Roberts, page 126).*

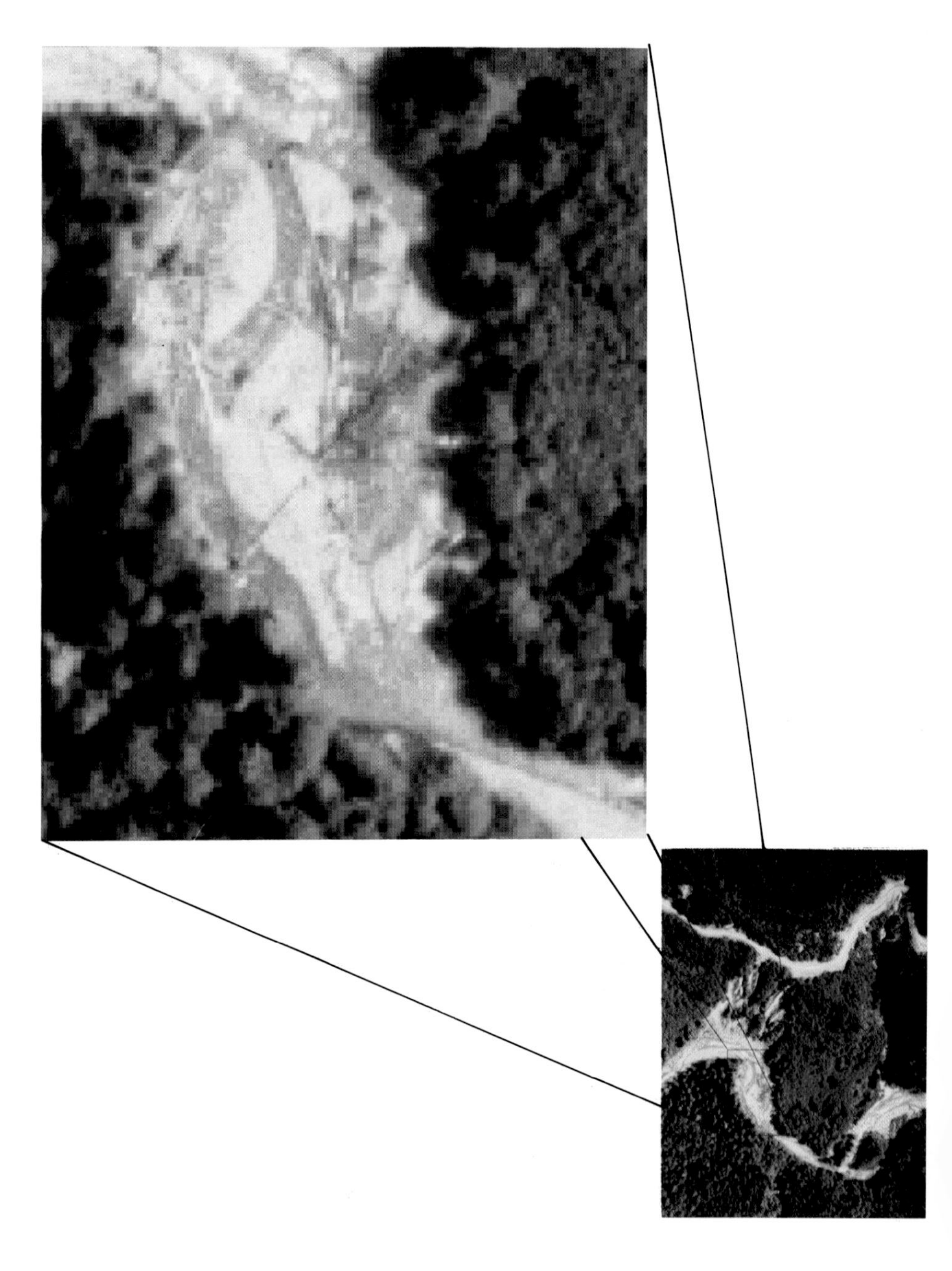

Plate 4 *Low altitude reconnaissance photography over Sowaqua Creek. This image shows an accumulation of large woody debris (Roberts, page 131).*

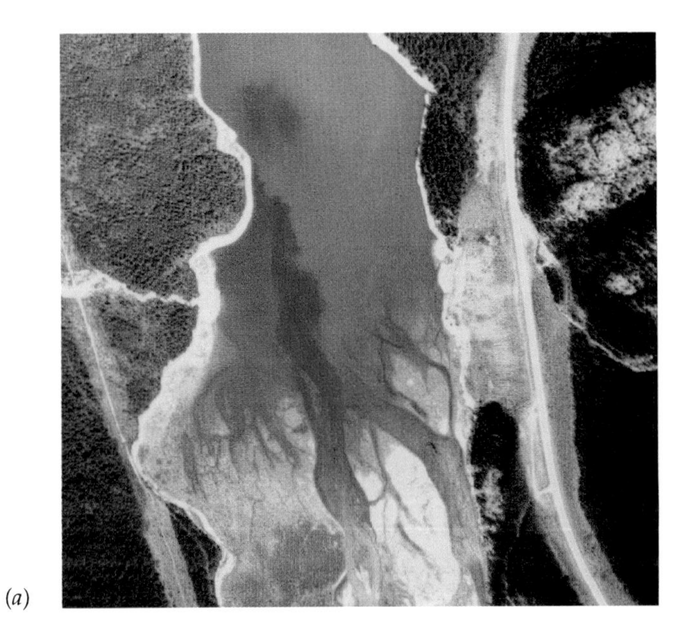

(*a*)

(*b*)

Plate 5 *Comparison between (a) normal colour and (b) colour infrared images of a portion of Barrier Lake in the Kananaskis Valley (Roberts, page 132).*

(a)

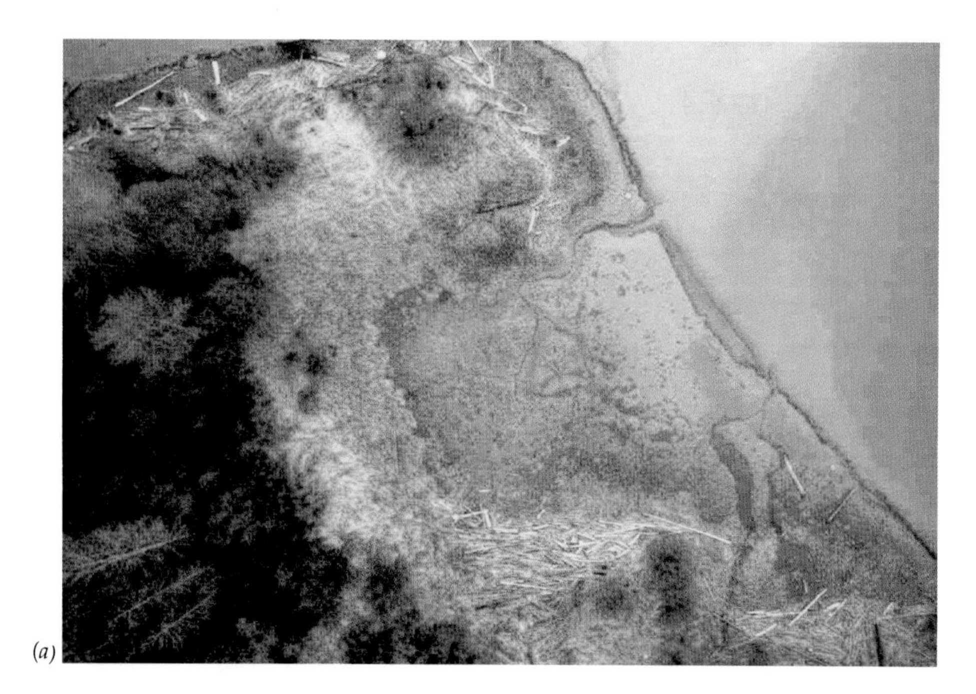

(b)

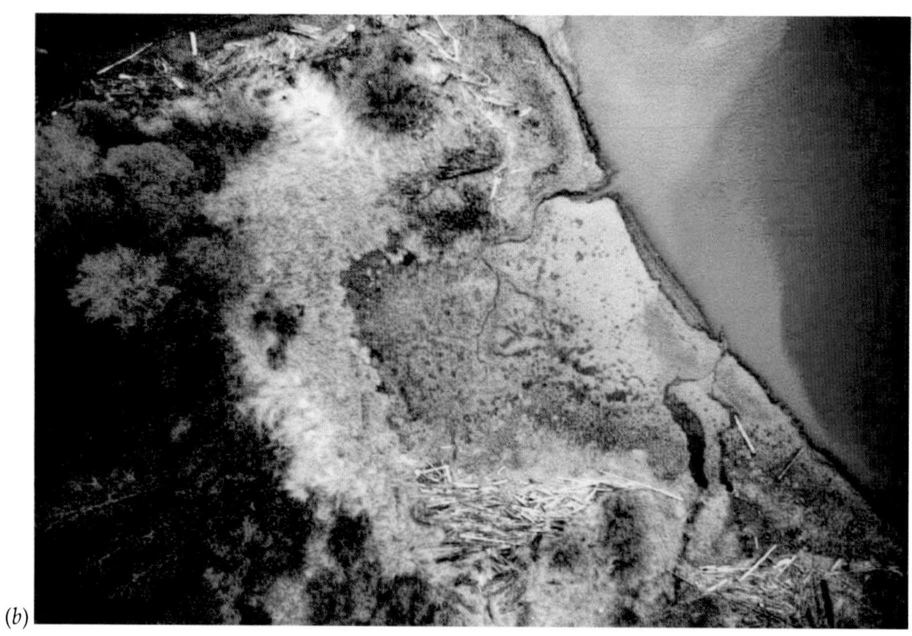

Plate 6 *Example of the benefits of colour infrared film over normal colour film in classifying vegetation: (a) normal colour and (b) colour infrared acquired over the Coquitlam River in March 1997 (Roberts, page 142).*

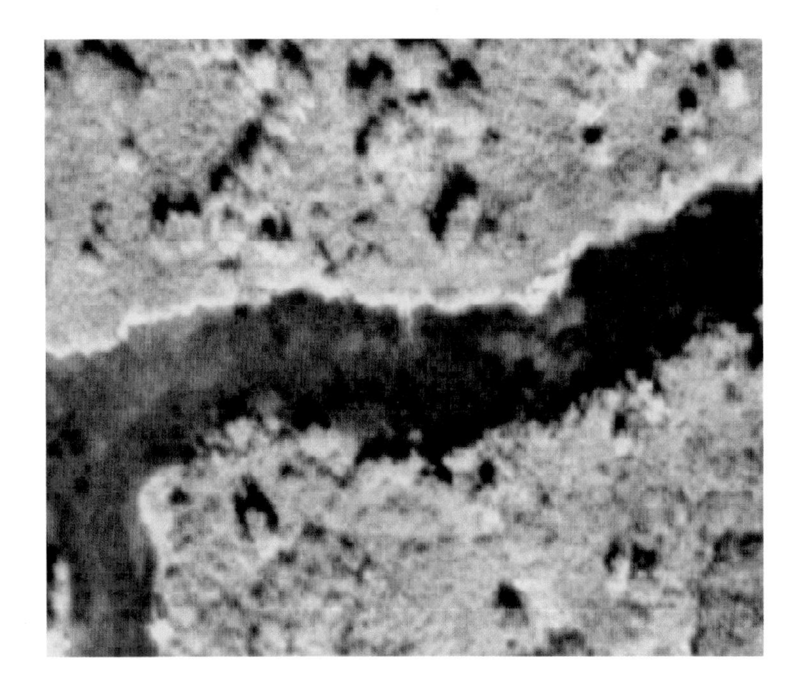

Plate 7 *Aquatic vegetation in the Horsefly River Delta (Roberts, page 144).*

Plate 8 *Submerged vegetation in the meandering section of the Horsefly River. The river has covered its banks. Submerged vegetation on the banks appears blueish while vegetation which is not submerged looks pink (Roberts, page 144).*

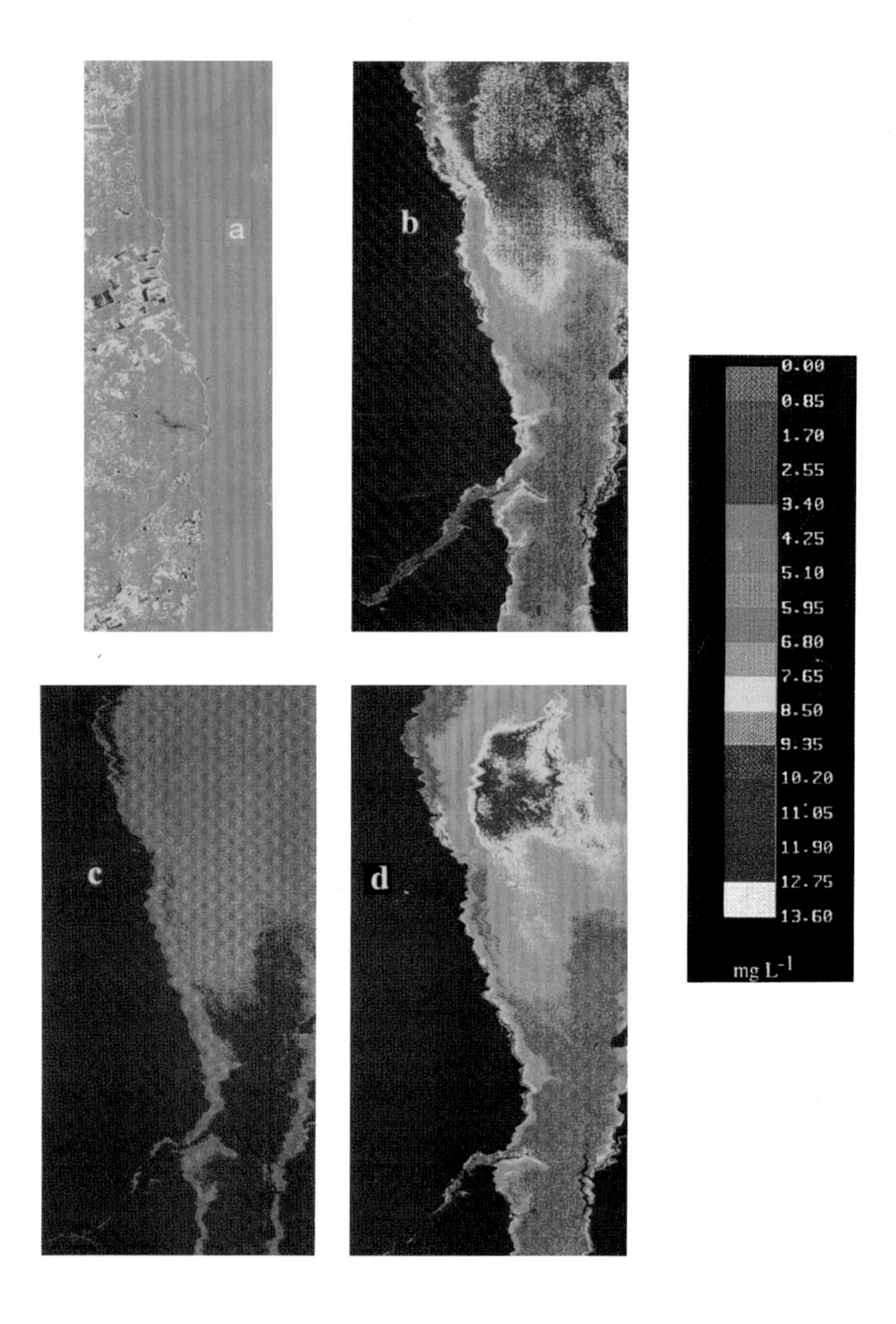

Plate 9 *Images of Southampton Water: (a) uncorrected image at 600 nm, (b) atmospherically corrected with land masked off, (c) water leaving radiance at 865 nm after atmospheric correction, (d) derived SPM concentrations.*

The carbon biogeochemical cycle: mangroves and the coastal zone

Jin-Eong Ong

Universiti Sains Malaysia

1 Introduction

Although carbon constitutes a mere 1% of the elements on Earth, it is literally the stuff of life. Carbon is perhaps one of the most basic elements of life on Earth. In its pure elemental state it can exist in different forms e.g. soft black graphite, or the hardest known substance, brilliantly clear diamond. Most carbon occurs as compounds like carbon dioxide, hydrocarbons (oil & natural gas) and limestone (calcium carbonate).

Carbon compounds change from one form to another basically via oxidation (e.g. burning fossil fuels and respiration) or reduction (e.g. the process of photosynthesis where carbon dioxide is reduced to carbohydrate). Energy (from the Sun) is captured in the reduction (photosynthesis) process and this energy is released during oxidation (e.g. in the process of respiration or during the burning of fossil fuels). Carbon is thus a vehicle for energy transfer. Carbon dioxide also readily dissolves in water to form bicarbonates and carbonates. So we see both biological transformations as well as chemical transformations. Limestone is relatively soluble so vast limestone hills or karsts can be completely dissolved with time. All these transformations from one form to another and back again is known as biogeochemical cycling, since biological, geological and chemical processes are involved and the elements do not disappear but just move around.

1.1 The carbon reservoirs

In biogeochemical cycles, there exist a number of pools or reservoirs. In the carbon biogeochemical cycle the biggest reservoir is in carbonate (limestone) rocks that make up mountains. The next biggest reservoir is dissolved carbon in the oceans (as carbonic acid, bicarbonates and carbonates). Then comes fossil fuels (peat, coal, oil and natural gas), carbon locked up in soils, carbon in biota (animals and plants), and finally, carbon dioxide and other gaseous forms of carbon (like methane) in the atmosphere.

The carbon pool in limestone rocks (75,000,000Pg, where 1Pg is 10^9 tonnes) is immense and is more than 3 orders of magnitude larger than the carbon pool in the oceans (depths 37,350Pg and surface 650Pg). Fossil fuels make up 5,000Pg, soil carbon (mainly derived from biomass) accounts for 1,200Pg and 750Pg occurs in the atmosphere. Terrestrial biota (mainly plants) make up 560Pg but marine organisms make up a mere 3Pg. Indications of the fluxes in the global carbon cycle are given in Figure 1.

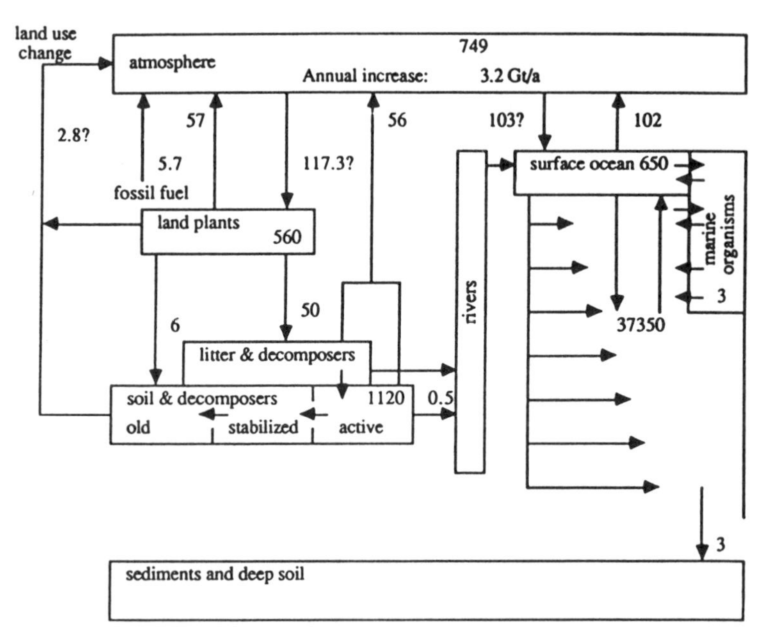

Figure 1 Carbon fluxes according to Jarvis and Dewar (1992).

1.2 Carbon fluxes

Fluxes refer to the two-way exchanges between one compound of carbon and another (e.g. carbon dioxide dissolving in seawater to form carbonic acid, bicarbonate and carbonate ions) or from one medium to another (e.g. atmosphere to oceans or to biota). If we look at the carbon fluxes (Figure 1) the main exchanges are between atmosphere-land and atmosphere-oceans. Only a tiny proportion goes from land to rivers into the oceans. Carbon dioxide (the major compound of carbon in the atmosphere) is thus the major currency as far as carbon exchange goes. Most of the atmosphere-land as well as the atmosphere-oceans fluxes are biologically mediated (through the processes of photosynthesis and respiration). Some 100Pg is exchanged between atmosphere and oceans with a net flow estimated at just 1 or 2Pg into the oceans. Exchange between atmosphere and land is in the same order as with the oceans and here again the net flow is again of the order of 1 or 2Pg (not taking into account fossil fuel combustion but taking into account land use change of about 3Pg). So in this budget we are trying to determine a very small net flow (around 1 or 2Pg) from in and out movements of around 200 Pg (two orders of magnitude higher than the net flow). Current estimates give an error band that is at least equal to the net flow. Without fossil fuel burning in the equation, the net flux is probably close to zero (as it probably was before industrial man).

If we look at the numbers in the global carbon cycle, it can be seen that the flow of carbon from land to the rivers and (we presume) eventually in the oceans is a mere 0.5Pg per annum. In comparison to inward and outward flux of around 100Pg per annum between the atmosphere and the oceans, this certainly appears small. On the other hand, the net flux between atmosphere and ocean is also only about 1Pg per annum from the atmosphere to the oceans. For cycles in equilibrium, we would expect net fluxes to be zero or very close to zero.

The net fluxes, although small, are the important figures to look at. For example, the in and out fluxes between the atmosphere and the lands and oceans is in the region of 200Pg per annum but a fossil fuel (combustion) input of 5Pg per annum has a very significant effect of increasing the atmospheric pool of carbon (despite approximately half the 5Pg disappearing into one sink or another). So the flux of carbon in the coastal zone is certainly not insignificant. Equally important is the fate of this carbon, i.e. whether most of it is oxidised and thus returned to the atmosphere or whether it is locked away in sediments. Also, together with the flow of carbon from the land via rivers there is also an increase in the flux of nutrients (nitrogen and phosphorus). We can ask whether this nutrient enrichment results in higher productivity, thereby resulting in the reduction of atmospheric carbon?

As we have already seen in the first article (Cracknell), the coastal zone is also where much of the human population is concentrated and thus a region of intense human activities. This results in major land use changes that affects the carbon biogeochemical cycle. Many coastal ecosystems like mangroves and peat swamp forests are significant sinks of carbon dioxide and modification of destruction of these systems could shift the balance in carbon flows.

Our understanding of the carbon biogeochemical cycle tells us that, in the short term, only about half of the approximately 8Pg per annum of carbon released into the atmosphere by human activities remain in the atmosphere. Thus there must be a sink or sinks somewhere that absorbs the other half. The rapid response of this absorption would suggest either chemical (e.g. the bicarbonate-carbonate pool of the oceans) or biological (e.g. increased photosynthetic uptake by plants) rapid processes. Nobody is certain where the sinks are but a number of investigations and initiatives are underway to determine this. A study of the carbon cycle is an important component of the IGBP's LOICZ programme (see chapter 1).

2 Atmospheric carbon dioxide

2.1 Recent Increase in Atmospheric Carbon Dioxide

The level of carbon dioxide in the atmosphere some 200 years ago (before the advent of industrial man) was about 275ppmv carbon dioxide. It had been at that level some 10,000 years before that. We know this from fossil records - from air trapped in polar ice. When snow falls and does not melt, as happens in polar regions, it traps air. The snow is compacted and turns to ice and the air that was trapped when the snow fell remains as air bubbles in the ice. So if different levels in the ice core can be aged then it is possible to melt that layer of ice to release the trapped air and then analyse this air sample for its carbon dioxide and methane (and other) concentrations. With this method it is possible to determine the concentrations of atmospheric carbon dioxide (and other gases) over thousands of years.

The Vostok ice core which goes to just over 2,000 metres deep and is equivalent to about 160,000 years of continuous record (e.g. Raynaud *et al.* 1993) is an excellent example of such an ice core. By looking at the oxygen isotopic ratio in the core, it is possible also to estimate the temperature at which the snow (which became incorporated into the ice core) formed. The Vostok ice core has been analysed for carbon dioxide, methane and stable oxygen isotope ratio (from which the surface temperature in the Antarctic, over where the Vostok ice core was obtained, can be estimated). It can be seen from the Vostok ice core records (Figure 2) that on a geological scale at least, there have been rather wild fluctuations in the carbon dioxide level in the atmosphere but that the level has been stable for some

10,000 years until recently. We thus know that changes in rates of transfer can occur from time to time and such changes can have widespread repercussions. The current hypothesis is that a huge meteorite hitting the Earth or an intense period of volcanic eruptions can reduce the intensity of the radiation from the Sun that reaches the surface of the Earth; this results in reduced photosynthesis and a build-up of carbon dioxide in the atmosphere. In the recent past, the Earth has not been hit by any large meteorite nor has there been intensive volcanic activity that had resulted in the extended blocking out of solar radiation.

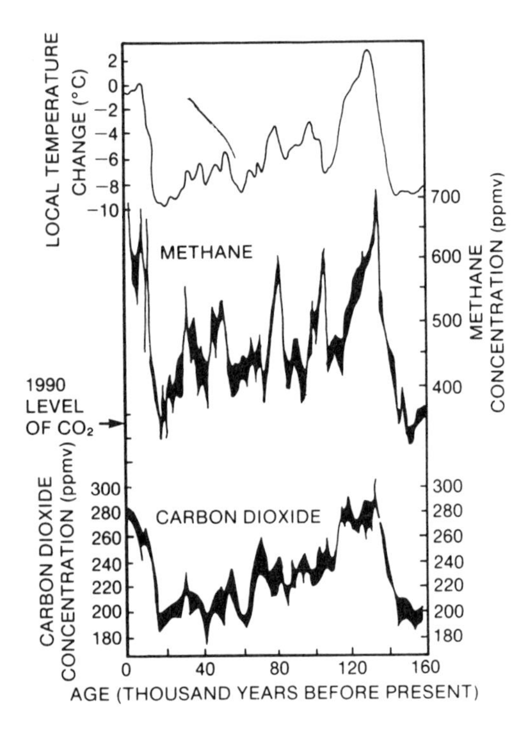

Figure 2 Fluctuating CO_2 levels from the Vostsok ice core, (IPCC 1990)

The level of carbon dioxide in the atmosphere has been increasing at a rapid rate in the past 200 years as a consequence of the advent of industrial man and the associated increase in the use of fossil fuels for energy. Early industrial man used mainly coal and the burning of coal produces more carbon dioxide than oil which is presently the main fuel together with natural gas (methane) which produces the least carbon dioxide of the three. Since we know the amount of fuel used each year, we have figures for the amount of carbon dioxide released into the atmosphere each year. The amount we add to the atmosphere from the burning of fossil fuels alone is more than enough to account for the increase in atmospheric carbon dioxide concentrations that we are seeing. We do not even have to take into account other sources (like carbon dioxide produced in the process of cement manufacture or land use changes). Only about half of what we oxidise can be accounted for in the increased concentration in atmospheric carbon dioxide. What happens to the other half? If we look at another time series of atmospheric carbon dioxide measurements, we may get a clue. We have this in the now famous Mauna Loa (Hawaii) data series where the level of carbon dioxide was carefully measured a number of times every year since 1958. These data (Figure 3) shows two main trends. First the mean concentration shows an increase from one year to the next. Secondly, there is a very clear intra-annual fluctuation with lower levels in the northern hemisphere summer and higher levels in the northern hemisphere winter. First, it clearly confirms the Vostok ice core data that demonstrate that the atmospheric carbon dioxide concentration is increasing. Secondly, the intra-annual seasonal fluctuation is a result of plants being involved in the process. Most of the world's temperate forests are located in the northern hemisphere. These forests (especially the deciduous ones) are only photosynthetically active during the warmer seasons when they absorb carbon dioxide from the atmosphere. During the colder season there is either no photosynthesis (deciduous forests) or minimal photosynthesis (evergreen forests) and so the draw down of carbon dioxide from the atmosphere by northern temperate forest plants is manifested in the intra-annual fluctuations in the Mauna Loa records. This again shows that the biological component plays a very important role in the carbon biogeochemical cycle.

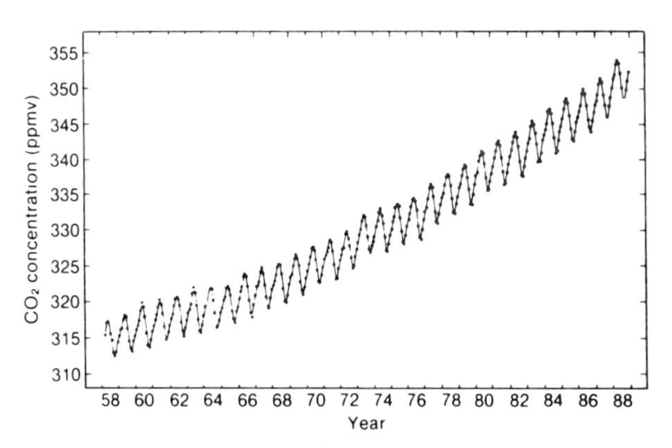

Figure 3 A recent CO2 time-series from Mauno Lao (Keeling et al. 1989)

So the carbon biogeochemical cycle presently has a capacity to absorb about half the anthropogenic (man-made) carbon dioxide. The questions are: how long can it continue to do this, and where are the sinks? Much research is being conducted to answer these questions.

An equally important question is what are the repercussions of increasing carbon dioxide levels in the atmosphere?

The global concern is that carbon dioxide is a very important greenhouse gas (the other greenhouse gases, other than water vapour, all combined contribute about the same amount of warming as the carbon dioxide) and its increase in the atmosphere could lead to the warming of the Earth. Whether this global warming is actually happening is still a subject of debate - mainly because there are many other considerations including numerous feedback loops (e.g. see Kerr 1997). Modellers have had a fantastic time with their global climate change models (GCMs). Unfortunately the spatial resolutions (typically 300km square grids) of all these GCMs are too coarse and, until recently, very few of them have taken into consideration the role of water vapour which is the most important of all the greenhouse gases. Even when water vapour is taken into consideration, cloud formation, for example, occurs at a scale smaller (a couple of orders of magnitude) than the present grid size used. Finally, but not least important, is the fact that some processes are not well understood.

The Vostok data, for example, gave a very convincing correlation between increase in concentrations of carbon dioxide and methane with increase in surface temperature. There is general agreement that the increase in temperature follows the rise in carbon dioxide but there is uncertainty if the increase in methane concentration is part of the cause or is a consequence of the temperature rise. On top of this, the rises in carbon dioxide and methane concentration and in temperature correlate very well with changes in sea level.

2.2 The greenhouse gases

Most of the radiation from thc Sun that falls on the Earth is in the visible range (0.4 - 0.7μm) because of the high temperature of the Sun (6,000K). Some of this radiation is reflected back into space by clouds and suspended particles as well as by the Earth's surface itself. This reflectivity is termed the albedo. The radiation that reaches the surface of the Earth warms up the Earth which, in turn, radiates this energy into space. But because of the lower temperature of the earth (255K) the radiation from the Earth is of longer (infrared) wavelengths (mostly from about 10-20μm) than that from the Sun. So-called greenhouse gases absorb energy in these longer wavelengths and so keep some of the Earth's radiated energy within its atmosphere. Carbon dioxide absorbs between 13 and 19μm and water vapour, the other major (and more important) greenhouse gas, absorbs strongly between 4 and 7μm. Most of the radiant energy that escapes from the earth (some 70%) is in the 7 - 13μm window. Some anthropogenic gases, like the stratospheric ozone layer which are responsible for depleting the chlorofluorocarbons (CFCs), absorb very strongly in this window (molecule for molecule the two most common CFCs have 10,000 times the warming effect of carbon dioxide, whereas methane on the other hand is only about 30 times more potent than carbon dioxide). Because there is an atmosphere on Earth that contains these greenhouse gases, the Earth becomes warmer than celestial systems (e.g. our Moon) that do not have this blanket of greenhouse gases.

Methane, nitrous oxide and ozone are some of the other naturally occurring greenhouse gases but these are also increasing as a result of anthropogenic activities. Methane is produced by anaerobic bacteria in wetlands (including rice fields). Methane is also trapped in permafrost and huge amounts can be released as a result of global warming. Increased use of nitrogenous fertilisers has also resulted in the increase of nitrous oxide in the atmosphere. The concentration of ozone has also increased in the troposphere (as opposed to in the stratosphere, where it is decreasing) from the burning of oil (e.g. photo-oxidation of automobile emissions).

2.3 Repercussions of increasing atmospheric carbon dioxide levels

Post-industrial man has added a very significant amount of carbon dioxide to the atmosphere the concentration of carbon dioxide has increased from about 280ppmv at the start of the Industrial Revolution to the present level of about 350ppmv. There is also little or no argument that increase in carbon dioxide concentration in the atmosphere is due mainly to the burning of fossil fuels. It is possibly aggravated by other activities like cement manufacture and land use change (mainly the conversion of forests to systems that do not absorb carbon dioxide). From here we move into an area of less certainty. Since the greenhouse gases retain heat, an increase in carbon dioxide concentration could, in itself a least (i.e. if feedback loops from other processes are not considered), lead to global warming. The consequences could be such things as expansion of the waters of the oceans, melting of glaciers and non-floating polar ice and a consequent rise in sea level (e.g. see Schneider 1992 and Lindzen 1990 for two different views of this situation).

We need to take one more step into the realm of the uncertain and this concerns the expected rise in sea level as a result of the increase in greenhouse gases. The predicted rate centres around 5mm per annum over the next 100 years or so (UNESCO 1990, IPCC 1990, IGBP 1992), but there is considerable uncertainty about this figure. The latest IPCC prediction (IPCC 1995, cited in Kerr 1997) has a confidence band as large as the mean.

3. Mangroves

3.1 Introduction

Mangroves dominate much of the usually low energy coasts of the tropics. These forests form the interface between land and sea. In the ever wet tropics, mangroves are probably the most productive systems, in terms of net productivity, as opposed to the often quoted high gross productivity of coral reef ecosystems. However, even where they dominate and are most luxuriant, in the wet tropics, mangroves comprise only approximately 2% of the total land area. Over the last thirty years or so, this coastal tropical ecosystem has been subjected to ever increasing human population and economic pressures. Ong (1982) estimated that the loss of mangroves in Malaysia was about 1% per year over the past 20 years. Since then the rate of destruction has probably increased. In Thailand, for example, most of the mangroves on the eastern coast have been reclaimed; mostly they have been converted to ponds for the culture of the tiger prawn, Penaeus monodon. Where human population pressures are minimal, the mangrove woodchips industry continues systematically to degrade huge areas of pristine luxuriant mangroves (Ong 1994). Small as they may be in area, the mangrove ecosystem is still a dominant and important tropical coastal ecosystem and its contribution as a carbon sink may be several times that of most other ecosystems (Ong 1993). Yet there are very few quantitative studies on mangroves.

In this section we present a description of a long term programme (started in the mid 1970s) to determine the fluxes of water, carbon, nitrogen and phosphorus in a tropical mangrove ecosystem. It must be emphasised that measuring fluxes from intertidal systems like mangroves and salt marshes is an extremely difficult task (e.g. see Nixon 1980). Despite our 20 years or so of study with intensive measurements and wide international collaborationthe problem is not yet solved.

3.2 Sea level change and the response of mangroves

The phenomenon of sea level change is rather complex and leaves many not directly in the field confused or perplexed. The perception of sea level change is very much dependent on where the measurement is made. If the measurement is made on sites where the land is rising faster (as in sites where glaciers have recently melted or are melting, thus reducing load on the land and letting it bounce back up) than the actual rising water level (as caused by an increased volume of the oceans), then what is seen is a relative fall in sea level. Glaciers are not the only thing that can cause the land to rise or sink. The Earth's plates are in constant motion so that on one side a plate can tilt downwards causing the opposite side to tilt upwards. So unless the rise or fall of sea level (as caused by volume change) is extremely rapid, relative sea level change is very much site dependent. So even if the oceans' volume is increasing (initially as a result of thermal expansion of water more than from increase in mass from melting polar ice caps), sea level rise will not be seen universally. In many areas land masses may be rising at a faster rate so a relative fall in sea level is still seen.

Here we look at the mangroves on the Malaysian side of the Straits of Malacca as a case study. We will look at the Straits of Malacca and in particular the mangroves on the Malaysian side of the Straits mainly because we have more readily available data. Since the last ice age, which reached its peak some 12,000 years ago, the sea level in the Straits of Malacca (Figure 4) rose to reach the present level around 7,000 years ago and continued until a peak of about 5m above the present by 5,000-4,000 years ago and in the last 4,000 years has fallen about 5m to the present level (Geyh *et al.* 1979). So, for the Straits of Malacca (located on the tectonically stable Sunda Shelf), relative sea level has been falling at a rate of around 1mm per year over the past 4,000 years. On the other hand, Peltier and Tushingam (1989) have estimated (from isostatically filtered tide gauge data) the global sea level change over the past 50 years to be +2.4±0.90mm per annum. What I like to emphasise is that the rate for the geophysical study is a mean over four thousand years and is located in the Straits of Malacca whereas the study from 40 tide gauges, spread over the world, is only over fifty years. If indeed the Straits of Malacca is on a tectonically stable plate, then the discrepancy in rates would be difficult to reconcile. Such is the uncertainty we have to contend with although this caveat is often not pointed out.

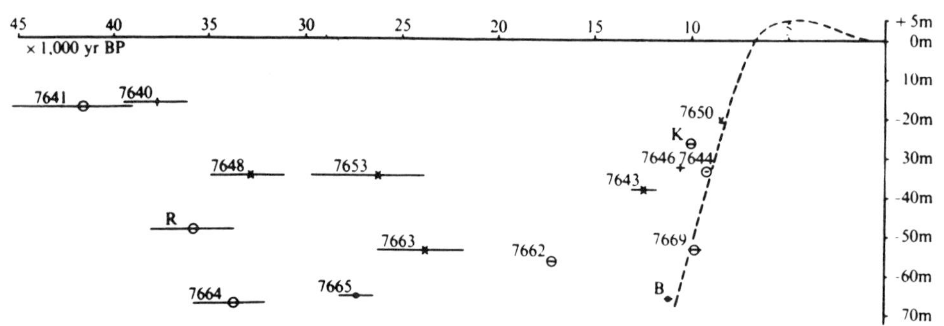

Figure 4 (Changing sea level in the Straits of Malacca, after Geyh *et al.* (1979); the numbers and letters against the points refer to the sources and sample identification.

The mangroves (at the same relative sea level as we know them today) have been very recently (4,500 years) subjected to a 2-3mm per year relative sea level rise (and an even greater rising rate of 10mm per annum earlier on) so we would expect that our present day mangroves will be able to handle a 5mm per annum predicted sea level rise. There is however one very major difference: post-industrial Homo sapiens. When the sea level rose 4,500 year or so ago, nature was able to operate unimpeded. Mangroves spread into the plains now occupied mainly by rice fields (it certainly was not rice then). Now, most of the land behind our mangroves are mainly rice fields dotted with towns and cities belonging to man and man has the technology to set up defences against this aqueous intrusion. The landward part of much of the Malaysian mangroves bordering the Straits of Malacca is already bunded to prevent saltwater intrusion and the land has been converted to agriculture and other uses. It is a simple matter of raising the level of these Earth bunds by a mere half a metre and we will be safe for the next 100 years. The mangroves on the seaward side will slowly drown and we would expect a change in species composition in those areas that allow mangroves to thrive.

Let us proceed to a less extreme scenario. Taken as a whole the mangroves on the Straits of Malacca are growing on accreting coasts. Macnae (1968) gives the vivid example of Palembang, a town on the Sumatran side of the Straits of Malacca which was still a river-mouth port 400 years ago but which is now 50km inland (accretion has moved at a rate of around 125m per annum). On the Malaysian side, the sedimentation rate is about 2mm per annum. This was during a period when there was no significant human perturbation of the environment. With the present activity of land clearing, sedimentation would probably be higher so it is reasonable to double the sedimentation rate to 4mm per annum which is not much different from the predicted 5mm sea level rise. In such a situation we will not feel the effects of the predicted sea level rise.

There are a few main points to be made from the above scenarios. The first is that no matter whether sea level rises or falls (at the rates we think we are seeing or generally predicting) the relative sea level at any particular site could be rising, falling or be stable depending on the nature of tectonic movement. If we understand this we are in a position to select the relevant sites that are stable, rising or sinking with respect to relative sea level. If we are interested in monitoring changes in the mangrove ecosystem, as some international organisations are, this is a necessary first step.

The second point is that should rapid relative sea level rise take place, there is very little likelihood of saving mangroves whose landward margins have been developed by man. It is vital that those responsible for the selection (or those who formulate the guidelines) of conservation sites for posterity fully understand this.

The third point is that for effective planning and management, it is vital to know if a particular site is stable, rising or sinking and so efforts should be directed to finding suitable methods for determining this.

So if we have to select only one site in the Straits of Malacca for conservation for posterity it would certainly be more rational to look at the Sumatran (Indonesian) side of the Straits where human activity is minimal and where there are still extensive areas of pristine mangroves as well as freshwater swamp forests and other natural ecosystem behind the mangroves. Potentially the mangroves here may migrate inland and survive as they did some 4,500 years ago. The question is then why the Indonesian people should be asked to contribute their mangroves for the world to keep in posterity.

3.3 Structure and primary productivity of a mangrove forest

This section is to give a background on the structure and physical environment encountered in a mangrove forest. The description here applies to a 20 year-old stand of Rhizophora apiculata Bl. dominated mangrove forest; it centres on various environmental parameters and the estimating of biomass and productivity of a stand of managed forest. It is meant to show that this is not as simple a task as might at first be expected and that there are still a number of technically difficult gaps to fill.

The study site is located near Kuala Sepetang in the Matang Mangrove Forest Reserve (100°36'E, 4°50'N) which covers an area of 40,000ha. This forest has been managed by the Perak State Forestry and since the Forestry Department keeps records on fellings, it is possible to estimate the age of any particular stand to within 2 years of the actual age.

This 20 year-old stand (in 1992) has been identified by the Forestry Department as a display area for the public to have better access to the mangroves and there is a walkway next to the site. The site is only inundated by high spring tides so comes under inundation class 4 of Watson (1928). Rhizophora apiculata grows best in inundation class 3 (inundated by all high tides) so this site is rather dry for R. apiculata.

3.4 Environmental parameters

Silicon light sensors (calibrated against a LI-COR underwater PAR (photosynthetically active radiation quantum sensor) and copper-constantan thermocouples were placed at different heights from just above the canopy to just above ground level using a 23m tall multi-platform scaffolding tower. These were connected to data-loggers to record 10-minute averaged readings continuously. Data were downloaded to portable personal computers daily.

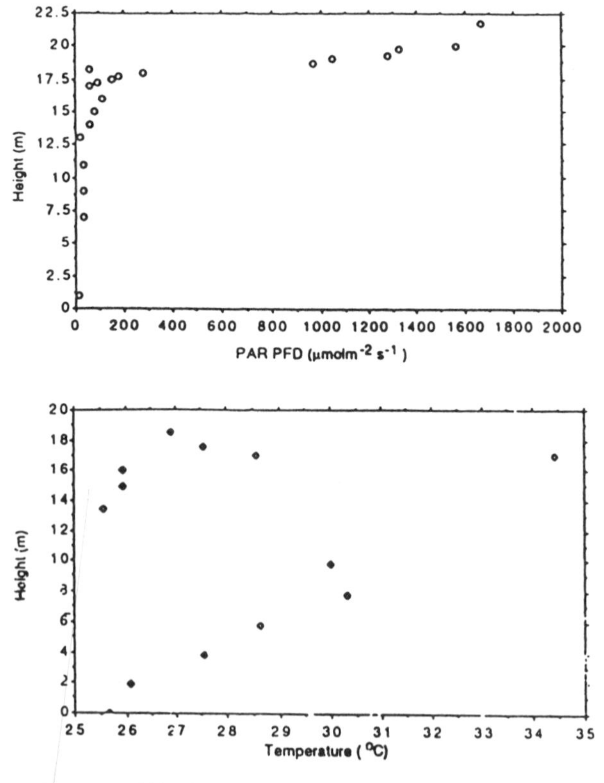

Figure 5 Midday light (PAR) and temperature profiles (Ong *et al.* 1995).

Figure 5 shows typical (noon) light (PAR) and temperature profiles through the canopy. Photon Flux Density (PFD) ranged from 1670mol $m^{-2}s^{-1}$ above the canopy to 14mol $m^{-2}s^{-1}$ a metre from the ground showing that only about 1% of the photosynthetically active radiation reaches the ground. There are no leaves below about 10m. Temperature ranged from 34.5°C at the canopy to 25.6°C just off the ground. There is considerable variability, especially for the group of sensors in the crown (above 10m). Soil salinity was measured by squeezing water from soil at different depths and measuring with a temperature compensated, hand-held refractometer (with a salinity scale). Mangrove shoot water potential was measured using a specially constructed high pressure bomb.

Figure 6 shows a diurnal plot of shoot water potential and light. It can be seen that the predawn potential was around -1.96MPa. This is approximately equivalent to the osmotic pressure of soil with a salinity of 18.0ppt (at a depth of 10cm at 0630hrs and 1900hrs). The shoot water potential decreased to -3.55MPa, flattened off in the middle of the day and started to increase again at about 1750 hrs reaching -2.50 MPa, not quite reaching the predawn potential suggesting that it takes a while for the plant to recover completely from water lost through transpiration.

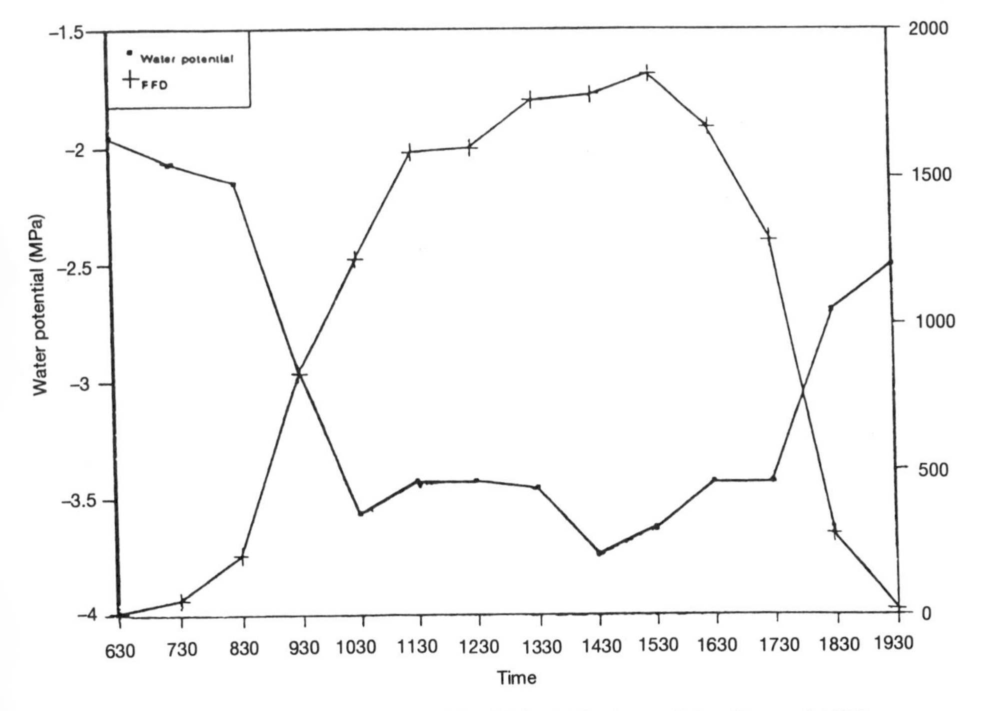

Figure 6 Diurnal shoot water potential and light (PAR) characteristics, (Ong et al. 1995).

3.5 Biomass & net productivity

Eight 10m x 10m plots (covering 20m x 40m) were established around the tower and the girth at breast height (GBH) of each tree was measured, noting the species and whether the tree was alive, dead or cut. The first measurement was made in June 1992 and the second measurement was made in September 1993.

The total above-ground weight, trunk weight, leaf weight and canopy (branches, twigs, buds, flowers and propagules) weight of every tree was calculated using the allometric regressions obtained by Ong *et al.* (1985) for Rhizophora apiculata in the Matang Mangrove Forest Reserve. These are:

W_{ag} (total above-ground weight) (kg)	0.0135	2.4243 cm GBH
W_{trunk} (total trunk weight) (kg)	0.0067	2.5414 cm GBH
W_{leaf} (total leaf weight) (kg)	0.0161	1.4363 cm GBH
W_{canopy} (total canopy weight) (kg)	0.0140	1.8453 cm GBH

The below-ground weight was calculated using the regression equation of Ong, Gong & Wong (unpublished). Total biomass was obtained by adding below-ground weight to total above-ground weight.

The method by which the coefficients in these relations were determined is illustrated as follows for there is a relation between the girth at breast height (1.3m) referred to as GBH and the weight of the whole tree (its trunk, branches, leaves, roots etc.). The way to obtain the allometric regression is to select some 30 trees covering a range from say 10cm girth to the biggest tree around (in our case some 90cm in girth). The girth of each tree is measured, the tree is cut down and its different components (trunk, branches, leaves etc.) weighed. A sample of each component is taken back to the laboratory and oven dried to obtain a wet weight to dry weight conversion. To convert dry weight to carbon a factor of 0.5 gives a reasonable estimate. A log transformation of both the GBH and the weight component gives a straight linear plot so that linear regressions may be obtained. The correlation is usually very good (e.g. r^2 of around 0.98 between GBH and total above-ground biomass). This was our first encounter with modelling: a good example of a simple stochastic model.

The tree density (live, dead or cut) of the 20m x 40m plot was equivalent to 2,425 stems per hectare. All were Rhizophora apiculata, apart from 5 small to medium size Bruguiera parviflora and a single very large (108.5cm girth at breast height) Bruguiera gymnorhiza so that non-Rhizophora apiculata species constituted just over 3%. 18.5% of the trees were dead or had been recently cut so the density of live Rhizophora apiculata trees in the plot was equivalent to 1,975 trees per hectare.

The size (GBH) and biomass distribution of Rhizophora apiculata trees in the plot is shown in Figure 7. The size of the Rhizophora apiculata trees (both living and dead) ranged from 9 to about 75.5cm. Most (85%) of the trees were between 20 and 50cm, with a peak (40%) at between 35 and 45cm. The mean GBH was 39cm. The canopy had an average height of about 21m. The smallest (14.9cm GBH) live Rhizophora apiculata tree weighed 10kg, the largest (75.5cm GBH) weighed 510kg and the mean biomass was 122 kg. About 70% of the trees were below 100kg but 30% of the bigger trees contributed to slightly more than half of the total biomass of the plot. There was a large (GBH=108cm) Bruguiera gymnorhiza tree weighing 1,488kg (above-ground biomass) about 3 times that of the biggest Rhizophora apiculata tree. This plot was left out of the calculations on productivity because of the possible bias that this one big B. gymnorhiza tree may cause.

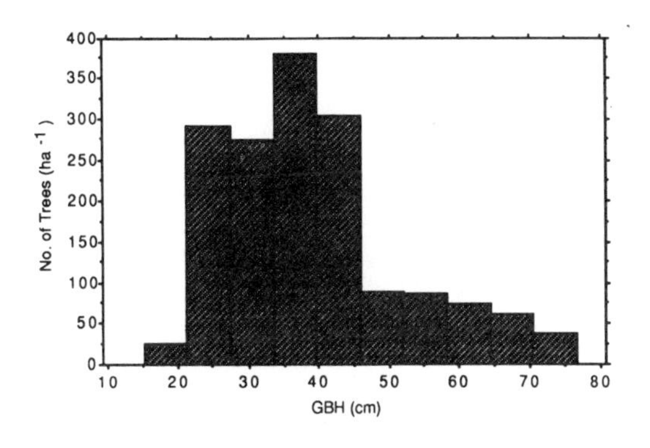

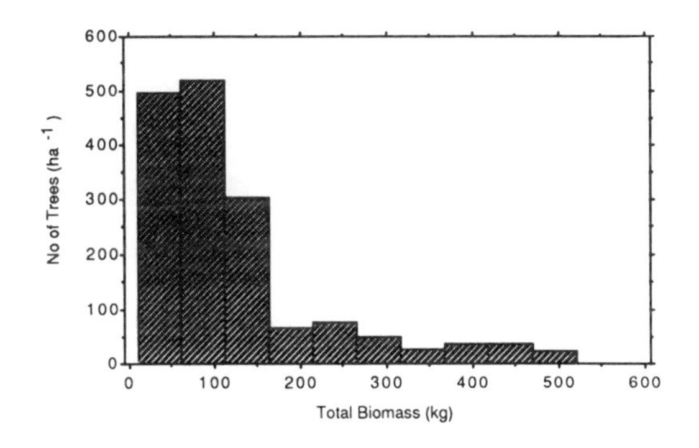

Figure 7 Tree size distribution, based on (a) girth and (b) biomass, after Ong *et al.* (1995).

The partitioning of biomass and of productivity in a typical 20-year-old stand of Rhizophora apiculata determined by Ong *et al.* (1995) is given in the following table (where tC ha^{-1} denotes tonnes of carbon per hectare):

	Biomass (tC ha^{-1})	Productivity (tC $ha^{-1}y^{-1}$)
Canopy - leaves	3.0	0.08 + 4.0
Canopy - branches (including fruits)	9.2	0.44 + 1.1
Trunk	84.5	5.56
Stilt roots	11.5	0.64
Roots	5.8	0.42 + ?
Total	114.0	7.14 + (5.1 + ?)

The two parts of some of the entries in the final column correspond to the productivity contributing to the standing biomass and that which is lost from the trees as litter (fallen leaves, small branches, fruits and flowers). There were no results available for the possible loss of organic carbon through the roots and this is indicated by the query. From the above figures one finds that, of the total biomass of 114.0tCha^{-1}, 74.1% was in the trunk, 15.2% in the roots (10.1% in stilt roots and 5.1% below-ground) and 10.7% in the canopy (2.6% in leaves and 8.1% in branches, twigs, buds, flowers and propagules). The figure for below-ground roots is low compared with other studies; see Clough (1992).

Net productivity consists of four components: increment in biomass per unit time, the turnover of litter (leaves, small branches, fruits and flowers) and fine roots, herbivory and the loss of dissolved organic materials from the roots. Most of the increment in biomass is in the trunk (78%) and this is followed by roots (15%); both stilt roots (9%) and below ground roots (6%). 7% (or about half the amount that went to roots) of the total annual increment in biomass went to the canopy (with only 1% going to leaf increment). Litterfall figures are those of Gong *et al.* (1984) for a 20 year-old in another site in the same Matang Mangrove Forest. Leaf litter productivity is 50 times standing leaf biomass increment. We do not have any data on fine root productivity but if it is anywhere near the 50 times standing biomass increment figure for leaves, then this would be a very important missing component. We do not have figures for herbivory but we have observed negligible herbivory during the 1 year and 3 months between the first and second measurement. In general, we have observed very little herbivory in Rhizophora apiculata and this may be because of its leathery leaves with its high phenolic content. We feel that it is safe to assume that herbivory did not decrease significantly the productivity estimate of this stand of Rhizophora apiculata. Since mangrove roots are tidally inundated, loss of organic carbon from roots could be significant (Clough, 1992). This is a problem that still needs investigating.

Although only about 1% of new standing biomass is in the leaves component, leaf litter turnover is 40% of total biomass increment. Turnover of trunk (perhaps some bark) and stilt root (above-ground) would be negligible but that of fine underground roots, for which we have as yet not been able to measure, may be very significant (perhaps as much as, if not more than leaf turnover). From the figures given above we see that the total net productivity in this 20 year-old stand of Rhizophora apiculata is at least (as we are as yet unable to measure fine root turnover) 12.24tCha^{-1}year^{-1}.

There is no estimate available anywhere of all the components on net productivity for mangroves (Clough 1992) and the figures reported here are as comprehensive as can be found for mangroves to date. The main gaps are for root turnover and for leaching of organic matter from the roots through tidal flushing. These are difficult gaps to fill, but nonetheless this needs to be done if we are to determine the role of mangroves as carbon sources or sinks.

3.6 Photosynthetic assimilation

Net photosynthetic assimilation as well as a number of related parameters like leaf temperature, stomatal water conductance, photosynthetically active radiation and relative humidity were measured using a LI-COR LI-6200 Portable Photosynthesis Meter. The LICOR LI-6200 measures net photosynthetic assimilation i.e. total carbon dioxide taken in by the leaf minus respiration (including photo respiration, should this occur) of the leaf. This is different from net productivity because respiration of all the other plant parts (like those for branches, trunk and roots) have not been accounted for. It is thus a measure of somewhere between gross and net productivity.

Measurements were made on three rosettes of leaves, two rosettes (1 and 2) of sun leaves and one rosette (3) of shade leaves, through the day (from about 0650 till 2000 hrs). The leaves in rosette 1 were measured 11 times, those in rosette 2 were measured 14 times and those in rosette 3 were measured 5 times in the course of a days measurements. There was thus a strong bias in sun leaves (measurements were made on shade leaves less than 5% of the time). The mean photosynthetic assimilation (from all the readings taken through the day) was about 6molm^{-2}s^{-1}. Net assimilation was plotted against photosynthetically active radiation, leaf temperature, stomatal conductance and intercellular carbon dioxide. One striking feature which was observed was the large scatter of the data points. It is not intended here to analyse or explain this data set in any detail as Cheeseman *et al.* (1991) have done this very adequately with a similar study on another mangrove Bruguiera parviflora. Despite the large scatter it is possible to discern that light saturation occurs at around 400molm^{-2}s^{-1} and net assimilation levels off at about 15molm^{-2}s^{-1}. Maximum assimilation recorded for the day was 23molm^{-2}s^{-1}. The light compensation point was around 50molm^{-2}s^{-1}. From the measurements made in the dark, a mean rate of just under 1.5molm^{-2}s^{-1} was obtained for dark respiration. There was also a large scatter with temperature but it can be seen that maximum net assimilation increased with temperature until about 38°C and decreased thereafter. The maximum leaf temperature recorded was 44°C.

3.7 Comparison

We have basically tried two approaches to finding an estimate of productivity.

The first is to estimate growth based on allometric regressions as well as the estimation of turnover of parts that are regularly shed (leaves and fine roots). It has already been noted that we are not yet able to obtain a reasonable figure for root turnover or for the possible loss of dissolved organic material through leaching from roots. The net productivity figures we have (annual increment in standing biomass plus small litter production) comes to 12.24tCha^{-1}yr^{-1} and if we assume that root turnover is about the same as canopy turnover then this gives a figure of around 17tCha^{-1}yr^{-1}. Unless leaching from roots is very high, we are very safe with a 5tCha^{-1}yr^{-1} error band and can confidently use the figure of 17±5tCha^{-1}yr^{-1} net productivity for this 20 year-old stand of Rhizophora apiculata. We are completely confident with the lower limit but the upper limit may well be higher if the loss of organic matter from roots is very significant. If we need to obtain an accuracy of better than the present 30%, then we need to try to measure directly the root turnover as well as the leaching from the roots.

The second approach is to use the gas exchange method. As we pointed out earlier, the measurements we made were net leaf assimilation and so we are getting a figure somewhere between gross and net productivity. If we know what the respiration of the leaf is (and we we have some measurements of dark respiration) then we will be able to make an estimate of gross productivity. To obtain net productivity we would need to have an estimate of respiration of the whole plant. This we have not been able to measure, but we can start by making a few order of magnitude approximations.

First, let us try to use the figures we have to see if we can arrive at some reasonable figures, based on a few approximations. Taking just the daylight hours, assimilation ranged from 0-23molm^{-2}s^{-1}. We will use the mean figure of 6molm^{-2}s^{-1} that we mentioned earlier. We use the figure of 1.5molm^{-2}s^{-1} for leaf respiration. So net productivity (not taking into account respiration of non-leaf tissues) comes to 4.5molm^{-2}s^{-1}. If we assume respiration of the non-leaf tissues to be the same as for leaves, then we are down to 3.0molm^{-2}s^{-1}. Using a leaf area index of 4 (based on our non-destructive measurements made on four 0.5m x 0.5m quadrants around our scaffolding towers), the figure for net productivity comes out as 11.4 tCha^{-1}yr^{-1}. This is close to the lower range of the 17±5tCha^{-1}yr^{-1} net productivity obtained using the allometric method. It must be pointed out that the mean assimilation of 6.0mol$m^{-2}s^{-1}$ may be an overestimate since only 5% of the measurements were made on shade leaves. On the other hand, the leaf area index of 4 may have been an underestimate. Also, we could have overestimated respiration (of non-leaf parts) and loss through leaching. What this means is that our accuracy with the gas exchange method is worse than the 5tC$ha^{-1}yr^{-1}$ confidence band we estimated with the allometric method. Perhaps the use of models that allow us to move from the leaf to the whole tree and stand level could improve this resolution. Such models already exist (e.g. the MAESTRO model of Jarvis *et al.* 1990). We have not tried these but are now in a position (in terms of the necessary data) to use our data on such a model.

3.8 A mass balance carbon and nutrient budget in a mangrove ecosystem

Mangroves are the main primary producers in many tropical estuarine ecosystems. Net primary production of such a system consists of a number of components: fine litter production, fine root turnover, herbivory, dead trees and tree growth. In mature (steady state) systems the production of dead trees is about equal to tree growth so that their standing biomass remains constant. The other components detach from the trees and are either buried (e.g. fine roots, litter and some dead trees), eaten by detritivores (e.g. sesarmid crabs feeding on leaf litter) or are carried away by tides (e.g. litter). These components are thus released and part or all may be exported from the system. If a budget is compiled then it is possible to obtain at least an order of magnitude estimate of plant biomass (= carbon) and nutrient flux. Gong and Ong (1990) have attempted this for the Matang Mangrove Forest in Malaysia.

3.9 Fluxes in the Merbok mangrove estuary

The primary site for the following studies is the Sungai Merbok mangrove estuary in Malaysia (5°40'N, 100°25'E). The physical characteristics of the site have been described by Ong *et al.* (1991). The estuary is some 30km long and its depth varies from 3-15m. Mangroves, dominated by Rhizophora apiculata and Bruguiera parviflora that grow up to about 30m, cover an area of about 5,000ha. The estuary is characterised by a 1.7m semi-diurnal tide with peak currents of 1.3ms^{-1}, and mean freshwater discharge of 20m^3s^{-1}. The estuary displays a pronounced neap-spring stratification-destratification cycle and the effective longitudinal dispersion is approximately 100m^2s^{-1}.

As water flows from rivers into estuaries, mixing with seawater and numerous chemical transformations occur. Apart from changes in chemical (e.g. redox) states, physical processes (like electrical charges) results in some of the chemical species settling out (e.g. through flocculation). Some of these chemical species are also taken up or transformed by living organisms. So numerous physical, chemical and biological processes occur between the head of an estuary and the mouth of an estuary. If these are not considered in material budgets that look at inputs from the riverine end and outputs at the marine end of the estuary then it may not be possible to balance the budget. Of course, if the flux measurements are made only at one point or a single cross-section (e.g. Kjerfve *et al.* 1981), such a budget is not affected by the processes.

The main aim of the whole exercise is to determine the fluxes of carbon and nutrients from the mangrove estuary. Initially this was linked to the question of the extent to which mangroves can be put to alternative uses without affecting the mangrove and adjacent capture fisheries but now (some 10 to 15 years later) the question is whether mangroves are a source or a sink of atmospheric carbon (Ong 1993).

3.10 Single cross-section budgeting

The Kjerfve method involved measuring current speed and direction and obtaining the covariance with the concentration of the scalar, at the mouth of the mangrove estuary. Despite data from 45 tidal cycles (4 to 9 stations, 3-5 depths at lunar hourly intervals), including a set with 31 continuous tidal cycles (to cover a spring/neap tidal cycle) estimates of water flux were about an order of magnitude higher than those based on calculations from rainfall, evapotranspiration and catchment area. The other important point was that a salt balance could not be obtained (Dyer *et al.* 1992). The reason for this is not clear but may be because the current measurements made were not accurate enough or there may be a systematic error in the current vanes we used (Simpson *et al.* 1997). We used simple current vanes, and trying to resolve a small residual from a huge tidal variation would require much more accurate, as well as possibly more intensive, measurements. There was still no solution 10 years down the hydrodynamics road but much had been learned about the estuary (Uncles *et al.* 1990, 1992, Ong *et al.* 1991, 1993).

Although this method appears to be successfully applied in the salt marsh estuary of North Inlet, South Carolina, there appear to be problems when applied to mangrove estuaries, not only in the Merbok Mangrove estuary, but also in the Ranong Mangroves in Thailand (Wattayakorn *et al.* 1990). The Thai study also applied the modelling method of Wolanski and Ridd (1986) with better success. Unfortunately this method is applicable only during the dry season when there is no freshwater flow and therefore cannot be applied to mangroves situated in areas where there is no dry season.

We are still hopeful and have just completed (in May 1997) another massive field study covering a two-week spring-neap tidal cycle (MERBOK 97). This time we used single point self-recording current meters (AANDERAA RCM 7s and General Oceanic Niskin winged meters) as well as a broad band 1200 MHz RDI Acoustic Doppler effect Current Profiler (ADCP). We are still involved in the task of processing and analysing the massive data set and we are looking forward to the outcome of our final attempt at solving the problem.

3.11 Numerical flow modelling

The more numerical fluid modelling approach has also been explored. There are now quite a number of 1-dimensional and 2-dimensional numerical flow models available that are not too difficult to apply but for systems like the Merbok Mangrove estuary where the water is not well mixed all the time, it is desirable, if not necessary, to consider 3-dimensional models. Unfortunately 3-dimensional models are complex and are too complicated for most biologists to handle. An attempt was made by Phang (1994) to apply a small data set from the Sungai Merbok mangrove estuary to the TRIM 3D numerical model (Cassuli and Cheng 1992 and Cassuli and Cattani 1993). The results suggested that this may be worth following up with better bathymetric and hydrographic data. It will be very useful to bear in mind the possible application to the TRIM model when designing the next exercise (using the ADCPs).

As with any model, unless the actual fluxes are known, calibration will be a problem. So without accurate estimates (with confidence limits) there will always be some doubt as to the reliability of the models or the predictions derived from the model.

In the mangrove ecosystem the mangrove trees are both the major primary producers and standing biomass. These plants take up nutrients like nitrogen and phosphorus from the estuarine waters. They also return these nutrients to the water via decomposition of litterfall, dead trees and fine root turnover. If the litter, parts of dead trees and fine roots get buried, then the estuary becomes a sink for carbon and mineral nutrients. This has to be taken into account in the input-output budget (often termed storage). In many places the plants are harvested, so this amount has to go to the output. A considerable amount of data (as well as a few assumptions) are required but the storage and harvest terms can be calculated (Gong and Ong 1990). Unfortunately this empirical approach does not have a predictive capacity. Alternatively, this process may be modelled using the modelling software tool STELLA. One particularly useful aspect of this tool is that one can use it to conceptualise how a particular system works and build very simple to very complex conceptual models. If the relevant data are available then it is possible to turn the conceptual model into a model that can be run. Two examples are given here.

Mangrove tree crop harvest model: STELLA

This is a simple example of what happens to the mangrove biomass if the thinning and harvest periods are changed. In Malaysia, many of the mangrove forests are managed for the production of timber. The management system varies from State to State. In the Merbok Mangroves the forest is harvested in patches of a few hectares over a 30-year rotation. So if there are 3,000ha of productive forest the annual coupe is 100ha i.e. 100ha of forest is harvested each year. In the Matang Mangroves (located in another State) - reputedly the world's best managed mangroves - a 30-year rotation is also used but here the trees are thinned at 15 years and at 20 years. About half the trees are removed at each thinning. Since the forests are managed and the forestry departments keep records of the coupes, it is possible to age the stands to within a year or two. Allometric equations (girth against biomass of different tree parts) were used to estimate the above- and below-ground biomass of different age stands. Litterfall data for different age stands were also available. From these it is possible to make very reliable estimates of the above-ground net productivity of these forests. One gap is that of fine root turnover (which could be almost as high as litterfall).

It is then possible to build a model of the change in forest biomass with time and see what happens during each of the thinnings and harvest (Figures 8a and b). From this, predictions can be made as to what happens if the thinning regime or the harvest cycle is modified. It was possible (without too much expenditure of time and effort) to run the biomass model (Radford 1990). Since data on the nutrient content of the trees are available, it would also be possible to run a model based on nitrogen and phosphorus and thus get a good idea of the transformation of inorganic to organic nutrients.

Complex model of water, salt and dissolved organic matter

This is a complex 3 compartment model looking at the flow of water, salt (conservative) and dissolved organic matter (non-conservative) through the estuary. There is not enough data to run this model. Figure 9 shows how complex the model can become. It also shows that by going through such an exercise, it is possible to know exactly what sorts of data are required so in this case the model will help in the conceptualisation as well as design of the experiment. It will save a lot of unnecessary collection of irrelevant data. We would thus strongly recommend the use of STELLA at the very start of a new project.

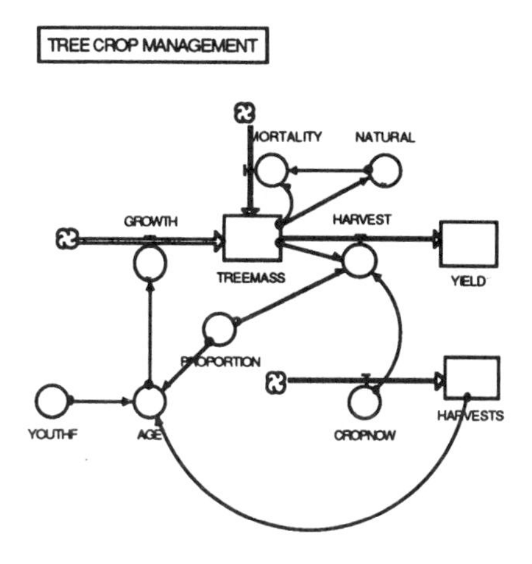

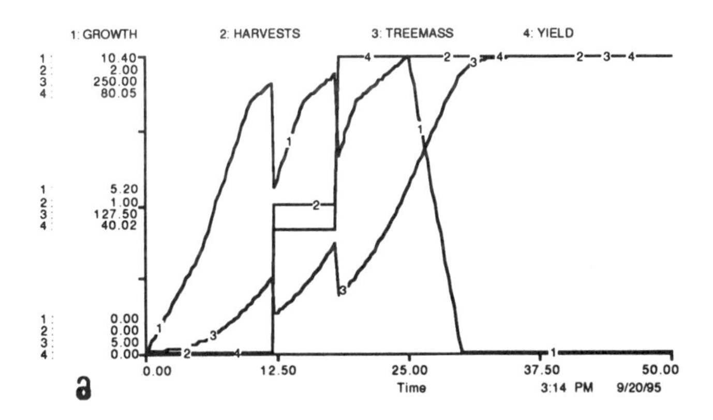

Figure 8 Mangrove Tree Crop Harvest Model STELLA II: (a) the model structure and (b) output.

3.12 The mangrove-atmosphere interface

Budgets cannot be complete if the atmospheric interface is not considered, so another aspect of this study (though more related to Biospheric Aspects of the Hydrological Cycle (BAHC) and Global Change and the Terrestrial Environment than to Land-Ocean Interactions in the Coastal Zone) is on fluxes from mangrove tree canopies. A number of sets of photosynthesis and related environmental and ecophysiological parameters in the mangrove canopy have been measured (Ong *et al.* 1995). This has reached a stage where there is a need to go into modelling (e.g. Jarvis *et al.* 1990). Such a model allows a certain amount of prediction.

A project is also being initiated on measuring carbon dioxide fluxes from mangrove canopies using the eddy covariance method. This is essentially based on the same principle as the Kjerfve method that we use for measuring fluxes in the water: it is a matter of aerodynamics and hydrodynamics. This is again a single point budgeting procedure with no predictive capacity.

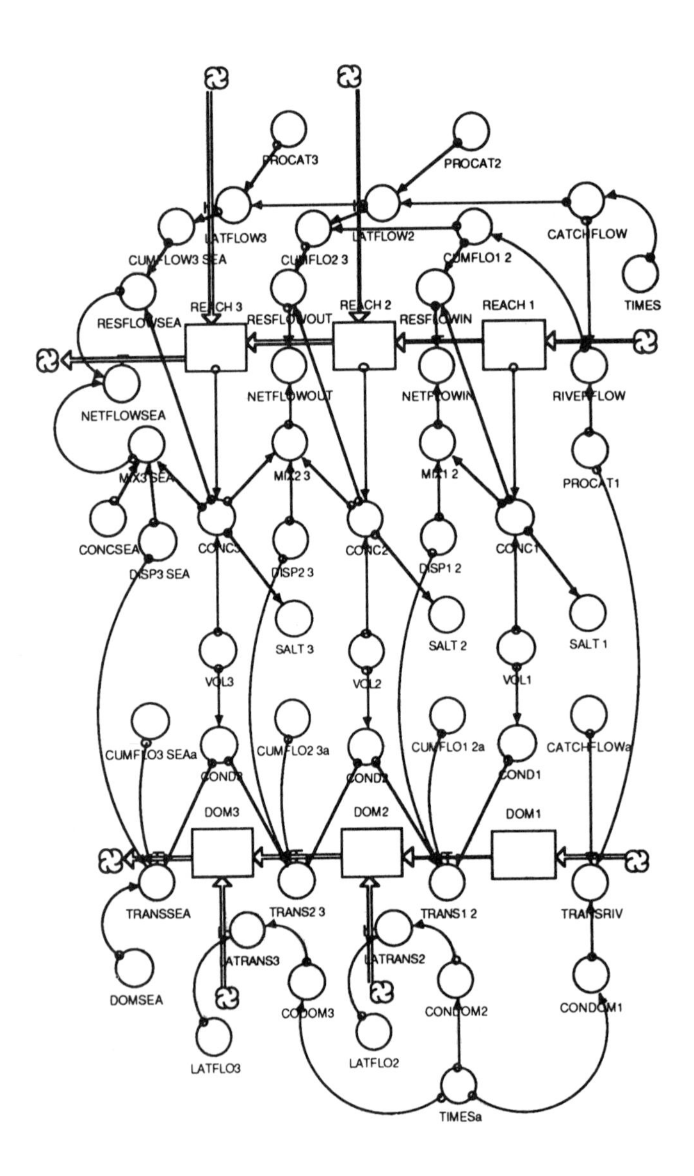

Figure 9 A model of water, salt and dissolved organic carbon in the Merbok Mangroves using STELLA II.

4. Conclusion

In our studies to try to close the carbon budget, we had to resort to a number of different methods and approaches. This is an absolute necessity: there is a need for some independent check. Apart from the approach, the ability to measure accurately or at least within known error limits is also extremely important. This is not as easy as it seems. First, there is often high natural variability. For instance, the measurement of total suspended solids (from which we can obtain the total suspended carbon by heating to 500°C in a muffle furnace for a few hours to drive off the organic carbon) is a simple procedure of filtering known volumes of water through pre-weighed glass fibre filters. A good calibrated balance ensures that measurements are made accurately but there is always the chance that there are small leaks in the filters. A larger problem is that natural variability can be high so that it becomes necessary to replicate. This can lead to thousands of samples for a 15 days time series which translates to hundreds of days of muffle furnace time since the capacity of our muffle furnace is only about a dozen per day! Determining dissolved organic matter is an even bigger problem. The instrument is complex and there is still a not quite settled controversy regarding the best method of combustion/oxidation. Conversion of organic carbon to carbon dioxide is allegedly more complete using catalytic converters. At some US$25,000 each we tend to keep our instruments as long as they are serviceable. Our instrument is only able to process 3 samples an hour and together with basic calibration we are limited to a maximum of 20 samples per working day. The alternative is to work round the clock. Even then we are limited to hundreds of samples per time series because samples can deteriorate with time. The practical problems are many and this is perhaps the reason very few people are physically involved in this type of study.

Models can be extremely useful tools but there are important caveats. They are often useful in indicating a trend. One has to be fully aware that until models are properly verified (i.e. no tweaking until it matches the data), their predictive usefulness is not only limited but can be dangerous. It is important to have a good understanding of the science involved before we can have a robust model. Models are only as good as the data available to support them.

Bibliography

GCTE (1994). IGBP in action: work plan 1994-998. IGBP Report No. 28.Stockholm

Gong W K and Ong J E, 1995, The use of demographic studies in mangrove silviculture, Hydrobiologia, **295**: pp 255-261

Proceedings of the UNESCO Asian Symposium on Mangrove environment: research and management, Universiti Malaya, Malaysia.

Gordon D C, Boudreau P R, Mann K H, Ong J E, Silvert W L, Smith S V, Wattayakorn G, Wulff F and Yanagi T, 1996, LOICZ Biogeochemical Modelling Guidelines, LOICZ/R&S/95-5, vi + 96 pp. LOICZ, Texel, The Netherlands.

IPCC Scientific Assessment 9, Sea level rise, WMO/UNEP, Cambridge University Press, Cambridge: pp 257-281

Japar Sidek Bujang, 1989, Studies on leaf litter decomposition of the mangrove, Rhizophora apiculata Bl. Ph.D. Thesis, Universiti Sains Malaysia, Penang. 322 pp

Nixon S W, Furnas B N, Lee V, Marshall N, Ong J E, Wong C H, Gong W K and Sasekumar A, 1984, The role of mangrove in the carbon and nutrient dynamics of Malaysian Estuaries. pp. 534-544. In : E.Soepadmo, A.N.Rao and D.J. Macintosh (eds), Proceedings of the Asian Symposium on "Mangrove Environment: Research and Management"

Schnack E and Pirazzoli P, 1990, Quaternary sea-level changes, Palaeogeography, Palaeoclimatology, Palaeoecology (Global and Planetary Change Section) **82**: 65-68

Von Caemmerer S and Farquhar G D, 1981, Some relationships between the biochemistry of photosynthesis and the gas exchange of leaves, Planta **153**: 376-397

References

Cassuli V and Cattani E, 1993, Stability, accuracy and efficiency of a semi-implicit method for three-dimensional shallow water flow, Computers and Mathematics with Applications **27**(4): pp 99-112

Casuli V and Cheng R T, 1992, Semi-implicit finite difference methods for three-dimensional shallow water flow, International Journal of Numerical Methods in Fluids **15**: pp 629-648

Cheeseman J M, Clough B F, Carter D R, Lovelock C E, Ong J E and Sim R G, 1991, The analysis of photosynthetic performance in leaves under field conditions: a case study using Bruguiera mangroves, Photosynthesis Research **29**: 11-22

Clough B F, 1992, Primary productivity and growth of mangrove forests. pp. 225-250 in: A I Robertson and D M Alongi (Eds.) Tropical mangrove ecosystems, Coastal and Estuarine Studies **41**, American Geophysical Union,Washington, DC

Dyer K R, Gong W K and Ong J E, 1992, The cross-sectional salt balance in a tropical estuary during a lunar tide and a discharge event, Estuarine, Coastal and Shelf Science **34**: 579-591

Geyh M S, Kudrass H R and Streif H, 1979, Sea-level changes during the late Pleistocene and Holocene in the Straits of Malacca, Nature **278**: 441-443

Gong W K and J E Ong, 1990, Plant Biomass and Nutrient Flux in a Managed Mangrove Forest, Estuarine, Coastal and Shelf Science **31**: 519-530

Gong W K, J E Ong, C H Wong and G Dhanarajan, 1984, Productivity of mangrove trees and its significance in a managed mangrove ecosystem in Malaysia, pp 216-225 in: Soepadmo E, A N Rao and D J Macintosh (eds)

Gribbin J, 1988, The greenhouse effect, Inside Science No. 13, New Scientist, **22** October 1988

Haron H A H, 1981, A working plan for the second 30-year rotation of the Matang mangrove forest reserve Perak, State Forestry Department, Perak. Rajan & Co. (Printers) Sdn. Bhd., Ipoh, Perak. 109 pp

IGBP, 1992. Global change: reducing uncertainties. The International Geosphere-Biosphere Programme, The Royal Swedish Academy of Science, Stockholm.

IPCC (Intergovernmental Panel on Climate Change), 1990, Climate Change

Jarvis P G, Barton C V M, Dougherty P M, Teskey R O and Massherder J M, 1990, MAESTRO. Pp. 167-178in : A. R. Kiester (Ed.) Development and use of tree forest response models. Acidic Deposition: State of Science and Technology Report 17. NAPP; Government Printing Office, Washington DC 20402-9325.

Jarvis P G and Dewar R C, 1993, Forest in the global carbon balance: from stand to region. In: J R Ehleringer and C B Field (eds) Scaling Physiological Processes: Leaf to Globe, Academic Press, San Diego

Keeling C D, Bacastow R B, Carter A F, Piper S C, Whorf T P, Heimann M, Wook W G, and Roeloffzen H, 1989, A three dimensional model of CO2 transport based on observed winds: 1. Analysis of observational data. In 'Aspects of climate variability in the Pacific and western Americas', edited by Petersen D H, *Geophysical Monograph* **55**, AGU, Washington, 163-236.

Kerr R A, 1997, Greenhouse forecasting still cloudy, Science **276**: 1040-1042

Kjerfve B, Stevenson B L H, Proehl J A, Chrzanowski T H and Kitchens W M, 1981, Estimation of material fluxes in an estuarine cross section: A critical analysis of spatial measurement density and errors, Limnology and Oceanography, **26**, 325-335

Lindzen R S, 1990, Some coolness concerning global warming, Bulletin of American Meteorological Society **71**: 288-299

Macnae W, 1968, A general account of the fauna and flora of the mangrove swamps and forests in the Indo-West-Pacific Region, Advancements in Marine Biology **6**: 73-270

Nixon S W, 1980, Between coastal marshes and coastal waters - a review of twenty years of speculation and research in the role of of salt marshes in estuarine productivity and water chemistry, In: P Hamilton and K B McDonald (eds), Estuarine Wetland Processes, Plenum Press, New York, pp 437-525

Ong J E, 1982, Mangroves and aquaculture in Malaysia, Ambio **11**: pp 252-257

Ong J E, 1993, Mangroves - a carbon source and sink, Chemosphere **27**: 1097-1107

Ong J E, 1994, The ecology of mangrove management and conservation, Hydrobiologia **295**: 343-351

Ong J E, Gong W K and Wong C H, 1985, Seven years of productivity studies in a managed Malaysian mangrove forest then what? pp. 231-223 in : K.N. Bardsley, J.D.S. Davy and C.D. Woodroffe (eds). Coasts and tidal wetlands of the Australian monsoon region. Australian National University North Australia Research Unit Mangrove Monograph No 1. Darwin, Australia.

Ong J E, Gong W K, Wong C H, Din Zubir Hj and Kjerfve B, 1991, Characterisation of a Malaysian mangrove estuary, Estuaries, **14**, 38 - 48

Ong J E, Gong W K and Uncles R J, 1993, Transverse structure of semidiurnal currents over a cross-section of the Merbok Estuary, Malaysia. Estuarine, Coastal and Shelf Science **38**: 283-290

Ong J E, Gong W K and Clough B, 1995, Structure and productivity of a 20 year-old stand of Rhizophora apiculata mangrove forest, Journal of Biogeography **22**: 417-424

Peltier W R and Tushingam A M, 1989, Global sea level rise and the greenhouse effect: might they be connected, Science **244**: 80 -810

Phang D, 1994, A numerical model study into the hydrodynamics of the Sungai Merbok Estuary. B. Eng. (Hons.) Thesis, Faculty of Engineering, University of Western Australia. 131 pp

Radford P J, 1990, Ecological modelling on personal computers. Paper presented at the Natural Environment Research Council and Institute of Mathematics and Its Application Conference on Computer Modelling in the Environmental Sciences. 10-11 April, 1990, Nottingham, England.

Raynaud D, Jouzel J, Barnola J M, Chappellaz J, Delmas R J and Lorius C, 1993, The ice record of greenhouse gases, Science **259**: 92 -933

Schneider S H, 1992, The climatic response to greenhouse gases, Advances in Ecological Research **22**: 1-32

Scurlock J and Hall D, 1991, The carbon cycle, Inside Science No. 51, New Scientist 2 November 1991

Simpson J H, Gong W K and Ong J E, 1997, The determination of net fluxes from a mangrove estuary system, Estuaries **20**: 103-109

Uncles R J, Ong J E and Gong W K, 1990, Observations and Analysis of a Stratification-Destratification Event in a Tropical Estuary, Estuarine, Coastal and Shelf Science **31**: 651-665

Uncles R J, Gong W K and Ong J E, 1992, Intratidal fluctuations in stratification within a mangrove estuary, Hydrobiologia **247**: 163-171

Watson J G, 1928, Mangrove forests of the Malay Peninsula, Malayan Forest Records **6**: 1- 275

Wattayakorn G, Wolanski E and Kjerfve B, 1990, Mixing, trapping and outwelling in the Klong Ngao Mangrove Swamp, Thailand, Estuarine, Coastal and Shelf Science **31**, 667-688

Wolanski E and Ridd P, 1986, Tidal mixing and trapping in mangrove swamps. Estuarine, Coastal and Shelf Science, **23**, 759-771

UNESCO, 1990, Relative Sea-level Change: A Critical Evaluation, UNESCO Reports on Marine Science, **54**

The Earth's radiation budget in coastal regions

Costas Varotsos and George Chronopoulos

University of Athens

1 Introduction

The Earth's radiation budget is a measure of all the inputs and outputs of radiative energy relative to the Earth's system. Due to atmospheric phenomena such as absorption and reflection from objects varying in size from dust particles to clouds, much of the incoming solar radiation into the Earth's atmosphere never reaches the ocean or the land surface. It is reflected back into space or retained in some way by the atmosphere and eventually processed into the system as a converted energy form. After being cycled, used, and converted to another energy form through water vaporisation and various gas vaporisation in the Earth's energy system, the outgoing radiation is released out of the atmosphere.

Short-wave radiation is the radiation flux resulting directly from solar radiation and has wavelengths of 0.3 to 4μm. Short-wave radiation received by an object may be the result of direct, diffuse, or reflected solar radiation. Direct radiation is highly directional and depends on the angle between the surface and the Sun. Radiation scattered and reflected by clouds has no specific direction and is called diffuse radiation. Additional reflection of short-wave radiation can occur from the ground surface or other objects. As with diffuse radiation, this terrestrial reflected radiation has no specific direction and is diffuse as well.

Rates of direct short-wave radiation may range from $150Wm^{-2}$ during very hazy or overcast periods to 900 or $1000Wm^{-2}$ during bright sunny periods. However, daily accumulated levels in Florida may range from 12 to $34MJm^{-2}$.

The reflectivity of a surface to short-wave solar radiation is known as the albedo of the surface. The higher the reflectivity of the surface, the higher the albedo. Typical values of albedo are: open water 0.05, dry soil (light colour) 0.32, woodland 0.16-0.18 and crop surfaces 0.15-0.26 (0.23 is often used for a complete canopy of green crops). The intensity of reflected radiation is the product of the albedo and the intensity of the incoming short-wave radiation.

Absorption of incoming solar radiation by water vapour, dust, and O_3 accounts for 16% of incoming solar radiation. Clouds account for 3%. The absorbed radiation helps to heat the clouds. Although this absorbed heat is not as influential as radiation which is absorbed by the ocean and land, it can help to fuel everyday plant and animal life. In terms of radiation which is reflected, air backscatters 6%, clouds reflect 20%, and the Earth's surface reflects only 4% of the total incoming radiation. The land or ocean will immediately receive 51% of the total incoming solar radiation. This number can be broken down into three major categories. The first is net surface emission which requires 21% of the total radiation, the second is sensible heat which takes in 7% of the total radiation, the third is evaporation/precipitation which accounts for 23% of the total radiation.

2 Radiation balance at the surface-atmosphere ininterface

The radiation balance in the system Earth's surface-atmosphere, B, consists of the following two main parts: the radiation balance of the Earth's surface, B_E, and the radiation balance in the atmosphere, B_A.

2.1 Calculation of the radiation balance of the Earth's surface

Denoting by:

Q_E the amount of incident direct solar radiation

q_E the amount of incident scattered sky radiation

A_E the albedo of the surface in a coastal zone (a function of Q_E and q_E)

$L_o\downarrow$ the amount of terrestrial radiation emitted by the atmosphere and intercepted by the Earth's surface

α the absorption coefficient of the surface for $L_o\downarrow$

$L_o\uparrow$ the amount of radiation emitted by the Earth's surface in the upward direction,

then the overall radiation balance of the Earth's surface (made up of the radiation gains and losses) may be described by the following equation:

$$B_E = (Q_E+q_E)(1-A_E) + \alpha L_o\downarrow - L_o\uparrow \quad (1)$$

Considering that $L_o\uparrow - \alpha L_o\downarrow = L_o$ (which is the effective radiation of the surface) the previous equation may be written as:

$$B_E = (Q_E+q_E)(1-A_E) - L_o \quad (2)$$

By using the appropriate albedo values this equation provides the radiation balance at a specific coastal zone.

2.2 Calculation of the radiation balance of the atmosphere

It follows from the above definitions that the overall balance of the atmosphere may be described by:

$$B_A = (Q_A+q_A) + \alpha L_o - (L_o\downarrow - L_\infty\uparrow) \quad (3)$$

where:

$L_\infty\uparrow$ is the radiation emitted by the atmosphere into space

Q_A is the amount of direct solar radiation

q_A is the amount of scattered sky radiation absorbed by the atmosphere.

Considering the transmissivity, P, of the atmosphere above a coastal zone for the amount of thermal radiation absorbed by the atmosphere, the quantity $\alpha L_o\uparrow$ may be written as $\alpha L_o\uparrow = (1-P)\, L_o\uparrow$ and thus the previous equation may be written as:

$$B_A = (Q_A + q_A) + (L_o\downarrow - L_\infty\uparrow) \tag{4}$$

This equation provides the radiation balance of the atmosphere above a coastal zone.

2.3 The radiation balance in the system surface-atmosphere

If A_A is the albedo associated with the back-scattering of the incident radiation by the atmosphere and A_C is the albedo of clouds, then the mean albedo of the Earth as a planet is $A_S = A_E + A_A + A_C$. By substituting A_E by the land or the ocean albedo then A_S gives the mean albedo of land or ocean. Thus the total radiation balance in the system surface-atmosphere in a coastal zone can be obtained by:

$$B = B_E + B_A = (Q_E + q_E)\,(1-A_E) + Q_A + q_A - L_\infty\uparrow \tag{5a}$$

or

$$B = Q_o\,(1-A_S) - L_\infty\uparrow \tag{5b}$$

where Q_o is the total flux of the solar radiation outside the atmosphere.

It is clear that from equation (4) that the term Q_A+q_A is small compared with $L_o\downarrow$ and $L_\infty\uparrow$. It is also clear that $L_\infty\uparrow$ is greater than $L_o\downarrow$, which means that $B_A<0$. We shall assume that the atmosphere is a uniform participant in the radiation balance, i.e. that B_A is independent of height. Therefore the annual variation of B_A is a function of geographical latitude. For instance, at northern latitudes B_A decreases between the equator and the latitude of 25°, increases until the latitude of 60° is reached, and decreases again with increasing latitude thereafter.

It should be noted from the above discussion regarding the calculation of the radiation balance in coastal zones, that it is evident that the albedo is the most crucial parameter. Therefore special attention has been given to the following:

- The term soil must be defined in terms of moisture content, grain size and the colour of the surface. The albedo of all kinds of soil decreases with increasing amount of moisture because the albedo of water is lower than that of soil.
- The albedo of a vegetation cover depends on its type and age.
- The albedo of snow depends on the amount of impurity it contains, on its surface roughness and on the angle of incidence.
- The albedo exhibits a diurnal and an annual variation. It is found to increase with decreasing elevation of the Sun.
- The albedo depends on the spectral composition of the incident radiation. The diurnal variation in albedo is only partly due to its dependence on the angle of incidence and the reflectivity of the surface. A further factor responsible for this variation is the fact that the incident radiation contains a higher proportion of short-wave radiation as the

elevation of the Sun increases. The effect of the wavelength dependence on albedo shows a trend towards an increase in albedo with decreasing elevation of the Sun.

- It is found that the maximum positive value of the radiation balance in the system Earth's surface-atmosphere (B) occurs at about noon, and the maximum negative value is found at night.

In the morning, the transition from negative to positive values of B is delayed by about an hour (it does not correspond exactly with sunrise). This effect is due to the fact that a finite time interval is necessary for the incoming radiation to compensate for losses caused by the effective radiation emitted by the surface. The time of the transition from positive to negative values of B precedes sunset by almost 1.5 hours. This effect is the consequence of the fact that the effective outward radiation exceeds the incoming radiation before sunset. The diurnal variation of B is asymmetric because it is lower in the afternoon than in the morning. This is due to the increase in the effective surface radiation in the afternoon.

3 Radiation balance and atmospheric constituents

The infrared absorption spectrum of the atmosphere is extremely complicated. In contrast with the main components of the atmosphere (nitrogen and oxygen) which do not absorb any thermal radiation, the variable components of the atmosphere such as water vapour, carbon dioxide, ozone, numerous oxides of nitrogen, and hydrocarbons, have innumerable absorption lines and bands in the infrared region of the spectrum.

3.1 Water vapour

Water vapour has the most intense absorption bands in the infrared region of the spectrum. The predominant role of water vapour in the absorption of thermal radiation was established towards the end of the last century. Special physical investigations have determined with adequate reliability the structure of the molecule of water vapour and established the origin of the individual absorption bands. It was found that in the unexcited state the water vapour molecule has the shape of an isosceles triangle. The emission and absorption of radiation by water vapour in the infrared region of the spectrum is caused by vibrational and rotational-vibrational transitions and in the far infrared region (λ>15μm) by purely rotational transitions. Three types of normal vibration of the water vapour molecule are known with the following wave numbers: 3670 cm^{-1}, 1675 cm^{-1} and 3790 cm^{-1}. All these types of vibration are active during absorption. The purely rotational spectrum arises from transitions of energy due to the rotation of the molecule.

The atmospheric thermal radiation, the principal part of which is at normal temperatures and located in the region between 4 and 40μm is almost completely absorbed by the atmosphere in the whole of this wavelength region, with the exception of the interval between 8 and 12μm, where the absorption by water vapour is insignificant. The wavelength interval 8-12μm is often referred to as a transparent window of the atmosphere or simply an atmospheric window.

3.2 Carbon dioxide and other polyatomic gases

As we have seen above water, vapour undoubtedly plays the predominant part in the absorption of long-wave radiation in the atmosphere. However, the atmosphere contains many other polyatomic gases having infrared absorption bands. They include carbon dioxide, ozone, a number of nitrogen oxides, hydrocarbons and some other compounds. With the exception of carbon dioxide all these gases are present in the atmosphere in very insignificant quantities and at considerable altitudes.

The molecule of carbon dioxide is linear. Intense absorption bands of carbon dioxide in the infrared region of the spectrum are due to the vibrational transitions. There are three types of normal vibrations of the carbon dioxide molecule, but only two of them are active in absorption. Two intense absorption band correspond to these basic frequencies, situated at about 4.3 and 14.7μm. Apart from these two main bands, there are also bands centred at wavelengths of 2.7 and 10μm and also a number of other weak bands.

The main absorption bands of ozone in the infrared region of the spectrum are caused by rotation-vibration transitions. The present knowledge of the structure of the ozone molecule is as yet inadequate. The identification of the infrared absorption bands of ozone is also insufficient at present. A general picture of the infrared absorption spectrum of ozone shows that there are clearly defined absorption bands centred at the following wavelengths: 2.7, 3.28, 3.57, 4.75, 5.75, 9.1, 9.65 and 14.1μm. Other absorption bands of ozone are overlapped by more intense bands of water vapour and carbon dioxide.

Apart from water vapour, carbon dioxide and ozone, the atmosphere contains a number of other gases which have absorption bands in the infrared region. Oxides of nitrogen (NO_2, N_2O, NO, N_2O_4) and a number of hydrocarbons (C_3H_8, C_2H_6, C_2H_4, CH_4) have the largest number of absorption bands in the infrared region of the spectrum. Absorption bands of sulphur dioxide, heavy water and a number of other substances have also been detected.

3.3 Atmospheric ozone

3.3.1 The role of ozone in the atmosphere

Although ozone is a minor constituent of the atmosphere it has played a decisive role in the evolution of the Earth. Since the ozone layer was formed in the stratosphere, life on the land has become possible under the protection of the ozone layer against harmful short-wave solar ultraviolet radiation. The impact of the biologically effective UV-B radiation (wavelength interval 280-320nm) on the biosphere (including such deleterious effects on humans as skin cancer, erythema and eye disorders) has motivated a necessity to study the spatial and temporal variability of the total ozone content with the use of ground-based measurements and satellite observations. Such a necessity has recently been enhanced by the discovery of the biological significance of tiny solar spectrum variations within Fraunhofer lines.

The increasing level of atmospheric pollution leads to opposite changes of ozone concentration in the stratosphere and in the troposphere. The transport of such ozone destroying gases as chlorofluorocarbons (CFCs) into the stratosphere with their subsequent conversion into active chlorine (Cl and ClO) results in destroying ozone molecules through heterogeneous photochemical reactions and the depletion of the ozone layer. Of special significance is atmospheric pollution produced by High-Speed Civil Transport aircraft (Weisentein et al. 1993). More than two decades ago an explosion of scientific interest in this problem took place (Crutzen 1971, Johnston 1971). Recent new studies have been stimulated by important achievements in the development of relevant simulation modelling.

Chandra et al. (1993) have emphasised the existence of the still significant disagreement between numerical modelling results and observations for the upper stratosphere expressed in the strong underestimation of calculated HCl concentrations but too large ClO concentrations at heights above 40 km. A new photochemical unit of the model has been developed which simulates such a redistribution of Cl_y family components which results in the formation of a much more substantial HCl reservoir and a decrease of the ClO reservoir. This makes results of computations more reliable; computed ClO/HCl concentration ratios are much closer to observations. An important role to achieve such results belongs to the use of the new channel of HCl + CO_2 formation through the reaction $ClO + OH \rightarrow HCl + O_2$ in addition to the formation of $Cl + HO_2$.

Photochemical reactions at the Earth's surface with the participation of nitrogen oxides (NO_x) and hydrocarbon compounds lead to the growth of surface ozone concentration which is harmful for the biosphere in general and for humans in particular (Borrell et al. 1993, Kondratyev et al. 1994, Kondratyev and Varotsos 1993, Varotsos and Kondratyev 1994, 1995a,b). In such large industrial cities as Mexico and Athens episodes of high surface ozone concentration have become a very serious danger for human health.

The variability of solar ultraviolet radiation is but a part of a very complex problem of the variability of ozone as an interactive component of the atmosphere which also plays an important role as a climatically significant gas. While stratospheric ozone is responsible for warming the stratosphere through the absorption of ultraviolet solar radiation, tropospheric ozone is a significant greenhouse gas (Kondratyev 1988). This results in a coupled ozone participation in the formation of global climate change (Kondratyev and Cracknell 1998). A newly opened aspect of ozone variability is an interaction between aerosols produced by volcanic eruptions, oceans (dimethylsulphide emission) as well as anthropogenic activities (SO_2 emissions and subsequent gas-to-particle conversion) and both tropospheric and stratospheric ozone (Kondratyev and Varotsos 1995). The relevant chains of interactions are equally important for global climate and ozone variabilities.

A special aspect of ozone and climate dynamics is its dependence on solar activity impact (Kondratyev 1989, Kondratyev and Cracknell 1998). This controversial problem is far from being solved, but it is quite probable that a potential mechanism of the solar activity impact on climate is being realised through the ozone layer in the stratosphere.

3.3.2 Production and depletion of atmospheric ozone

The formation of the ozone layer was a very important stage in the evolution of the Earth because the ozone layer provides a reliable protection of life against harmful solar ultraviolet radiation. Total ozone content in the atmosphere on a seasonal time scale is controlled by both the transport of ozone within the stratosphere and photochemical processes. Kaye (1993) summarised this global pattern. Net ozone is produced in the tropical upper stratosphere. This ozone is transported polewards and downwards, particularly in autumn and winter. In the lower stratosphere, the lifetime of ozone is long during spring and winter. As ozone is transported downwards it begins to accumulate at high latitudes. The transition from winter to summer circulation in the stratosphere dramatically slows the accumulation of ozone and ozone decreases due to increased photochemical destruction in summer. Schoeberl (1993) has shown by using TOMS records that the transport towards the north is greater and extends more polewards than in the southern hemisphere. In the stratosphere the planetary wave activity is responsible for most of the north-south transport of ozone and heat. Since the southern hemisphere lacks the significant orographic features found in the northern hemisphere, planetary waves are weaker. As a consequence, the winter antarctic stratosphere is on average more than 10degK colder than the arctic stratosphere, the antarctic polar vortex lasts longer and the winter accumulation of ozone is weaker. This principal global dynamical pattern and its hemispheric differences are the basic frame for the variation of the vertical ozone distribution caused by dynamical effects as well as by chemical and radiative processes.

The global pattern of the ozone layer is finally produced by photochemical and catalytic reactions as well as by dynamical transport (Brasseur and Solomon 1984, World Meteorological Organization 1992, Kaye 1993). Ozone (O_3) is formed by the UV photolysis of molecular oxygen (O_2) followed by the recombination of atomic oxygen (O) with O_2:

$$O_2 + h\nu\,(UV) \rightarrow O + O$$

$$O + O_2 + M \rightarrow O_3 + M$$

where M indicates an air molecule. Although mixing ratios of ozone rarely exceed 10 ppmv in the stratosphere, its total amount comprises some 90% of the atmospheric ozone. The absorption of UV radiation by stratospheric ozone leads to heating of the stratosphere when the energy of radiation is converted to kinetic energy of air molecules (M*) by the recombination of O and O_2:

$$O_3 + h\nu\,(UV) \rightarrow O + O_2 \text{ (UV absorption)}$$

$$O + O_2 + M \rightarrow O_3 + M^* \text{ (heating of stratospheric air).}$$

This heating of stratospheric air causes increasing temperature with altitude. Therefore the stratosphere becomes more stable with respect to vertical mixing, and large vertical gradients may occur in trace constituents. They can only be mixed vertically by the global dynamics associated with large-scale planetary waves, or by meso-scale wave disturbances of the stratosphere, e.g. rising vertical motion in the tropics balanced by descent at high latitudes.

While the stratospheric production of ozone is only based on the photolysis of oxygen, it is mainly destroyed by a large number of catalytic chemical reactions involving free radical species containing either hydrogen, nitrogen, chlorine, or bromine atoms (Brasseur and Solomon 1984, World Meteorological Organization 1992, 1995). The general two-step reaction chain is:

$$O_3 + X \rightarrow XO + O_2$$

$$O + OX \rightarrow X + O_2\,.$$

The most important radicals X include hydroxyl (OH), nitric oxide (NO), chlorine (Cl), and bromine (Br). Their tropospheric sources are water (H_2O) and methane (CH_4) for OH, nitrous oxide (N_2O) for NO, methyl chloride (CH_3Cl) and anthropogenic CFCs for Cl, as well as methyl bromide (CH_3Br) and halocarbons (halons) for Br. But some ozone is also directly destroyed by a process involving only oxygen by the following reaction chain:

$$O_3 + h\nu\ (UV, visible) \rightarrow O + O_2$$

$$O + O_3 \rightarrow O_2 + O_2\ .$$

The natural and anthropogenic sources for these-ozone destroying radicals are found in the troposphere. Transported into the stratosphere, these chemically inert gases are photolysed by solar UV radiation and reactive atoms are released. Most ozone production and destruction occurs in the tropical upper stratosphere, where UV amounts are largest. Photodissociation of ozone, however, extends lower into the stratosphere and to higher latitudes than production does. So transport processes play an important role in controlling the distribution of ozone, whose production and destruction are not in local balance. The separation between production and destruction and the global transport are yielding to the net vertical profile of stratospheric ozone and its latitudinal variation. Any change in global transport would change the ozone distribution significantly.

The reason for separation between the production and destruction regions is that ozone production is driven by UV with wavelengths below 240 nm, while ozone loss typically involves production of atomic oxygen which occurs at longer wavelengths in the UV range (220-350 nm) and to a lesser extent in the visible range (400-800 nm). Since longer wavelength photons penetrate deeper into the atmosphere, there is a difference between the altitudes for production and destruction.

The catalytic ozone loss processes are in equilibrium with a large number of reactions between these catalytically active constituents themselves. Reservoir gases are formed by these reactions. They are unreactive towards ozone. The main reservoir gases are HNO_3, HCl, and $ClONO_2$. They are formed by the following reactions:

$$OH + NO_2 + M \rightarrow HNO_3 + M^*$$

$$ClO + NO_2 + M \rightarrow ClONO_2 + M^*$$

$$OH + NO + M \rightarrow HONO + M^*\ .$$

Because the reservoir gases are only slowly photolysed by UV and have a lifetime of the order of days to weeks, they can build up in large concentrations. Therefore the equilibrium between the catalytic radicals and the photolysis of reservoir gases specifies another control for the net ozone loss and subsequently the vertical ozone distribution. Additionally in the polar stratosphere at extremely low temperatures, heterogeneous chemical reactions at the surfaces of aerosol particles and polar stratospheric cloud particles can change the chemical composition and yield unusual losses in ozone there.

Thus the large number of chemical processes as well as the transport of source gases from the troposphere into the stratosphere, the transport of stratospheric species back into the troposphere, and the transport and mixing of chemicals within the stratosphere are all critical for understanding the spatial distribution of ozone and its temporal changes for a wide range of time scales. Ozone in the polar atmosphere is basically controlled by all of these processes. So detailed studies of the ozone layer have to refer to observations with a sufficient vertical resolution (Gernandt et al. 1997). On the other hand the mid-latitude ozone distribution is strongly affected by polar stratospheric processes which control the ozone exchange to the mid-latitudes. Thus an understanding of the processes in the polar regions and of the transport of air through polar regions to the mid-latitudes is very important for understanding and predicting the mid-latitude ozone loss.

3.3.3 Impact of solar activity on the ozone layer

The impacts of solar activity on the atmosphere and climate are still a subject of controversy (Kondratyev and Cracknell 1998). There is, however, no doubt that solar irradiance variability influences the ozone layer.

On the basis of the analysis of the Nimbus-7 TOMS data, Chandra (1991) has shown that the solar UV related change in total ozone content over a solar cycle is about 1.5%, which may be attributed to about 6% change in the solar UV flux near 200 nm. This estimate is also consistent with the solar UV changes in the total ozone content over a time scale of 27-day solar rotation. In a later paper Chandra and McPeters (1994) have analysed the combined Nimbus-7 solar backscattered UV and NOAA-11 SBUV/2 data, covering a period of more than a solar cycle (about 14 years), to assess the response of ozone in the stratosphere to the UV variations. The principal conclusion of this study is that about 2% change in TOC and about 5-7% change in ozone mixing ratio in the upper stratosphere (0.7 to 2hPa) may be attributed to the change in the solar UV flux over a solar cycle. A very important result of this analysis is that in the upper stratosphere, where photochemical processes are expected to play a major role, the observed solar cycle variation of ozone is significantly larger than is inferred either from photochemical models or from the ozone response to the 27-day solar UV modulation (Table 1). The relatively low observed sensitivity at the 10hPa level could be caused by measurement error, since ozone measurements in this height range were seriously affected for 1-2 years by aerosol clouds injected into the atmosphere during the El Chichon and Mount Pinatubo volcanic eruptions. Causes of this disagreement are not clear as yet.

Jackman et al. (1993) have pointed out that very large solar proton events, such as occurred from 19-27 October, 1989 are predicted to produce short-lived increases in HO_x and long-lived increases in NO_x species which can both lead to ozone destruction. Computations made by Jackman et al. with the 3D chemistry and transport model to simulate the distribution of NO_x and ozone after the solar proton events showed that ozone and NO_x behaviour for two months after the October 1989 solar proton events was characterised by significant interhemispheric differences which are qualitatively consistent with SBUV/2 ozone observations. The ozone depletion in the northern hemisphere was much more substantial (12% in the latitude band 60^0 - 80^0N at 4hPa) than in the southern hemisphere (only 1%) due to differences in atmospheric circulations (the amount of HO_x produced in both hemispheres is similar). The 3D model predictions (without chlorine chemistry) gave -18% (northern hemisphere) and -8% (southern hemisphere).

Pressure (hPa)	0.5	0.7	1.0	2.0	5.0	10.0	TOC
Sensitivity:							
Observed	0.66	0.72	0.88	0.99	0.33	0.11	0.27
Modelled	0.29	0.33	0.39	0.56	0.59	0.44	0.27

Table 1. The ozone sensitivity to 11 year solar cycle: observed and modelled values (Fleming et al. 1994). TOC denotes thetotal ozone content.

4 Applications of the radiation balance in coastal regions

4.1 Atmospheric physical processes and the greenhouse effect

As has been suggested by Kondratyev and Moskalenko (1984) the greenhouse effect, G, may be defined by the difference between thermal emission from the surface, E, and outgoing longwave radiation, F:

$$G = E - F \tag{6}$$

Assessments of global/annual average values on the basis of satellite observations led to the following figures: $E = 390\text{Wm}^{-2}$; $F = 235\text{Wm}^{-2}$, so that $G = 155\text{Wm}^{-2}$.

The clear-sky greenhouse effect, G_A, may then be defined as

$$G_A = E - F_{clear} \tag{7}$$

with global/annual average value $F_{clear} = 265\text{Wm}^{-2}$ (Harrison et al., 1990) so that we have $G_A = 125\text{Wm}^{-2}$. Although it may seem a paradox, the cloud longwave radiative forcing is thus equal to only 30Wm^{-2}.

Following Raval and Ramanathan (1989) let us introduce normalised quantities to remove the strong temperature dependence:

$$G' = E/F_{clear} \tag{8}$$

There is a very simple possibility to assess qualitatively the influence of atmospheric temperature and humidities on the clear-sky greenhouse effect. Obviously,

$$F_{clear} = \varepsilon_A \,\sigma\, T_A^{\,4} + (1 - \varepsilon_A)\,\sigma\, T_S^{\,4} \tag{9}$$

where ε_A is atmospheric emissivity, σ is the Stefan-Boltzmann constant, T_A is the mean temperature and T_S is the mean surface temperature. Now, from equation (8), we have

$$G' = T_S^{\,4}/(\varepsilon_A \,\sigma\, T_A^{\,4} + (1 - \varepsilon_A)\,\sigma\, T_S^{\,4}). \tag{10}$$

Or it is possible to introduce a parameter g (Raval and Ramanathan 1989):

$$g = \varepsilon_A\,[1 - (T_A/T_S)^4] \tag{11}$$

Hence:

$$G' = 1/(1 - g) \tag{12}$$

and

$$G' = 1/[1 - \varepsilon_A(1 - (T_A/T_S)^4)]\,. \tag{13}$$

Here it is assumed that the emissivity ε_A is proportional to the total column moisture content, w. Although it is a clear simplification, equations (11) and (13) illustrate the essential result that there is a separation between the effects of the atmospheric humidities, as measured by w and represented by ε_A in these equations, and the temperatures, as represented by T_A. As is seen, the increase of w and, consequently, T_A leads to the enhancement of G or g, while the T_A increase results in the decrease of temperature contrast between the atmosphere and the surface and, thus, to the G' decrease. It is quite obvious that the roles of w and T_A have to be geographically specific.

4.2 The sea-surface temperature

Webb et al. (1993) have pointed out that since the temperature field in the tropics is comparatively homogeneous, the dominating role in the greenhouse effect variability belongs to w, which is controlled by the sea surface temperature. The opposite situation takes place in middle and high latitudes where air temperatures are highly variable and therefore exert much more powerful control over G' than at low latitudes.

The analysis of satellite data on the Earth radiation budget (Earth Radiation Budget Experiment) and the total column moisture (Special Sensor Microwave/Imager) as well as the results of radiative transfer simulations made by Webb et al. (1993) have confirmed the conclusions of the simplified qualitative assessment. At low latitudes the clear-sky greenhouse effect varies mainly due to w changes (which, on the other hand, are controlled by the sea surface temperature) and seasonal variations are small. In contrast, at middle and high latitudes both G' and w exhibit strong seasonal variations. The clear-sky greenhouse effect variation is controlled by the seasonal changes in atmospheric temperatures, which are strong enough to overcome the opposing effect of the moisture variations. There are strong seasonal variations of the greenhouse effect which has the maximum in winter (when the surface-atmosphere temperature difference increases and leads to the enhancement of the greenhouse effect, in spite of small moisture content) and the minimum in summer. The combined impact of both temperature and column moisture results in the formation of the meridional profile of the greenhouse effect which is characterised by the decrease of G' towards the winter pole at a much slower rate than it does towards the summer pole.

Hallberg and Inamdar (1993) have pointed out a special role of water vapour as a greenhouse gas (see also Kondratyev 1956, 1969, 1988, Kondratyev and Moskalenko, 1984). In this context they discussed the so called super greenhouse effect defined, in accordance with the earlier definition given by Raval and Ramanathan (1989), as a situation in which atmospheric greenhouse trapping increases more rapidly with spatially increasing sea surface temperature than the infrared surface emission (spatially increasing means that if one were to move from a region of lower sea surface temperature to a region of higher sea surface temperature, one would generally find a large atmospheric greenhouse effect at the new location).

Under clear sky conditions the observed values of the atmospheric greenhouse effect rise abruptly with spatially increasing sea surface temperature at sea surface temperatures above roughly 298K. The derivative $dG_A/dT_S \approx 8 Wm^{-2}K^{-1}$ becomes even larger than the change in surface emission $dE/dT_S = 6.12 Wm^{-2}K^{-1}$ at 300 K, i.e. the loss of infrared energy to space (in the absence of clouds) is reduced in regions of high sea surface temperature. This is the phenomenon which is called the super greenhouse effect. Its existence has been discovered on the basis of satellite observations in various parts of the World's oceans.

On the basis of the analysis of numerical modelling results and satellite observations Hallberg and Inamdar (1993) have shown that four processes contribute to the formation of the super greenhouse effect, but the principal contribution belongs to water vapour continuum absorption and thermodynamically controlled increases in water vapour concentration at constant relative humidity with increasing atmospheric temperature.

Besides, such processes are significant as the increase of atmospheric moisture content (in the upper and middle troposphere in particular) over the warmest sea surface temperature, while the atmospheric vertical temperature profile is increasingly unstable. Regions with these high sea surface temperatures are also increasingly subject to deep convection, which suggests that convection moistens the upper and middle troposphere in regions of convective relative to nonconvective regions, resulting in the super greenhouse effect. Hallberg and Inamdar (1993) note that for the explanation of the observed super greenhouse effect the impact of dynamic processes also has to be taken into account.

Gutzler (1993) has analysed uncertainties in climatological tropical humidity profiles to demonstrate that they are very important for assessments of the greenhouse effect. An important potential consequence of the enhanced greenhouse effect is global climate warming. In connection with the establishment of potential mitigation measures against global warming Clerbaux et al. (1993) have calculated Global Warming Potentials for 10 alternative hydrohalocarbons, which may be used as substitutes for chlorofluorocarbons CFC11 and CFC12, on the basis of new measurements results for infrared cross sections of the substitutes. Table 2 illustrates relevant results (global warming potentials were calculated relative to CFC11 as a reference gas). As can be seen from the table, for three compounds (HCFC22, HCFC142b, HFC125) the global warming potentials are higher than unity for a 5-year to 10-year period. Moreover, these compounds maintain a relatively strong global warming potential for a very long time. Thus, it may be concluded that the problem of mitigation measures has not been solved as yet.

Halocarbon		Radiative Forcing per		Global warming potential (years)						
Name	Lifetime	molecule	kilogram	5	10	20	50	100	200	500
CFC11	57.0	1.00	1.00	1.00	1.00	1.00	1.00	1.00	1.00	1.00
HCFC22	14.3	0.85	1.35	1.19	1.06	0.86	0.56	0.41	0.35	0.34
HCFC123	1.5	0.90	0.81	0.25	0.13	0.07	0.04	0.03	0.02	0.02
HCFC124	6.0	0.94	0.95	0.67	0.50	0.32	0.17	0.12	0.10	0.10
HCFC141b	9.7	0.66	0.77	0.63	0.53	0.39	0.22	0.16	0.14	0.13
HCFC142b	21.1	0.82	1.12	1.04	0.97	0.86	0.64	0.50	0.43	0.11
HCFC225ca	2.4	1.08	0.73	0.32	0.19	0.10	0.05	0.04	0.03	0.03
HCFC225cb	6.8	1.31	0.88	0.65	0.50	0.34	0.18	0.13	0.11	0.11
HCF125	33.9	0.91	1.04	1.01	0.98	0.93	0.82	0.71	0.64	0.62
HCF134a	13.1	0.78	1.06	0.92	0.81	0.64	0.41	0.29	0.25	0.24
HCF152a	1.5	0.50	1.03	0.31	0.17	0.09	0.05	0.03	0.03	0.03

Table 2. Lifetimes (years), radiative forcings, and global warming potentials relative to CFC11.

4.3 The climate change in the context of the atmospheric composition

Recent years have been marked by a number of attempts to assess the reliability of existing datasets which were used for empirical analyses of climate changes. For instance, Van den

Dool et al. (1993) have compared for this purpose a time series of 43 years of observed monthly mean surface air temperature at 109 sites in the 48 contiguous United States to monthly mean air temperature specified from hemispheric gridded 700-hPa heights. Such a check of mutual data consistency showed that in both datasets cooling (of about 0.5^0C) from 1951 to about 1970 and subsequent warming (also by 0.5^0C), that continues to the present time, took place. This allows one to conclude that interdecadal temperature changes considered are probably real. Another important conclusion made by Van den Dool et al. (1993) is that a comparison of the full set of 109 stations to a clean subset of 24 has indicated that the influence of common problems in surface data (station relocation, urbanisation, etc.) is quite small. This conclusion is in contradiction, however, with the results of the detailed analysis of urban bias in air surface temperature of China's northern plains accomplished by Portman (1993), which indicates that despite past efforts to remove the effects of the urban heat islands from land-surface datasets, large urban biases may still remain. Thus, the problem of observation errors still deserves serious attention.

Madden and Meehl (1993) have analysed the bias in the global mean temperature estimated from sampling a greenhouse warming pattern with the current surface observing network. The principal conclusion is that the observations prove adequate to estimate the globally averaged temperature change associated with the pattern of CO_2 warming from a general circulation model with a bias whose absolute value is generally less than 2%. The calculated pattern of climate change was obtained as a result of a 60-year run of the global coupled atmosphere-ocean NCAR R15 model (4.5^0 latitude by 7.5^0 longitude resolution) with a linear increase in atmospheric CO_2 of 3.3 ppm yr^{-1}. The CO_2 signal is positive all over the globe with the largest values exceeding 1^0C north of 60^0N. The global average (last 30 years of the 60-year run) CO_2 signal is 0.808^0C.

Madden et al. (1993) have concluded that for global temperature maps generated by a long (120 ensemble members) perpetual January general circulation model run, the empirically estimated mean-square deviation between perfectly and imperfectly spatially sampled temperatures varies from 0.050^0C^2 before the turn of the century to 0.002^0C^2 after 1950. Corresponding RMSD are 0.224^0C and 0.045^0C. Since 1950 imperfect spatial sampling is less of a problem than imperfect temporal sampling. Some initial calculations to assess sampling errors for real observational data have shown that in case of 1000-hPa temperatures for the time period 1982-1987 spatial variance is near 2^0C^2 for every month of the year. Madden et al. (1993) believe that such a small spatial variance in combination with averaging the twelve months data will result in real errors in annual means which are smaller than those obtained from numerical modelling results.

Although existing observational networks produce reliable enough information and historic data series for the previous century which are more or less representative from the climatic viewpoint, an important fact is that (Karl et al., 1993b) over the past decades observational networks have been designed for a variety of purposes but rarely have been designed to detect and monitor climate change and variations. Therefore the problem of data quality is still urgent. Karl et al. (1993b) have pointed out in this connection the following principal aspects of managing climate data from weather observing systems:

- long-term homogeneous databases
- resolution of datasets for various time and space scales

- information about the observing systems, data collection systems and data reduction algorithms, broadly defined as metadata (important information about the data)
- enhancing weather observing systems to reduce the uncertainties about how climate has (or is) changed and varied.

A number of recent global climate diagnostics studies have arrived at several unexpected conclusions. For instance, although most global climate models expect the strongest impact of the CO_2 rise in the high latitudes (see details in Kondratyev and Grassl, 1993) neither systematic change of minimum and maximum temperatures nor an overall warming has been observed in the Arctic over the last 50 years or so (Kahl et al. 1993). The analysis of tropospheric temperature trends in the Arctic during the time period 1958-1986 made by Kahl et al. (1993) has shown that absolute trends of 3^0C/30 yr or higher were found, with both cooling and warming tendencies observed in the four layers considered: 850-700, 700-500, 500-400, 400-300 hPa. The majority of the trends, however, are not statistically significant at the 90% confidence level. Therefore Kahl et al. (1993) have concluded that "greenhouse-induced warming is not detectable in the arctic troposphere for the 1958-1986 period". This is, of course, a rather surprising conclusion, since northern high latitudes have always been considered as the zone of maximum chances to detect a greenhouse climatic signal. Such a conclusion is especially important because it was made on the basis of the most reliable observational data for three recent decades.

Another climatic surprise has been a discovery of asymmetric trends of daily maximum and minimum temperature. Karl et al. (1993a) have pointed out that monthly mean maximum and minimum temperatures for over 50% (10%) of the northern (southern) hemisphere landmass, accounting for 37% of the global landmass, indicate that the rise of the minimum temperature has occurred at a rate three times that of the maximum temperature during the period 1951-1990 (0.84^0C versus 0.28^0C). The decrease of the diurnal temperature range is approximately equal to the increase of mean temperature. The asymmetry is detectable in all seasons and in most of the regions studied. Table 3 data illustrate observed trends of diurnal temperature range. For this part of the globe (37%) the rate of the decrease in the diurnal temperature range (-1.4^0C/100 years) is comparable to the increase of the mean temperature (1.3^0C/100 years).

A very long record of maximum and minimum temperatures for the Klementinum Observatory in Prague (Czech Republic) shows an increase of the diurnal temperature range from the early to the mid-twentieth century, with a substantial decrease since about 1950. The increase coincides with the increase of mean temperature since the turn of the century and the decrease occurs when the mean temperature reflects little overall change.

Although the variability of diurnal temperature range is affected by many factors, Karl et al. (1993a) have shown that two variables are most important, namely cloud amount and cloud height. The decrease in the daily temperature range is partially related to increases of cloud cover. More complete identification of various causes requires further studies. It is clear, however, that "the direct radiative effect of increasing CO_2 alone is unlikely to explain the current trends".

Kukla *et al.* (1994, unpublished) have summarised the results of a "Minimax Workshop", which confirmed conclusions concerning the decrease of diurnal temperature range made earlier as well as the conclusion that this decrease is due to the rise of the night-time temperatures. Although, in accordance with earlier findings, it was confirmed that the decrease in daily temperature range is related closely with the increase in cloud cover, this is not always the case. The undeniable fact is however that the cloud cover is indeed increasing in most places where the decrease of diurnal temperature range has been reported and that the preferential increase of night-time temperatures is in line with such development. According to Parungo et al. (1994) the cloud cover increased between 1952 to 1981 by 2.3% in the northern hemisphere and 1.2% in the southern hemisphere. The daytime A_S and A_C between 30^0 and 50^0N latitude increased by 27%.

Seasons:	MAX	MIN	Diurnal range
Northern Hemisphere (50%)			
D-J-F	1.3	2.9	-1.5
M-A-M	2.0	3.2	-1.3
J-J-A	-0.3	0.8	-1.1
S-O-N	-0.4	1.3	-1.7
Annual	0.5	2.0	-1.4
Southern Hemisphere (10%)			
D-J-F	1.6	2.2	-0.6
M-A-M	1.7	2.5	-0.8
J-J-A	1.0	1.3	-0.4
S-O-N	0.8	2.1	-1.3
Annual	1.3	2.0	-0.8
Globe (37%)			
D-J-F	1.3	2.9	-1.6
M-A-M	1.9	3.1	-1.2
J-J-A	-0.2	0.8	-1.1
S-O-N	-0.3	1.4	-1.7
Annual	0.7	2.1	-1.4

Table 3. Trends of temperature (^{0}C/100yr) for annual and three-month mean maximum (MAX), minimum (MIN), and diurnal temperature over the period 1951–1990 Percent of the land area covered for the northern and southern hemisphere and the globe is denoted within parenthesis. D-J-F denotes the period ofDecember, January, February, etc.

Since global climate models do not predict the increase in cloud amount over land, it may be assumed that the increase of cloud cover (frequency and density of clouds) could be due to an impact of atmospheric pollution.

Of course, surface air temperature should be not the only parameter to be used for climate diagnostics. The problem, however, is that time series for other parameters are too short for climatological analysis. In fact, only sea surface temperature data merged with surface air temperature data have been extensively used for the analysis of global climate changes. Deser and Blackmon (1993) discussed the results of the analysis of four components of the climatic system, sea surface temperature, surface air temperature, surface wind and sea level pressure, applying the empirical orthogonal functions technique. The results show that the patterns of sea surface temperature and surface air temperature change between the time periods of 1900-1929 and 1939-1968 and demonstrate that the climate warming was concentrated along the Gulf Stream east of Cape Hatteras and may have been a result of altered ocean current. Warming also occurred over the Greenland Sea and the eastern subtropical Atlantic.

A number of efforts have been undertaken to analyse available data on precipitation. Bauer and Schluessel (1993) discussed combined datasets on rainfall, total water and water vapour over sea from polarised microwave simulations and Special Sensor Microwave/Imager (SSM/I) data. The analysis of the global oceanic precipitation from the Microwave Sounding Unit (TIROS-N satellites) during 1979-1991 on a 2.5^0 grid shows that peak annual rainfall (5600 nm) occurs in a quasi-stationary portion of the Intertropical Convergence Zone (ITCZ) over the eastern Pacific, while peak monthly rainfall (over 900 mm) occurs in the north-eastern Bay of Bengal in June. Comparisons with other data (both conventional and satellite data) revealed several important differences from existing ocean rainfall climatologies which interpretation allows to judge about higher reliability of the MSU data. It is important, however, that at the 2.5^0 gridpoint level the correlation between two satellites' monthly anomalies (MSU and GOES data for 81 months) is generally above 0.8 in the tropics, reaching 0.99 in the central Pacific.

In connection with the problem of the antarctic polar ice sheets dynamics Bromwich and Robasky (1993) studied changes of the precipitation rate over the Antarctic which appears to have increased by about 5% over a time period spanning the accumulation means for the 1955-1965 to 1965-1975 periods. During the same time period over Greenland the precipitation rate decreased by about 15% since 1963 with secondary increase over the southern part of the ice sheet starting in 1977. At the end of the 10-year overlapping period, the global sea level impact of the precipitation changes over Antarctica dominates that for Greenland and yields a net ice-sheet precipitation contribution of roughly $-0.2 mmyr^{-1}$. These conclusions should be considered as tentative, however.

Important studies of climate changes have been accomplished with the use of satellite information on sea ice and snow cover dynamics. In the context of global warming Karl et al. (1993a) have examined contemporary large-scale changes in solid and total precipitation and satellite-derived snow over the North American continent. They show that snow cover extent over the last 19 years decreased up to 6×10^5 km^2 related to a 0.93^0C (0.33^0C) increase in North American (northern hemisphere) temperature. Over the last two decades the decrease in snow cover during winter (December-March) has largely occurred through reduced frequency of snow cover in areas that typically have a high probability of snow on the ground with little change in the frequency of snow cover in other areas. Similar characteristics were observed during spring (April-May) in areas with high snow cover probability except for an expansion of the snow-free regions. Anomalies in these two seasons dominate the interannual variability (nearly three-fourth of the variance) of snow cover.

Karl et al. (1993a) have emphasised that the apparent unprecedented global warmth of the 1980s was accompanied by a retreat of the mean annual North American snow cover, a 10% increase in annual Alaskan precipitation, a significant decrease (-7%) in annual snowfall over southern Canada (while the total precipitation remained above normal) and a more than twofold increase in the variance of the ratio of frozen to total precipitation over the contiguous United States, where also precipitation has significantly increased (2%-3% per decade) during the last four decades, but on a century time scale the increasing trend is not yet statistically significant. Karl et al. (1993a) have cautioned against oversimplification of the relation between precipitation and snow cover extent. An increase of precipitation with enhanced greenhouse gases may result both in the retreat of snow cover (in case of rainfall) and in the snow cover expansion (in case of snowfall). Therefore, a priori, the early evolution of future changes of snow cover is uncertain in high latitudes as the climate warms. Thus a careful investigation of the climatology of snowfall and snow cover is necessary. This is equally true for ice sheets.

A very important source of information is satellite measurements of the Earth's radiation budget, especially in the context of the atmospheric greenhouse effect dynamics. This problem has been discussed in detail by Kondratyev et al. (1994).

4.4 Aerosol cooling, greenhouse warming and phytoplankton

In connection with the problem of aerosol cooling versus greenhouse warming, a great deal of attention has been attracted by the hypothesis of dimethylsulphide (DMS) climatic impact which is based on the assumption that phytoplankton could affect cloud albedo by producing DMS, that the latter is a precursor to aerosols and cloud condensation nuclei and that cloud albedo could in turn effect the productivity of the phytoplankton thus creating a climatically significant feedback cycle (see Kondratyev and Cracknell, 1998).

Lawrence (1993) has undertaken an empirical analysis of the strength of this phytoplankton-dimethylsulphide-cloud-climate cycle by considering available data on the relations between individual components of the feedback and developing an empirical model of the cycle as a whole which allows an assessment of the strength of the cycle to regulate climate thermostatically. The feedback considered by Lawrence (1993) includes three components:

- the coupling between phytoplankton, DMS and cloud condensation nuclei,
- the effect of changes in cloud condensation nuclei levels on albedo, incident irradiance at the surface and surface temperature,
- the response of phytoplankton to changes in incident irradiance and seawater temperature.

The first and third components of the cycle have been derived primarily from observations of their respective chemical, physical and biological constituents. The second component was based on the estimations that a 30% increase of cloud condensation nuclei in the area covered by marine stratiform clouds would cause a decrease in the global average surface temperature of 1.3^0C, as well as on the conclusion that enhanced sulphur emissions should cause roughly the same magnitude of cooling via the direct aerosol backscatter effect as via the indirect cloud albedo effect.

Calculations made by Lawrence (1993) for several different situations gave results which indicated that the marine biogeochemical cycle could be playing a significant role in the global climate change (for example, in aerosol cooling versus greenhouse warming). The climate is extremely sensitive to the changes in aerosol and cloud condensation nuclei levels. Lawrence (1993) has pointed out that studies in two areas in particular are needed to reduce uncertainties, (a) phytoplankton ecology and the concomitant DMS production and (b) the relation between DMS emissions and aerosol particle density.

Important studies of biogenic sulphur aerosol in the arctic troposphere have been made on the basis of long-term observations from 1980 to 1990 of aerosol methanesulphonate, sulphate, sodium and other related chemical species as well as a shorter time series of aerosol sulphur isotopic composition at Alert, Northwest Territories, Canada. The principal aim of the studies was to determine the contributions of sea salt, biogenic sources and pollution to aerosol SO_4^{2-}. The utilisation of two different processing techniques gave similar values of methanesulphonate/biogenic SO_4^{2-} mass ratio within 0.2 to 0.9 for the summer months and much lower in October to March (<0.08). The analysis of the isotopic composition data show that aerosol SO_4^{2-} in summer is 25 to 30% biogenic, 1 to 8% sea salt and the rest (62 to 74%) anthropogenic in origin. At other times of year it is <14% biogenic, 1 to 8% sea salt and the rest anthropogenic in origin.

Various aspects of anthropogenic aerosol and the climate problem have been studied recently. An important finding has been that the satellite data (low-level cloud albedo) demonstrate enhanced cloud albedo near the coastal boundaries where sulphate concentrations are large. Similar trends are absent over ocean regions of the southern hemisphere that are far away from anthropogenic sulphate sources.

Kaufman et al. (1991), have assessed the impact of fossil fuel and biomass burning on climate. They have pointed out that although coal and oil emit 120 times as many CO_2 molecules as SO_2 molecules, each SO_2 molecule is 50-1100 times more effective in cooling the atmosphere (through the effect of aerosol particles on cloud albedo) than a CO_2 molecule is in heating it. Kaufman et al. (1991) concluded that the cooling effect from coal and oil burning presently range from 0.4 to 8 times the heating effect. Within this large uncertainty, it is presently more likely that fossil fuel burning causes cooling of the atmosphere rather than heating. Biomass burning associated with deforestation, on the other hand, is more likely to cause heating of the atmosphere than cooling since it aerosol cooling effect is only half that of fossil fuel burning and its heating effect is twice as large. For a doubling in the CO_2 concentration due to fossil fuel burning, the cooling effect is expected to be 0.1 to 0.3 of the heating effect.

In a later publication Kaufman and Chou (1993) have concluded (using a conservative approach) that the sulphate aerosol induced cooling can presently counteract 50% of the CO_2 greenhouse warming. Their assessment also showed that a complete cessation of SO_2 emissions will result in a warming surge of 0.4^0C in the first few years after the elimination of the SO_2 emission. Table 4 data illustrate the results of numerical modelling for two extreme IPCC scenarios for the CO_2 growth: BAU (business as usual: the strongest increase of CO_2 concentration) and D (minimum increase of CO_2 concentration).

Kaufman and Chou (1993) believe that the uncertainties in the SO_2-induced radiative forcing and climate cooling is a factor of 3-4. According to the business as usual scenario a potential of offsetting the CO_2-induced warming is equal to 60% at the present time and 25% by 2060. Hunter et al. (1993) have studied seasonal, latitudinal and secular variations in temperature trend with the purpose of identifying the influence of anthropogenic sulphate. These data indicate that pronounced minima in the rate of temperature increase in summer months in northern hemisphere mid-latitudes are consistent with the latitudinal distribution of anthropogenic sulphate and changes in the rate of SO_2 emissions over the industrial area.

Emission	Change from 1900 to 1980			Change from 1900 to 2060		
Scenario	NH	SH	Global	NH	SH	Global
BAU scenario						
CO_2 only	0.31	0.29	0.30	2.04	1.73	1.89
$CO_2 + SO_2$	0.01	0.21	-0.20	1.31	1.54	1.43
SO_2-induced cooling	0.30	0.08	0.19	0.73	0.19	0.46
Scenario D						
CO_2 only	0.31	0.29	0.30	1.13	0.98	1.06
$CO_2 + SO_2$	0.01	0.21	0.11	0.89	0.89	0.89
SO_2-induced cooling	0.30	0.08	0.19	0.24	0.09	0.17

Table 4. The CO_2- and SO_2-induced changes in the surface air temperature (^{0}C) for years 1980 and 2060 for the northern hemisphere (NH) and southern hemisphere (SH). BAU is 'business as usual', *ie* max increase, and scenario D is a minimum increse.

To summarise, one can say that all existing empirical and theoretical assessments of aerosol climatic impact are still highly speculative. As far as numerical modelling results are concerned, it is highly surprising that they have ignored aerosol absorption. The observations made within the CAENEX programme long ago showed that under various conditions studied aerosol absorption of shortwave radiation in the troposphere was, on the average, equal to water vapour absorption. Therefore it is very important to undertake new field measurement of the chemical and physical properties of atmospheric aerosol particles.

Undoubtedly, the aerosol versus greenhouse gases climate change problem is very important. There are many reasons to believe that aerosol contribution to climate formation is comparable with the impact of greenhouse gases. The interval for current climate forcing due to anthropogenic sulphate ranges from -1 to -2 Wm^{-2}; estimates by Rodhe (1994) has led to the interval from -0.2 to -1 Wm^{-2} versus 2-2.5 Wm^{-2} greenhouse warming.

However, the conclusion of Parungo and Hicks (1993), sounds persuasive "because of the strong role of cloud dynamics, the independent sulphate aerosol concentration effects on cloud microphysics, cloud amount and rainfall are difficult to assess".

It is worthwhile emphasisingat more attention in the context of aerosol and climate problem should be devoted to the analysis of the results of the impact of volcanic eruptions on climate (Kondratyev 1984) as well as to the climatic consequences of the Arctic haze.

4.5 The coastal zone color scanner

For most regions of the world, the colour of the ocean is determined primarily by the abundance of phytoplankton and their associated photosynthetic pigments. As the concentration of phytoplankton pigments increases, ocean colour shifts from blue to green. Taking advantage of this change, NASA developed the Coastal Zone Color Scanner (CZCS) which was launched on the Nimbus-7 satellite in October 1978.

Simple, semi-empirical equations can be used to estimate the concentration of chlorophyll-a and its degradation products from satellite measurements of backscattered sunlight at three wavebands centred at 443, 520, and 550 nm, covering the blue and green regions of the spectrum. These radiances are not merely reflected from the sea surface, but are derived from sunlight that has entered the ocean, been selectively absorbed, scattered and reflected by phytoplankton and other suspended material in the upper layers, and then backscattered through the surface. This approach permits quantitative estimates of phytoplankton pigment concentrations within the uppermost tens of metres of the open ocean, and within somewhat lesser depths in coastal waters.

Sunlight backscattered by the atmosphere contributes 80-90% of the radiance measured by a satellite sensor at these key wavelengths. Such scattering arises from dust particles and other aerosols, and from molecular (Rayleigh) scattering. However, the atmospheric contribution can be calculated and removed if additional measurements are made in the red and near-infrared spectral regions (e.g. 670 and 750 nm). Since blue ocean water reflects very little radiation at these longer wavelengths, the radiance measured is due almost entirely to scattering by the atmosphere. Long-wavelength measurements, combined with the predictions of models of atmospheric properties, can therefore be used to remove the contribution to the signal from aerosol and molecular scattering.

With this approach, the CZCS measured reflected sunlight at 443, 520, 550, 670, and 750 nm with a spatial resolution of about 1 km across a swath 2200 km wide. An additional thermal-infrared spectral channel, at 11.5μm, was included as well to permit concurrent measurements of sea-surface temperature. During its $7^1/_2$ years lifetime (Oct.78 - June 86), CZCS acquired 68,000 images, each covering up to 2 million square kilometres of ocean surface, thereby laying the foundation for systematic studies of ocean colour from space.

Suspended sediment, detritus, and pigments other than chlorophyll also affect the spectrum of backscattered sunlight. When reflectance by these substances contributes a relatively high fraction of the total signal, as in the cases of some coastal waters, the semi-empirical equations used to estimate phytoplankton concentrations from CZCS measurements yield unreliable results. Research is underway to develop equations that can distinguish detrital, sediment, and phytoplankton signatures in observations of coastal waters. The application of these new equations requires measurements in more spectral channels than were provided by CZCS; future sensors provide the required capabilities.

References

Bauer M and Schluessel P, 1993, Rainfall, total water, ice water and water vapor over sea from polarised microwave simulations and special sensor microwave imager data, *Journal of Geophysical Research* **98** 20737-20760.

Borrell, P M, Borrell, P, Cvitas, T, and Soiler, W (Eds.), 1993, Photooxidants: Precursors and Products. Den Haag, SPB Academic Publishers

Brasseur, G, and Solomon, S, 1984, Aeronomy of the Middle Atmosphere, Dordrecht, Reidel

Bromwich D H and Robasky, F M, 1993, Recent precipitation trends over the polar ice sheets, *Meteorology and Atmospheric Physics* **51** 259-273.

Chandra, S, 1991, The solar UV related changes in total ozone from a solar rotation to a solar cycle, *Geophysical Research Letters* **18** 837-840

Chandra, S, Jackman, Ch. K, Douglass, A R, Fleming, E L, and Cosidine, D B, 1993, Chlorine catalized destruction of ozone: implications for ozone variability in the upper stratosphere, *Geophysical Research Letters* **20** 351-354.

Chandra, S, and McPeters, R D, 1994, The solar cycle variation of ozone in the stratosphere inferred from the Nimbus-7 and NOAA-11 satellites, *Journal of Geophysical Research* **99** 20665-20672

Clerbaux, C, Colin, R, Simon, P C and Granier, C, 1993, Infrared Cross Sections and Global Warming Potentials of 10 Alternative Hydrohalocarbons, *Journal of Geophysical Research* **98** 10491-10497

Crutzen, P J, 1971, Ozone production rates in an oxygen, hydrogen, nitrogen oxide atmosphere, *Journal of Geophysical Research* **76** 7311-7327

Deser, M, and Blackmon, M L, 1993, Surface climate variations over the North Atlantic ocean during Winter: 1900-1989, *Journal of Climate* **6** 1743-1753

Fleming, E L, Chandra, S, Jackman, C H, Considine, D B., and Douglass, A R, 1995, The middle atmospheric response to short and long term solar UV variations: Analysis of observations and 2D model results, *Journal of Atmospheric and Terrestrial Physics* **57** 333-366

Gernandt, H, von der Gathen, P and Herber, A, 1997, Ozone change in the polar atmosphere. In Atmospheric ozone dynamics: observations in the Mediterranean region, edited by C. Varotsos, NATO ASI Series I, Vol. 53, Berlin, Springer, pp. 73-100.

Gutzler D S, 1993, Uncertainties in climatological tropical humidity profiles: Some implications for estimating the Greenhouse effect, *Journal of Climate* **6** 978-982

Hallberg R and Inamdar, A, 1993, Observations of seasonal variations in atmospheric greenhouse trapping and its enhancement at high sea surface temperature, *Journal of Climate* **6** 920-931

Harrison E F, Minnis, P, Barkstrom, B R, Ramanathan, V, Cess R D, and Gibson, G G, 1990, Seasonal variation of cloud radiative forcing derived from the Earth Radiation Budget Experiment, *Journal of Geophysical. Research* **95** 18687-18703

Hunter, D E, Schwartz, S E, Wagener, R and Benkovitz, C M, 1993, Seasonal, latitudinal and secular variations in temperature trend: influence of anthropogenic sulfate, *Geophysical Research Letters* **20** 2455-2458

Jackman, C H, Nielsen, J E, Allen, D S, Cerniglia, M C, McPeters, R D, Douglass, A R, and Rood, R B., 1993, The effects of the October 1989 solar proton events on the stratosphere as computed using a three-dimensional model, *Geophysical Research Letters* **20** 459-462

Johnston, H S, 1971, Reduction of stratospheric ozone by nitrogen oxide catalysts from supersonic transport exhaust, *Science* **73** 517-522

Kahl W, Serreze, M C, Stone, R S, Shiotani, S, Kisley, M and Schnell, R C, 1993, Tropospheric temperature trends in the Arctic 1958-1986, *Journal of Geophysical Research* **98** 12825-12838.

Karl, R, Groisman, P Ya, Knight, R W and Heim, R R, 1993a, Recent variations of snow cover and snowfall in North America; relation to precipitation and temperature variations, *Journal of Climate* **6** 1327-1344

Karl, R, Quayle, G and Groisman, P Ya, 1993b, Detecting climate variations and change: New challenges for observing and data management systems, *Journal of Climate* **6** 1481-1494

Kaufman, R S and Chou, M-D, 1993, Model simulations of the competing climatic effects of SO_2 and CO_2, *Journal of Climate* **6** 1241-1252

Kaufman, R S, Fraser L and Mahoney, R L, 1991, Fossil fuel and biomass burning effect on climate - Heating or cooling, *Journal of Climate* **4** 578-588.

Kaye, J A, 1993, Stratospheric chemistry, temperatures, and dynamics. In Atlas of Satellite Observations Related to Global Change, edited by R J Gurney, J L Foster and C L Parkinson, CUP, pp. 41-57

Kondratyev K Ya, 1956, Radiative Heat Exchange in the Atmosphere, Oxford, Pergamon Press.

Kondratyev K Ya, 1969, Radiation in the Atmosphere, New York, Academic Press.

Kondratyev, K Ya, 1984, Volcanoes and Climate, World Climate Paper-54, Geneva, WMO

Kondratyev K Ya, 1988, Climate Shocks: Natural and Anthropogenic, New York, Wiley

Kondratyev, K Ya, 1989, The International Geosphere-Biosphere Program: the role and place of sun-atmospheric interrelationships, *Geofisica Internacional* **28** 453-466

Kondratyev, K Ya, and Cracknell A P, 1998, Observing global climate change, Taylor and Francis, London

Kondratyev, K Ya, Danilov-Danilyan, V I, Donchenko, V K and Losev, K S, 1994, Ecology and Politics, St. Petersburg, Academic Publishers (in Russian).

Kondratyev K Ya and H Grassl, 1993, Global Climate Change in the Context of Global Ecodynamics, St. Petersburg, PROPO (in Russian)

Kondratyev, K Ya, and Moskalenko, N I, 1984, Greenhouse Effect of the Atmosphere and Climate. Moscow, VINITI (in Russian)

Kondratyev, K Ya, and Varotsos, C A, 1993, Total ozone depletion at St. Petersburg, *Doklady of the Russian Academy of Sciences* **331**, 622-624 (in Russian).

Kondratyev, K Ya, and Varotsos, C A, 1995, Atmospheric ozone variability in the context of global change, *International Journal of Remote Sensing* **16** 1851-1881

Lawrence, A, 1993, An empirical analysis of the strength of the phytoplankton-dimethylsulfide-cloud-climate feedback cycle, *Journal of Geophysical Research* **98** 20663-20674

Madden D and Meehl G A, 1993, Bias in the global mean temperature estimated from sampling a greenhouse warming pattern with the current surface observing metwork, *Journal of Climate* **6** 2486-2489

Madden D, Shea, D J, Branstator, G W, Tribbia, J J and Weber, R O, 1993, The effects of imperfect spatial and temporal sampling on estimates of the global mean temperature: Experiment with model data, *Journal of Climate* **6** 1057-1066

Parungo J F and Hicks, B, 1993, Sulfate aerosol distributions and cloud variation during El Niño anomalies, *Journal of Geophysical Research* **98** 2667-2675

Parungo J F, Boatman A, Sievering, H, Wilkinson, S W and Thicks, B B, 1994, Trends in global marine cloudiness and anthropogenic sulfur, *Journal of Climate* **7** 434-440

Portman, D, 1993, Identifying and correcting urban bias in regional time series: Surface temperature in China's northern plains, *Journal of Climate* **6** 2298-2308

Raval, A, and Ramanathan, V, 1989, Observational determination of the greenhouse effect, *Nature* **342** 758-761

Rodhe, H, 1994, Global distributions of atmospheric sulphur compounds and their potential impact on climate. In EUROTRAC Symposium-94. Transport and tranformation of pollutants in the troposphere. Abstracts of lectures and posters, Garmisch-Partenkirchen, 11-14 April, 1994, 224

Schoeberl, M R, 1993, Stratospheric ozone depletion. In Atlas of Satellite Observations Related to Global Change, edited by R J Gurney, J L Foster and C L Parkinson, Cambridge, University Press, pp. 59-65

Van den Dool, H M, O' Lenic, E A and Klein, W H, 1993, Consistency check for trends in surface temperature and upper-level circulation, *Journal of Climate* **6** 2288-2297

Varotsos, C A, and Kondratyev, K Ya, 1994, Athens environmental dynamics: from a rural to an urban region, *Optics of the Atmosphere and Ocean*, 7 3-17

Varotsos, C A, and Kondratyev, K Ya, 1995a, Interrelationship between solar ultraviolet radiation and total ozone content: observations in Greece, *Optics of the Atmosphere and Ocean* **8** 608-613 (in Russian)

Varotsos, C A, and Kondratyev, K Ya, 1995b: The tropospheric pollution and the solar ultraviolet radiation, *Optics of the Atmosphere and Ocean* **8** 614-618 (in Russian)

Webb M, Slingo, A and Stephens, G L, 1993, Seasonal variations of the clear sky greenhouse effect: The role of changes in atmospheric temperatures and humidities, *Climate Dynamics*, **9** 117-129

Weisenstein, D K, Ko M K W, Rodriguez, J M, Sze, N-D, 1993, Effects on stratospheric ozone from high-speed civil transport: sensitivity to stratospheric aerosol loading, *J Geophysical Research* **98** 23133-23140

World Meteorological Organization, 1992, Scientific Assessment of Ozone Depletion: 1991, Report 25 of the Global Ozone Research and Monitoring Project, Geneva, WMO

World Meteorological Organization, 1995, Scientific Assessment of Ozone Depletion: 1994, Report 37 of the Global Ozone Research and Monitoring Project, Geneva, WMO

Near-shore bathymetry and side-scan sonar

Silke Wewetzer and Robert Duck

University of Dundee

1. Introduction

The problems of locating objects underwater have intrigued and challenged investigators for generations. Since light and radar waves are too rapidly attenuated within the water column, sound is the only useful medium in the sea for obtaining information over distances or depths. This acoustic equipment is called SONAR (SOund Navigation And Ranging).

According to Fish and Carr (1990) underwater acoustics as a research discipline began in 1826 when Daniel Colladon measured the velocity of sound in water on Lake Geneva. He positioned two boats 16km apart and fastened a large trumpet, fitted with a membrane that would respond to underwater sound, on the first boat. From the second boat he suspended a bell underwater and on deck a pan of flash powder and a small flare. The bell and the flare were set off simultaneously. The theory was that the light of the powder would travel instantaneously while the sound of the ringing bell would take some time to travel that distance through the water. Colladon, in the first boat, watched for the flash and started a stopwatch which he stopped when he heard the sound about 10 seconds later. Colladon's method may seem crude but his calculated value for the velocity of sound of 1435ms^{-1} is very close to the generally accepted values which lie between 1470 and 1540ms^{-1} depending on the temperature, salinity and pressure of the water.

More sophisticated sonar systems have been developed in this century especially during wartime when the need arose to detect and track submarines and surface vessels. However, the systems were found to have a far wider range of applications. Acoustic remote sensing of the sea floor includes bathymetric surveying (echo-sounding) which provides information on the water depth and usually forms a basis for other kinds of marine survey and acoustic imaging (side-scan sonar) which is most often used to determine the geomorphology of the bed and detect obstacles. Both techniques provide a rapid means of assessing the major geological and sedimentological characteristics of an area. Even though there have been continuous improvements in acoustic equipment and techniques, the basic principles have remained the same. The survey results are still subject to the same constraints imposed by water depth, sea state and the weather, as well as the human operator. In the nearshore environment the major difficulty is the shallow water depth since it limits the size of the survey vessel. The sea state and the weather are of great importance in any marine survey since they can prevent the successful operation of geophysical equipment being towed behind the survey vessel as well as increasing the ambient background noise levels.

Any measurement or observation made must be located with reference to a defined map co-ordinate system. Accurate position-fixing is essential to make it possible to return to a previously identified site and to compile observations made at different sites into maps and charts showing their correct spatial relationships (McQuillin and Ardus 1977).

2. Basic principles of acoustic surveying techniques

All acoustic surveying systems function with three basic components.

1. An energy source or transducer that emits acoustic pulses at specific power and frequency levels.
2. One or more receivers that pick up the acoustical echoes after they are reflected back from the seabed.
3. A recording instrument that converts the reflected acoustical signals to a more permanent record (Williams 1982).

These geophysical remote sensing techniques do not provide a direct measurement of distance but measure the two-way travel time ($2t$) of a pulse of acoustic energy from a vessel to the sea bed and back, and convert this to water depth (d) according to

$$d = v_S\, t \quad (1)$$

As mentioned, the velocity of sound (v_S) is dependent on the temperature, salinity and pressure of the water; for approximate calculations $v_S = 1500\text{ms}^{-1}$ (Ingham 1975).

A transducer generates acoustic energy which is propagated through the water. When the acoustic wave reaches a target, parts of the wave will be reflected and others refracted. Some of the reflected wave will, after further attenuation, reach the receiver as an echo. The strength of this echo depends on the power of the original transmission as well as its propagation through the medium and its reflection from the target. Although acoustic pulses are less easily attenuated in water than light or radar waves, energy is still attenuated with increasing distance. The attenuation is generally related to the frequency and wavelength of the equipment with low frequencies being used over long working ranges and high frequencies over short working ranges as shown in Table 1. Sonar frequencies commonly lie in a range between 50kHz and 500kHz (Fish and Carr 1990).

Frequency (Hz)	Wavelength (m)	Distance (km)
1×10^2	15	$\geq 1 \times 10^3$
1×10^3	1.5	$\geq 1 \times 10^2$
1×10^4	1.5×10^{-1}	10
2.5×10^4	6×10^{-2}	3
5×10^4	3×10^{-2}	1
1×10^5	1.5×10^{-2}	6×10^{-1}
5×10^5	3×10^{-3}	1.5×10^{-1}
1×10^6	1.5×10^{-3}	5×10^{-2}

Table 1: Two-way working ranges of modern sonar systems (Fish and Carr 1990).

3. Echo-sounder

The most elementary and widely used sonar search technique is the vertical beam sonar usually called the echo-sounder. It measures the time interval between transmission of an outgoing pulse and detection of the reflected sea bed return, which is converted into water depth, and gives a display of the sea bed profile. Because the vertical beam configuration yields information only about that area of the sea bed directly beneath the vessel this technique has restricted applications.

3.1 Basic components

An echo-sounder record is built up by a series of sweeps of a stylus across recording (electro-sensitive) paper. The recorder motor drives the stylus belt and the paper drive rollers, thereby providing the two axes of depth and ship travel for the sea bed profile. The forward movement is represented by the movement of the paper trace. The stylus moves from top to bottom with each outgoing acoustic pulse and represents the depth measurement by marking the trace at the instant of transmission and at the receipt of the returned echo. Many modern forms of echo-sounders use video displays which have the disadvantage of not giving a permanent record unless they record the data digitally and offer a play-back facility. The record on dry paper is produced by burning away the surface of the paper to expose black graphite beneath as the stylus current passes through the paper. The width of a paper record is naturally limited and the range of scales offered is confined within these dimensions, so that the paper width might represent a depth range of 0–10m or of 0–1000m (Ingham 1975). Both the continuous paper and video record are periodically marked with lines relating to the ship's position fixes, recorded simultaneously in order to determine the relation between the ship's position and the echo-sounding record.

3.2 Calibration

A survey echo-sounder should be calibrated at least at the beginning and/or end of survey operations. This is usually done by employing the "bar check". An iron bar is lowered successively in steps of 1m beneath the transducer suspended from two lines. The corresponding depths are recorded and read off from the record. If any errors occur, the stylus speed and the zero settings can be adjusted to eliminate them. This should also include the velocity settings depending on the survey environment since the speed of sound varies in salt and freshwater. In a tidal environment the recorded depth will vary according to the height of the tide at the time. Therefore, all depths must be reduced to a constant datum level which involves the recording of tidal level readings throughout the survey and adding or subtracting the adjustments as appropriate before plotting the measured depths.

3.3 Factors affecting the record

There are some factors which should be considered when interpreting echo-sounder records. Because the beam of the echo-sounder spreads out as a cone, the most direct sound path to the bed may not be vertically under the transducer. This may cause errors when looking specifically at the slope of the bed although it will result in the recording of the minimum depth. In rough weather conditions ship motion due to the wind and waves also cause errors in depth recordings. Air bubbles in the water, fish with gas-filled swim bladders and seaweed with gas-filled bladders cause the reflection of the acoustic pulse and the seabed return is lost. Sometimes "ghost echoes" above the seabed can be observed. These are usually produced by layers of fluid mud or other high concentrations of suspended sediment, but they are only observed when high frequency transducers are used, since low frequencies penetrate through these layers and are reflected only by higher density substrates.

3.4 Applications of echo-sounding

Vertical beam sonar is most useful for simple bathymetric surveys when specified depth values, measured from the echo-sounder trace, and corresponding position fixes are plotted onto a map. These depth values are usually joined by contour lines to produce a bathymetric chart. Echo-sounding is also applied when the position of an object of interest has been located by another technique, like side-scan sonar, and the elevation of the object above the bottom as well as the size has to be determined. This can include measurements of dunes where the best results are achieved by running transverse to the crestal orientations. Dunes produced in predominant tidal flows have crests which extend across the direction of flow so that transverse sounding lines might easily miss the deepest or shallowest parts of the bedforms and the characteristic dune-shape might escape identification from the contoured soundings (Ingham 1975). These restricted limitations of the echo-sounder are overcome by the side-scan sonar technique which can cover a large area in a short time.

4. Side-scan sonar

Side-scan sonar systems were developed on the basis of the echo-sounder when researchers began to turn the transducers on their sides to look at a series of echoes along the bottom. To maximise the coverage obtainable per traverse, dual channel systems were developed scanning to both the port and starboard side of a vessel simultaneously, hence doubling the effective coverage and reducing the operating time to survey an area. Although this was developed primarily to detect man-made objects (especially submarines), its ability to locate features with positive and negative relief such as rock outcrop, dunes, channels, wrecks and to distinguish between major sediment types has made it an important underwater remote sensing technique (Duck and McManus 1985).

4.1 Basic principles

Like the echo-sounder the side-scan sonar is an active sonar system which transmits sound and records the returning echo. At the heart of the system is the transducer which converts the oscillating electric field produced by the transmitter into a mechanical vibration which is then transferred into the water as the sound pulse. The sound travels away from the transducer through the water until it strikes the sea bottom or an object. Only a fraction of the outgoing sound is scattered back to the receiving transducer (Johnson and Helferty 1990). The detected energy is amplified and presented on some kind of display, *i.e.* a paper recorder or video, which is called a sonograph.

The main features which distinguish side-scan sonar from most other forms of sonar are: sideways look, dual channel, towed body and narrow horizontal but wide vertical beam angle. To decouple the sonar from the ship's motion the transducers of a side-scan sonar system are mounted in a streamlined, hydrodynamically balanced body called a tow fish.

The system recorder processes the echo of the transmitted acoustic signal. The stronger the echo, the darker the mark on the record. The sonograph is built up by a succession of scan lines giving a continuous record. In a dual-channel system both channels are printed onto the same roll of recording paper showing the survey line in the middle and printing the echoes of the port and starboard transducers to the left and right. The paper advance is related to the forward motion of the fish while the range is shown across the width of the paper. Since the recording process is instantaneous, the opportunity is given to take a position fix as soon as the contact appears bearing in mind that the vessel's position is not the position of the tow fish.

4.2 Geometry of a side-scan sonar record

The geometry of side-scan sonar is the key to understanding and interpreting sonographs. Since the sonar measures and displays the ranges of reflectors from the transducers housed in the tow body at some depth below the water surface everything must be referred to this position (Klein 1985). It is important to remember that the technology is based upon the elapsed time between the outgoing pulse and the returned reflection from a target (Fish and Carr 1990).

Since the same transducer emits and receives the pulse, the first transmitted pulse will be a strong signal and therefore produce a dark mark on the record of the port and starboard channels (Figure 1). This signal is known as the trigger pulse (Fish and Carr 1990). Depending on the depth position of the tow fish in the water column, the first recorded echo may be either the first water surface return or the first sea bottom return (Klein 1985). The white area between the dark marks of the trigger pulse and the first return is known as the water column. The first surface return is usually a good indication of the tow fish depth which can be scaled using the scale marks printed by the recorder (depending on the system, distances between the scale lines are 15m, 25m or more). The first bottom return is almost always a strong reflection and a good indication of tow fish height (Fish and Carr 1990). The first bottom return will be followed by returns from the sea floor at successively greater distances from the tow fish. The recorder will print or display them at correspondingly increasing distances. These distances do not represent the true range across the sea bed but the slant range from the tow fish to the various features on the bottom. Large objects projecting above the sea floor will prevent sound from insonifying the bottom beyond the object, thus producing an acoustic shadow. This will leave a white patch on the record and its width and position relative to the fish can be used to calculate the object's height (Flemming 1976). However, a simple target will not appear identical if viewed from different angles or track line distance or if insonified with a different acoustic system at a different frequency (Johnson and Helferty 1990).

4.3 Across-track (range) resolution

Resolution is the ability to resolve multiple targets as distinct and separate on the sea floor. The range resolution is mainly determined by the pulse length as well as the forward motion of the sonic footprint (distance or thickness of the acoustic pulse when it has reached the sea floor). In resolving two objects, one closer and one farther from the tow fish, a shorter pulse will first insonify the closer target and then travel beyond it to the farther one recording them as two distinct targets. A long pulse might encompass both objects at the same time and record them as one target. An acoustic pulse forms an arc in the water resulting in a larger footprint in the near ranges and a smaller footprint in the far ranges (Fish and Carr 1990). This implies that the across-track resolution is better farther away from the sonar.

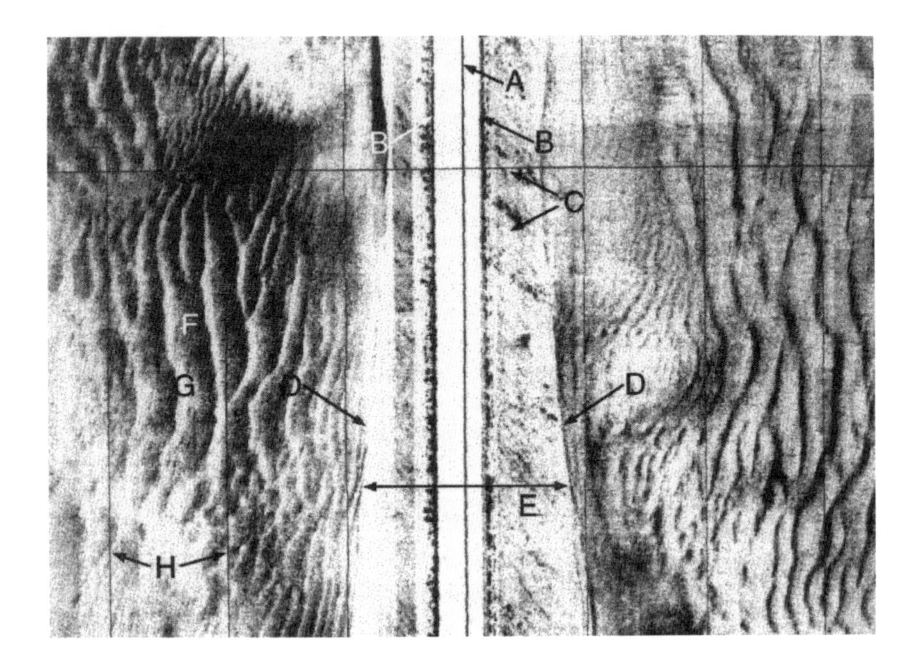

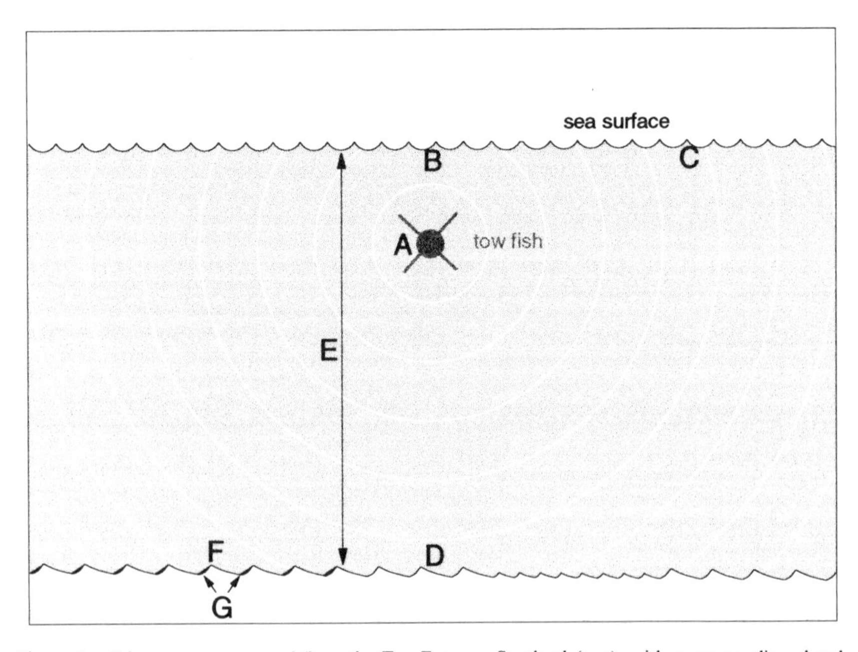

Figure 1: Side-scan sonar record from the Tay Estuary, Scotland (top), with corresponding sketch (bottom; Fish and Carr 1990) describing the conditions under which the record was made. A: trigger pulse, B: first surface return, C: sea clutter, D: first bottom return, E: water column, F: dune, G: shadow, H: 15m scale marks.

4.4 Along-track (transverse) resolution

The along-track (or transverse) resolution is the minimum distance between two objects parallel to the line of travel that are displayed separately on the sonar. This minimum distance is equivalent to the beam width (which widens with distance from the tow fish) at any particular point (Flemming 1976). The transverse resolution depends on the chosen range setting and the horizontal beam width. If two objects are separated by a distance which is less than the spread of the sonar beam at that range they will appear as one object. If the same two objects were in the closer range where the beam is narrower they would be resolved as two separate targets. Hence, the along-track resolution is better at close range.

4.5 Time Varied Gain (TVG)

Absorption, spreading and scattering of the acoustic pulse weaken the strength of the returning signal. Since the display of the sonar data should look alike for any given bottom type over a chosen range, the returned signal must be amplified to counteract the occurred losses over the travelled distance. The TVG correction assumes a constant speed of sound in water and an appropriate reduction in the returning echoes with range.

4.6 Interference

The reflection of the transmitted pulse is not the only sound received and recorded. Noise in terms of underwater measurement is the unwanted part of the signal (Fish and Carr 1990). Much of it is recognisable as interference patterns. Noise sources fall into two categories: self-made and ambient. Self-made noise sources are due to other acoustic instruments such as echo-sounders and sub-bottom profilers as well as the ship's engines. Ambient noise sources include wave motion at the sea surface including wave breaking and rain, as well as biological noise created by marine animals.

Another false trace, less common with towed fish transducers, in shallow water conditions, is the Lloyd's Mirror effect. This appears only in exceptionally calm weather when the water surface can act as a mirror and results in dark and light interference fringes (Morang and McMaster 1982; Werner 1982). These Lloyd's Mirror bands run approximately parallel to the ship's track across the sonograph. Towed fish transducers reduce this effect since only the weaker side lobes of the beam reach the water surface.

4.7 Record interpretation

Skill in interpreting sonar records can come only from experience. But even the most experienced operator can sometimes make mistakes. Interpretation remains a thoroughly qualitative and tedious process. For accurate interpretation the operator must use the entire record as well as any available data recorded from previous surveys such as ground observations from divers, video recordings, sediment samples etc. for calibration.

The side-scan method produces a plan view of the shape and texture of the surface of the sea floor (Belderson *et al.* 1972). However, the side-scan sonar image is not a representation of how the sea floor would look if the water were somehow removed. Instead it is a graphical presentation of how the sea floor interacts with acoustic energy. This conversion from how the sea floor ''sounds'' to how our models tell us the sea floor should look, can be a major pitfall for the interpreter of the images (Johnson and Helferty 1990).

4.8 Distortion

Features recorded by side-scan sonar are not normally presented in their true proportions. The main factors causing distortions are tow fish instabilities in the water column, the variations in survey speed affecting the along-track direction and the slant range distortion affecting the across-track direction. These cause major difficulties in the interpretation of side-scan sonar records. A solution to the problem of scale correction for distorted sonographs is provided by using a photographic technique which cannot correct for heading variations or slant range effects but only for distortion in the along-track direction or by the application of digital image processing techniques if the signals are recorded in digital form.

Speed distortion

Distortion will occur in the along-track direction due to variable ship speeds during the survey. Slow traverse speeds produce visual distortions on uncorrected sonographs so that objects will appear larger. Travelling too fast on the other hand minimises the opportunity for reflections to be detected and decreases the dimensions of targets in the direction of traverse (Flemming 1976) (Figure 2). According to Flemming no distortion occurs at about 2–3 knots which has been confirmed by Duck and McManus (1987).

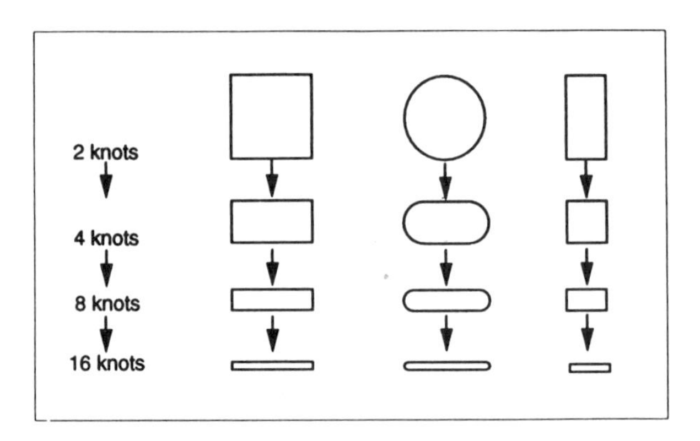

Figure 2: Distortion effects on some common shapes parallel to the line of travel caused by various ship speeds (after Flemming 1976).

Slant range distortion

As noted earlier the side-scan recorder shows the slant range from the tow fish to the various features on the sea bottom and not the true range. If an object is detected and it is required to make a precise measurement of its size and position relative to the ship, corrections for slant range distortion should be applied. This can be done by using the Pythagorean theorem with the tow fish height (H_f) forming one side of the right-angled triangle, the measured slant range (R_s) the hypotenuse and the horizontal offset (R_h) the third side (Figure 3):

$$R_h = \sqrt{R_s^2 - H_f^2} \qquad (2)$$

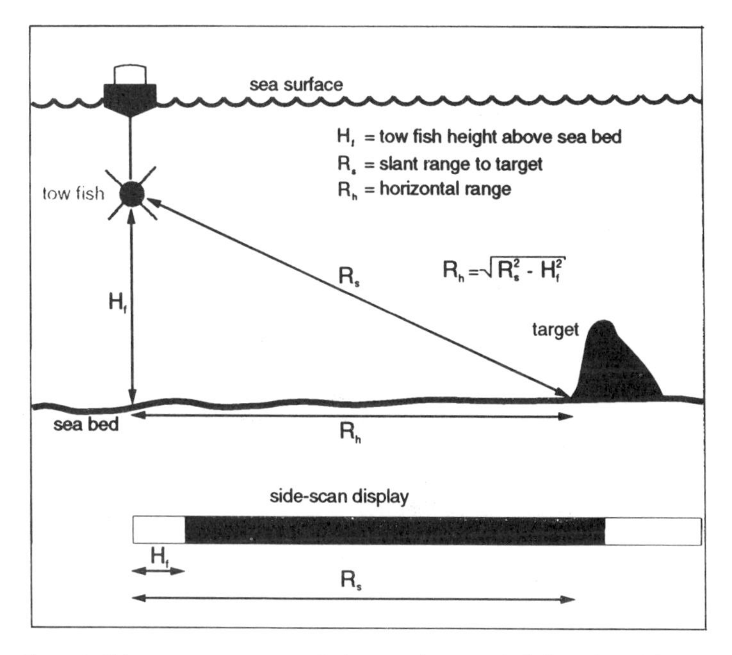

Figure 3: Side-scan sonar geometry for horizontal range calculation (after Klein 1985).

4.9 Acoustic shadows and backscatter

Acoustic shadows are of extreme importance in the interpretation of sonographs. Their position, shape and intensity give clues to the actual conditions of the insonified sea bed. Objects such as rocks, dunes or ship-wrecks will cast a clear, harsh shadow while other targets such as a gentle upward localised slope will cause only a light shadow on the record. The causes of lighter areas on a sonar record are grouped into three general categories (after Fish and Carr 1990).

1. Shadow zones that have been blocked from the sonar beam by an acoustically opaque object.
2. Areas of topography that provide less backscattering of the sonar beam.
3. Areas that are oriented in such a way as to provide less backscatter, such as an area inclined away from the tow fish.

The geometry of the sonar slant range (Figure 4) can be used to determine the height of a target as follows:

$$H_t = \frac{L_s \times H_f}{L_s + R_s} \tag{3}$$

where H_t is the target height, L_s is the shadow length, H_f is the height of the tow fish above sea bed and R_s is the slant range to the target.

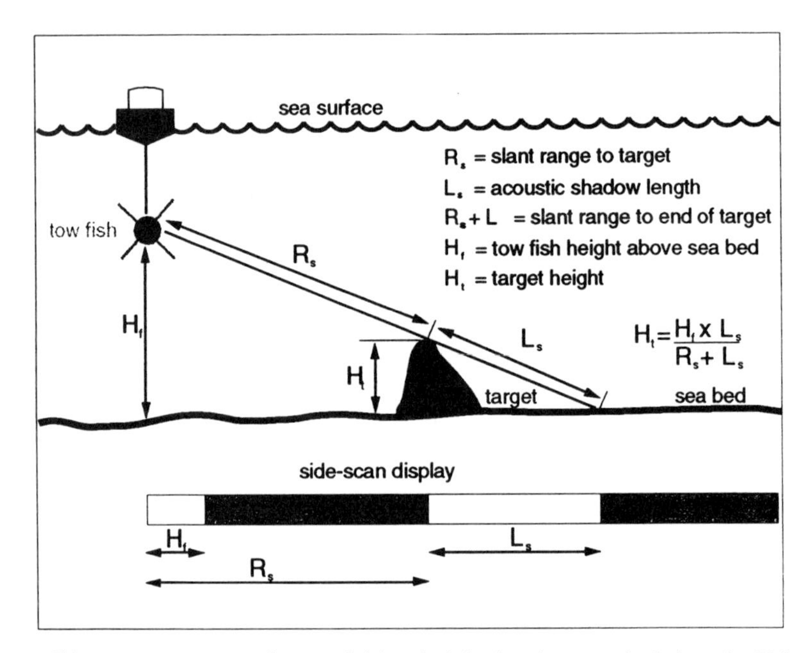

Figure 4: Side-scan sonar geometry for target height calculation based on acoustic shadow after Klein 1985).

The same formula can also be applied for target depth calculations if the target is a negative feature. This is possible only if the sonograph shows a shadow zone followed by a zone of strong reflection. However, it is not possible to calculate target depth on the basis of a low tonal colour zone. Therefore, height calculations for negative targets are possible only if the steep slope is closer to the transducer and the gentle slope farther away hence producing the required acoustic shadow for the measurements.

It may be difficult to determine the exact position of the target casting the shadow so that care must be taken. Side-scan sonar geometry can also prevent a target from casting a shadow on the sea bed. This may occur if the target is in the water column at a similar height to that of the tow fish so that the shadow would be beyond the displayed range. Also targets above the tow fish in the water column will be displayed on the record but will not cast a shadow on the sea bed.

4.10 Geology and topography

Side-scan sonar is an ideal tool for determining the configuration of the sea bed over a large scale. When the sonar signal is transmitted, part of it is attenuated and scattered as it travels through the water. Further energy losses occur when the pulses strike the sea floor. The intensity of the returning signal will determine the tonal shading on the sonograph and largely depend on the composition and porosity of the sediment on the sea bed as well as the bottom topography (McQuillin and Ardus 1977).

According to Williams (1982), relations between tonal intensity and sediment characteristics are not yet fully understood but, in general, the denser and coarser the sediment, the greater the reflectivity and consequently the darker the tone on the sonograph. Hence, outcrops of rock will have a very high reflectivity and will appear darkest on sonographs, gravel patches will be darker than sand and sand will be darker than fine-grained muddy sediments (Flemming 1976). Williams (1982) has found an inverse relation between acoustical impedance and sediment porosity. High porosity sediments like clay have a relatively low impedance and low reflectivity and therefore appear light on a sonograph, whereas low porosity material such as well packed medium sand has a higher impedance and higher reflectivity and therefore appears as darker tones.

Topographic features of the sea floor also affect the reflectivity and may cause effects on the record similar to changes in sediment composition and porosity. The slopes of dunes and ridges facing the transducer reflect sound waves better than surfaces lying obliquely to the sound beam and will consequently result in darker tonal intensities on sonographs (Flemming 1976). Depending on their height they may also produce a pronounced acoustical shadow on the far side where no signal is reflected and a white area is produced (Williams 1982). The patterns produced by the strong reflection and the acoustic shadow can reveal not only the presence of dunes but also their shape and asymmetry, thereby providing evidence of the direction of sediment transport – a technique widely applied by various scientists (McCave and Langhorne 1982; Goedheer and Misdorp 1985, Wewetzer 1997).

5. Conclusion

Even though side-scan sonar has become a widely employed instrument in marine sciences it is not always possible to distinguish between natural phenomena and anthropogenic structures. Therefore the technique should only be used as a reconnaissance tool and in no way as a substitute for on-site investigations by divers (Duck and McManus 1987). However, the technique has revealed the true extent of the complexity of surficial sediment relationships. It has also demonstrated the inadequacy of areal bottom classification based only on spot sediment samples (Bennett *et al.* 1992). In conclusion, side-scan sonar surveys, in conjunction with ground sampling, have proved to be a cost-effective method for acquiring remotely sensed data of the sea bed.

References

Belderson R H, Kenyon N H, Stride A H and Stubbs A R, 1972, Sonographs of the sea floor - A picture atlas, Elsevier Publishing Company (Amsterdam)

Bennett R H, Li H, Richardson M D, Fleischer P, Lambert D N, Walter D J, Briggs K B, Rein C R, Sawyer W B, Carnaggio F S, Young D C and Tooma S G, 1992, Geoacoustic and geological characterization of surficial marine sediments by in situ probe and remote sensing techniques (in CRC Handbook of geophysical exploration at sea, edited by Geyer R A, CRC Press (Boca Raton, Florida), pp 295 - 350

Duck R W and McManus J, 1985, A sidescan sonar survey of a previously drawn-down reservoir: a control experiment, International Journal of Remote Sensing, **6**, pp 601 - 609

Duck R W and McManus J, 1987, Sidescan sonar applications in limnoarchaeology, Geoarchaeology, **2**, pp 223 - 230

Fish J P and Carr H A, 1990, Sound underwater images - a guide to the generation and interpretation of side scan sonar data, Lower Cape Publishing (Cataumet, Massachusetts)

Flemming B W, 1976, Side-scan sonar: A practical guide, International Hydrographic Review, **53**, pp 65 - 91

Goedheer G J and Misdorp R, 1985, Spatial variability and variations in bedload transport direction in a subtidal channel as indicated by sonographs, Earth Surface Processes and Landforms, **10**, pp 375 - 386

Ingham A E, 1975, Sea surveying, John Wiley & Sons (London)

Johnson H P and Helferty M, 1990, The geological interpretation of side-scan sonar, Reviews of Geophysics, **28**, pp 357 - 380

Klein, 1985, Side Scan Sonar Record Interpretation, Klein Associates, Inc., Salem, New Hampshire

McCave I N and Langhorne D F, 1982, Sand waves and sediment transport around the end of a tidal sand bank, Sedimentology, **29**, pp 95 - 110

McQuillin R and Ardus D F, 1977, Exploring the geology of shelf seas, Graham & Trotman (London)

Morang A and McMaster R L, 1982, Nearshore bedform patterns along Rhode Island from side-scan sonar surveys - reply to Friedrich Werner, Journal of Sedimentary Petrology, **52**, pp 679 - 680

Werner F, 1982, Nearshore bedform patterns along Rhode Island from side-scan sonar surveys - discussion, Journal of Sedimentary Petrology, **52**, 674--677

Wewetzer S F K, 1997, Bedforms and sediment transport in the middle Tay Estuary, Scotland: a side-scan sonar investigation, University of St. Andrews (St. Andrews)

Williams S J, 1982, Use of high resolution seismic reflection and side-scan sonar equipment for offshore surveys, U.S. Army Corps of Engineering, Coastal Engineering Technical Aid

An introduction to underwater light processes

Tim Malthus

University of Edinburgh

1. Introduction

Remote sensing offers considerable potential for contributing to our understanding of the dynamics of coastal waters. One particular area where a significant contribution can be made with this technology is in characterising water colour, where the wide synoptic coverage afforded by remote sensing can help further knowledge of the distribution of suspended sediments, circulation patterns around variable coastlines, and of the production of such regions through the mapping of chlorophyll concentrations (including algal blooms). Nevertheless, while there are numerous publications demonstrating the utility of remote sensing for its application to coastal regions, the optical properties of the coastal zone remains poorly understood compared to our knowledge of the optical properties of the open oceans. Since there are more factors which influence the optical regime in coastal waters, a working knowledge of the principal underwater optical properties is vital in order to understand the nature of reflectance from coastal areas and in applying remote sensing for investigating coastal water quality. This chapter is an introductory review of underwater optics in the coastal zone and of methods for measuring underwater optical conditions.

2. The importance of underwater optical properties

Morel and Gordon (1980) defined the following three different approaches by which measurements of spectral irradiance from water could be used to estimate the concentrations of water quality parameters using remote sensing.

- **Empirical approaches** based on the development of statistical relations between measured spectral values and measured water quality parameters. The limitations of this approach are that causal relations between the parameters are not necessarily implied, and algorithms developed in this manner are frequently not generally applicable to other data sets. Nevertheless, empirical methods are by far the most common methods employed for algorithms for the coastal zone (e.g. Bagheri and Dios 1990).

- **Semi-empirical approaches** when the spectral characteristics of the water quality parameters are known, more appropriate wavebands or combinations of wavebands can be used in the development of relations (e.g. Topliss *et al.* 1990). However, like empirical approaches, the coefficients from any such relations may only apply to the data from which they were derived, with each new application requiring recalibration.

- **Analytical approaches** - in this approach the optical properties of the water column are physically related to subsurface reflectance $R(0-)$, and more importantly, vice versa. By inverting model approaches the concentrations of water quality parameters can be optimally retrieved from remotely-sensed data, with multi-temporal applicability.

Clearly, the preferred approach for the successful monitoring by remote sensing of features in dynamic coastal zones over time requires a rigorous application along the lines of analytical methods. Knowledge of the underwater optical properties is clearly vital for such an approach to proceed.

3. Water surface effects

Light impinging on a smooth water surface will undergo one of two effects. It will either be reflected from the surface itself back out into the atmosphere or it will pass across the air-water interface into the water, being refracted in the process. For the purposes of this paper, the surface reflected component is of little value, although it may be an unwanted signal in remotely-sensed imagery. We are more interested in that fraction of light which passes into the water, interacts with it and then may be reflected back across the interface to be detected by a sensor.

Refraction can be calculated according to Snell's law: $n_a \sin\phi_a = n_w \sin\phi_w$, where n_a and n_w are the refractive indices for air and water respectively, and ϕ_a and ϕ_w are the angles of incidence or refraction in air and water respectively.

The refractive index of a medium is defined as the ratio of the velocity of light in a vacuum to the velocity of light in that medium. Ordinarily, n_a is usually taken to be equal to 1 and for most purposes n_w can be regarded as being 1.338 (although it is affected by both water temperature and salinity, Plass and Kattawar 1972):

$$\frac{\sin\varphi_a}{\sin\varphi_w} = \frac{n_w}{n_a} = 1.338$$

The implications of refraction at the air-water interface are that, for a flat sea surface, the whole of the hemispherical irradiance from the atmosphere which passes across the interface is compressed into a cone of underwater light with a half angle of 48.8°. This phenomenon also has implications for reflected radiance. Any backscattered light travelling upwards and striking the surface at angles greater than 48.8° will be totally internally reflected; it will not penetrate the surface. Similarly, the flux contained within the solid angle below the surface will be spread out because of the refraction above the surface when it passes across the interface.

The effects of surface roughness

The surface of the sea is almost never flat; wind driven waves will have a major effect on the ability of light to pass across the air-water interface. The effect of wind roughening is generally to widen the solid angle through which light will penetrate, i.e. some light will penetrate the water at angles greater than 48.8° degrees. The presence of slicks or whitecaps will further modify the light field in different ways from that of wave action (Estep and Arnone 1994, Gordon and Wang 1994). Oil slicks, apart from having a dampening effect on wave action will cause higher reflectance in certain regions of the spectrum.

4. Attenuation in water - Beer's Law

Light entering a body of water undergoes **attenuation**, the loss of light from the incident beam (the processes which contribute to attenuation are discussed in greater detail below). Irradiance is assumed to decrease with depth in an exponential manner, in accordance with the following equation (Beer's Law):

$$I_z = I_0 \exp\{-K_d z\} \quad (1)$$

where I_z and I_0 are the irradiances at depth z and just below the surface, and K_d is the downwelling vertical attenuation coefficient for the light. The attenuation coefficient can be found from the slope of the regression between depth and the logarithm of irradiance. The attenuation coefficient is a useful parameter for comparing the relative clarity of different bodies of water. The higher the K_d value, the greater the amount of light attenuated. Table 1 lists a number of attenuation coefficients for different oceanic and coastal water bodies. Generally, coastal waters have K_d values that are higher than those of the open oceans. Estuarine waters generally have higher attenuation coefficients again.

	Water body	K_d (PAR)(m^{-1})
Oceanic waters		
	Sargasso Sea	0.03
	Gulf Stream, off Bahamas	0.08
	Pacific Ocean, 100 km off Mexico	0.11
Coastal waters		
	Bjornafjord, Norway	0.15
	North Sea, Netherlands	0.41
	North Sea, Dogger Bank	0.06 - 0.15
Estuarine waters		
	Elms Dollard, border of Netherlands and Germany	1 - 7
	Shannon, Ireland	0.35 - 8.6

Table 1. Attenuation coefficients calculated from measurements of downwelling photosynthetically active radiation (PAR, 400 to 700 nm) for a range of marine water bodies (selected from Kirk 1994a).

5. Underwater optical properties

The optical properties of seawater were defined by Preisendorfer (1961) who identified two principal groups of optical properties.

- **Inherent optical properties**, whose magnitudes depend only on the optically important substances which comprise the aquatic medium through which the light penetrates, and
- **Apparent optical properties**, which are dependent not only on the optical water quality parameters in situ but also on the particular properties of the ambient light field in the water at the time of measurement (e.g. as a result of changes in solar elevation and atmospheric inference).

We shall start with a discussion of inherent optical properties first:

6. Inherent optical properties

Light penetrating through the water column undergoes one of two processes, which contribute to the attenuation of light with depth. These are the processes of absorption and scattering. These are defined with reference to a parallel beam of monochromatic light interacting with an infinitesimally thin layer of medium, perpendicular to the light beam. Of the light that is not absorbed by the medium, most is transmitted but some is scattered, but mostly in a forward direction.

Absorption represents the loss of photons from the light stream by components in the medium. The fraction that is absorbed, divided by the thickness of the layer is called the absorption coefficient, a. The fraction of the light beam that is scattered, divided by the thickness of the layer, is called the scattering coefficient, b. Thus, the beam attenuation c represents the summation of the processes of absorption and scattering ($c = a + b$). The units of a, b, and c are metre $^{-1}$.

There is one final but important optical property, the volume scattering function. While b defines the proportion of light that is scattered, it takes no account of the direction in which a photon may be redirected. The volume scattering function represents the angular distribution of the scattering flux. Volume scattering $\beta(0)$ (units $m^{-1}sr^{-1}$) is formally defined as the radiant intensity in a given direction from a volume element, illuminated by a parallel beam of light per unit of irradiance on the cross-section of the volume, and per unit volume. The shape of $\beta(0)$ is important in determining the evolution of light at each particular depth and in describing the spatial distribution of the radiative field. Changes in its shape can be expected in marine waters due to the variation in the relative proportions of molecular and particulate scattering.

The scattering coefficient is naturally related to the volume scattering function in that it represents the integral of the volume scattering function over all directions

$$b = 2\pi \int_0^{\pi} \beta(0) \sin\theta \, d\theta \tag{2}$$

In addition, and with remote sensing in mind, it is appropriate to distinguish between forward and backwards scattering components, denoted b_F and b_B, respectively.

Volume scattering is a difficult process to measure requiring a device that is capable of measuring light scattered in directions from 0 to 180° to determine the angular distribution. The main problem is that the majority of light is scattered very close to the forward direction (0–5°), close to the path of the illuminating beam. Consequently, very few measurements have been made of the shape of the process for natural waters. The most commonly used volume scattering functions are those determined by Petzold (1972).

Figure 1 illustrates the contrast between measurements of normalised volume scattering in Atlantic Ocean water near the Bahamas ($b = 0.037m^{-1}$) and turbid water in San Diego harbour ($b = 1.583m^{-1}$). Note that back-scattering (90–180°) constitutes a much greater fraction of scattering in the case of the clear ocean water than for the more turbid water. Note also that, particularly in the forward directions, the shapes of the two functions are basically similar. These curves may be taken as reasonably typical for clear oceanic water and for moderate to highly turbid waters, respectively (Kirk, 1994a).

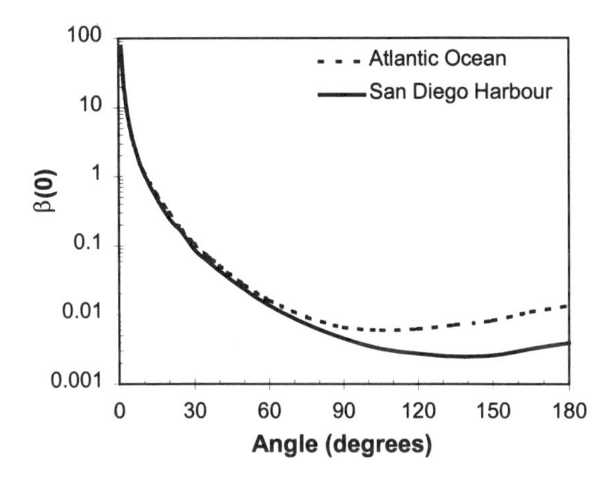

Figure 1. Normalised volume scattering functions [$\beta'(0)$] for clear oceanic and turbid waters (Petzold 1972).

Volume scatteringby pure water has a different shape from those in which particulate material may be found (Figure 2). This figure shows that β(0) of pure seawater is a minimum at 90° and rises symmetrically towards greater or lesser angles. Thus, the function is much more symmetrical in angular distribution than for particulate-dominated waters and explains the greater backscattering evident in the oceanic water in Figure 1. Morel and Gentili (1991) demonstrated that molecular backscattering may be significant in waters with low particulate concentrations. For the majority of coastal waters, however, we can expect that particulate volume scattering will dominate.

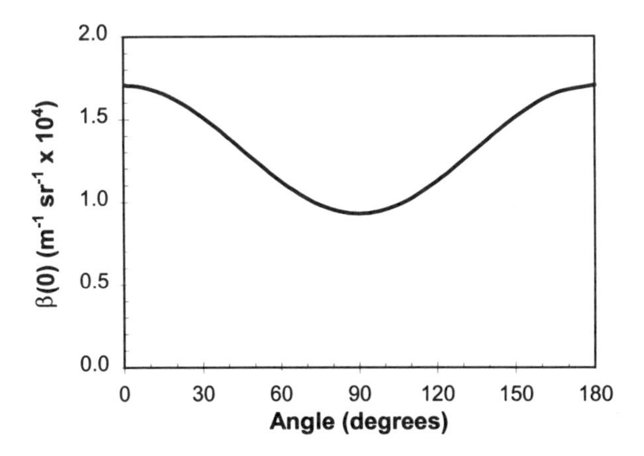

Figure 2. Volume scattering function [$\beta(0)$] for pure water.

7 Spectral variation of absorption and scattering processes

There are four principal components in the water column which contribute to absorption and scattering processes. These have been termed the optically important components of the water column and may be listed as:

- pure water,
- dissolved yellow substances (DYS),
- phytoplankton (the living photosynthetic and particulate component) and
- non-living organic and inorganic particulate matter.

Each of these components can have a marked effect on the spectral penetration of light with depth, and hence on the spectral nature of reflectance. The inherent optical properties of the optical components of each of these components will be discussed in turn.

7.1 The inherent optical properties of pure water

A good set of measurements of the optical properties of pure water was made by Hakvoort (1994), although numerous other sets of measurements are available (e.g. Smith and Baker 1981; Palmer and Williams 1974). The absorption properties of water are shown in Figure 3. These clearly show that water absorbs weakly in the blue region of the spectrum rising beyond approximately 550nm into the red region with absorption shoulders apparent at 600 and 660nm. As the wavelength increases into the near infrared, absorption by water increases significantly. It is for this reason that wavelengths beyond about 720nm are of little use in remote sensing of ocean colour.

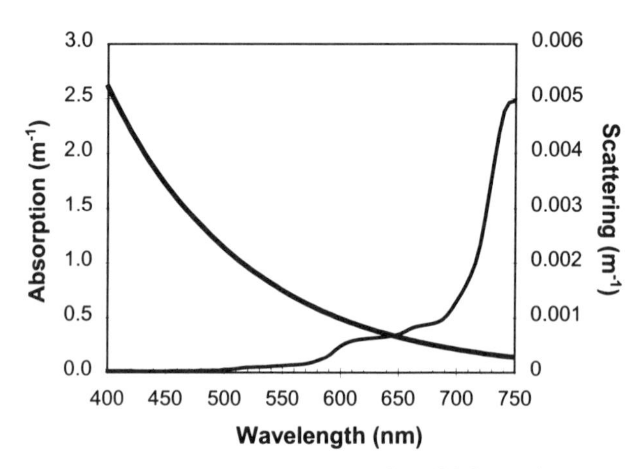

Figure 3. Inherent optical properties of pure water; scattering, thick line (Hakvoort 1994).

Scattering by water, by density fluctuation (both Rayleigh and Einstein-Smoluchowski scattering), varies with wavelength according to an inverse power:

$$b_w = \lambda^{-4.32}$$

Thus, scattering is most intense in the blue region of the spectrum which, combined with low absorption in this region, explains why the clearest oceanic waters appear blue in hue (Figure 3). Seawater scatters around 30% more intensely than pure water. The volume scattering function for pure water is shown above (Figure 2).

7.2 The inherent optical properties of dissolved yellow substances (DYS)

Dissolved yellow substances (DYS) refer to the natural breakdown products of plant matter, specifically to humic substances (mostly comprised of humic and fulvic acids), which are brown/yellow in their colour. Concentrations of such substances in water can be significant in coastal waters as a result of the breakdown of products (from phytoplankton and macro-algae) within the water but also from the input of humic matter of terrestrial origin from land runoff. In the water optics literature, humic substances have been given a range of different names by different researchers (e.g. gilvin, gelbstoff, dissolved yellow substances and dissolved aquatic humus).

The optical properties of humic substances are easily determined using a spectrophotometer on a water sample which has been filtered through a fine pore filter (0.2μm) to remove the particulate material. A typical spectrum is shown in Figure 4. This shows that absorption by DYS is highest in the blue region of the spectrum, falling in an exponential manner with increasing wavelength. Absorption is minimal in the red region of the spectrum. Bricaud et al. (1981) showed that the shape of the spectrum of DYS absorption for most marine waters can be determined from the relations:

$$a_{\mathrm{DYS},\lambda} = a(\lambda_0)\exp\{-0.014(\lambda - 440)\} \qquad (3)$$

where $a(\lambda_0)$ is the absorption coefficient at a reference wavelength, usually 440nm.

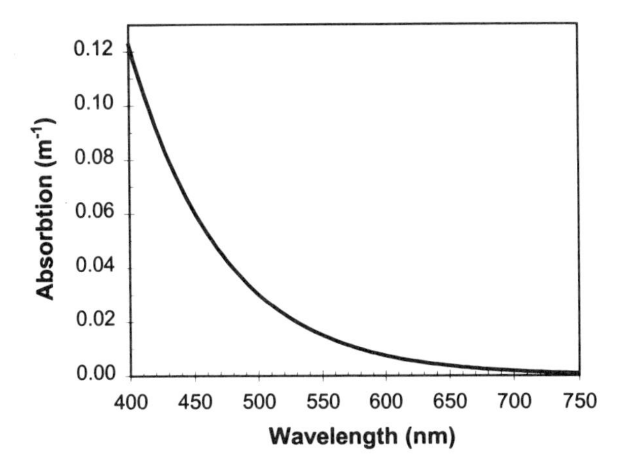

Figure 4. A typical example of the spectral absorption properties of dissolved yellow substance.

If concentrations of DYS are significant in a water column their absorption may 'compete' with phytoplankton for light due to strong absorption in the blue region of the spectrum, where chlorophyll a and other pigments also absorb. Field measurements of the relative concentrations of DYS should therefore be made to substantiate satellite or aircraft remotely-sensed data, especially in coastal regions where aquatic humus concentrations may be high.

7.3 The inherent optical properties of phytoplankton

The principal components of phytoplankton that contribute mostly to absorption are the photosynthetic pigments consisting of the chlorophylls, carotenoids and biliprotiens. Different taxonomic phytoplankton groups contain different combinations of pigments (Prezelin and Boczar 1986). It is important to note that the absorption properties of photosynthetic pigments when measured in intact cells (*in vivo*) can be markedly different from those measured when the pigments are extracted in a solvent.

The absorption properties of phytoplankton and other particulate matter are more difficult to measure than dissolved absorbing components, requiring a spectrophotometer modified such that the detector is capable of collecting light scattered by the turbid sample medium (usually achieved using an integrating sphere or similar turbid sample holder accessory). Another method is to measure directly the absorption of particulate matter collected on a filter pad. A typical average spectrum for marine phytoplankton absorption (a_{PH}) is shown in Figure 5. This shows the predominant regions of absorption in the blue and red regions of the spectrum where the main pigments (chlorophylls) absorb. Absorption is typically lowest in the 550–650nm region. Other shoulders in the spectrum are the result of other pigments, which effectively widen the spectrum of light over which the phytoplankton are capable of absorbing light for photosynthesis.

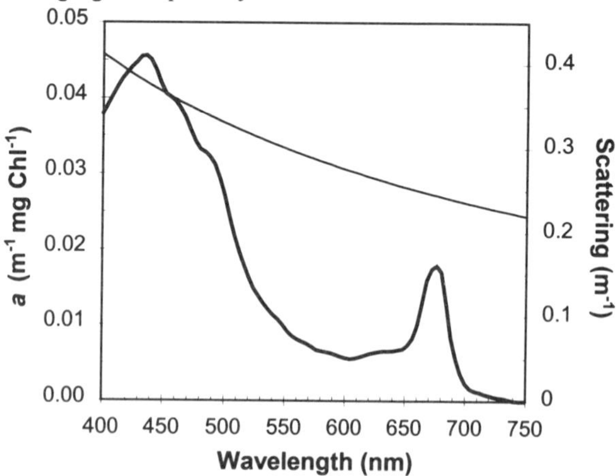

Figure 5. Absorption and scattering curves for phytoplankton. The absorption curve is an average of a number of spectra from different marine species (Morel 1988). The scattering curve is that predicted from Equation [4].

Separate absorption spectra for four different phytoplankton species are shown in Figure 6. These show that absorption properties of different algal classes can be markedly different, as a result of differences in pigment content and other environmental factors under which the phytoplankton have been growing (e.g. light intensity, nutrient status, etc.).

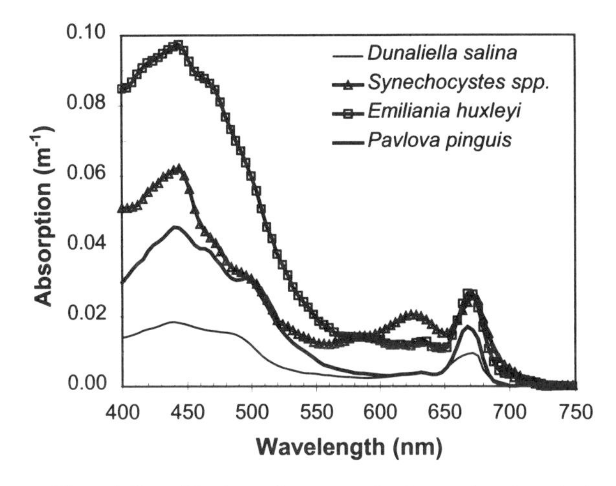

Figure 6. Chlorophyll specific absorption spectra for four marine phytoplankton species.

Compared to the wavelength-dependent scattering of water, scattering by particulate matter appears to be relatively insensitive to changes in wavelength since, with particulate diameters greater than the wavelengths of visible light (typically greater than 2μm), Mie scattering theory operates. This is supported by evidence from clear open oceans which have $b(\lambda)$ varying as λ^{-1} rather than $\lambda^{-4.32}$, indicating the influence of particulate material (Morel, 1973).

Fewer measurements are available for the spectral scattering properties of cultured and natural phytoplankton assemblages. These generally show that, with some variations, scattering by particulate matter (both phytoplankton and tripton) varies inversely with wavelength. Gordon and Morel (1983) gave the equation

$$b_p = \left(\frac{550}{\lambda}\right) 0.3\,\mathrm{Chl}^{0.62} \tag{4}$$

where b_p is scattering by particulates and Chl is the chlorophyll concentration. This equation describes the inverse relationship between particulate scattering and wavelength and also shows that scattering in clear oceanic waters (i.e. where suspended sediments are negligible) can be non-linearly related to pigment concentration.

7.4 Non-living organic and inorganic particulate matter

Absorption by inanimate particulate matter (a_t) is probably the least studied area of underwater light properties because of the difficulties in being able to separate non-living and living particulate components physically in real samples. Most measurements of particulate absorption that are reported include both the living and non-living fractions combined. The optical properties of either fraction can only be inferred in natural situations where it is possible to be sure that one or other component is by far the dominating influence in a particular water sample.

Bricaud and Stramsky (1990) reported a method for the retrieval of algal and non-algal components from absorption spectra of particulate matter. This approach utilises the fact that phytoplankton absorption spectra at 380nm and 505nm typically have spectral ratios close to unity. Thus, particulate absorption spectra showing an increase towards shorter wavelengths have significant amounts of non-algal absorption. A series of simultaneous equations could be solved to extract the living and non-living fractions of the particulate absorption. For ocean waters, Bricaud and Stramsky (1990) found that the spectral dependency of the absorption by non-living (detrital) material was similar to that for aquatic humus (equation (3), above). Further validation of this technique is required.

Under normal circumstances, it might be considered that inanimate particulate matter does not absorb light strongly, but that it may scatter to a significant degree. Scattering by this component is largely considered to be of a similar nature to phytoplankton.

8. Optical measurements and apparent optical properties

Light attenuation is most frequently measured using a flat cosine collecting sensor either lowered through the water column pointing upwards or downwards (thus measuring irradiance over the hemisphere). Broad-band irradiance is most often determined, most frequently in the 400-700nm range, defined as the photosynthetically active radiation (or PAR) range. Attenuation coefficients derived from such measurements are frequently used as the basis against which to compare remotely-sensed reflectance measurements (e.g. Gould and Arnone 1997). Kirk (1994b) provided a method for estimating the average broad-band inherent optical properties from series of downwelling and upwelling irradiance measurements. K_d has been described as a quasi-inherent optical property since it varies only weakly through the day with changes in solar elevation.

Scalar irradiance is the measurement of radiance distribution over *all* directions, both upwelling and downwelling and is usually measured with a spherical collector. Scalar irradiance is considered to be superior to upward and downwelling measurements particularly for determining light available for uptake by phytoplankton for photosynthesis as a population of phytoplankton cells will absorb light equally from all directions. Attenuation derived from such measurements (K_o) have been shown to be similar to K_d measured with a flat sensor.

Secchi Disk transparency is measured by lowering a flat white disk through the water column and noting the depth (Z_{SD}) at which it disappears. Measurements of Z_{SD} can be empirically related to, and therefore represent a (sometimes inaccurate) means to estimate K_d, since it varies inversely with the sum $(c + K)$. The optical properties of the Secchi Disk are in fact complicated and were reviewed by Tyler (1968) and most recently by Preisendorfer (1986). Davies-Colley *et al.* (1993) reported that the measurement of Z_{SD} may be considered a more quasi-inherent apparent optical property than K_d as it can be shown to be less dependent on ambient lighting conditions. Algorithms relating Z_{SD} and remotely-sensed reflectance have been reported by Topliss *et al.* (1988) amongst others.

Scattering measurements are more difficult to measure directly. Instruments that can be used to estimate scattering include transmissometers, designed to measure c; if measuring in near infrared wavelengths where the absorption by all constituents except water is negligible, then $b(\lambda_{NIR}) = c(\lambda_{NIR}) - a_w(\lambda_{NIR})$. Relative scattering measurements at fixed angles can be made using nephelometers, although the technique has some limitations (Davies-Colley *et al.* 1993).

Optical depth is defined as $OD = zK_d$ (which has no units). Thus, in a coloured turbid water with a high K_d, a given optical depth will correspond to a much shallower actual depth than a clear colourless water with a low K_d. The euphotic depth, defined as the depth at which light is reduced to 1% of the surface intensity corresponds to $OD = 4.6$.

Spectral measurements of the irradiance field using spectro-radiometers. For PAR measurements, deviations from the ideal of Beer's law can occur because the spectral distribution of the light field may change significantly with depth. Most accurate measurements of the underwater light field are performed using spectro-radiometers, which are instruments which are capable of measuring underwater optical parameters at a number of discrete wavebands across the spectrum of interest. In the example from Stagnone Lagoon, Sicily (Figure 7), the presence of organic colour in the water leads to the rapid attenuation of light at the blue end of the spectrum. Similarly, with water absorbing strongly at longer wavelengths, the penetrating light is increasing constrained to green wavelengths, peaking around 570nm.

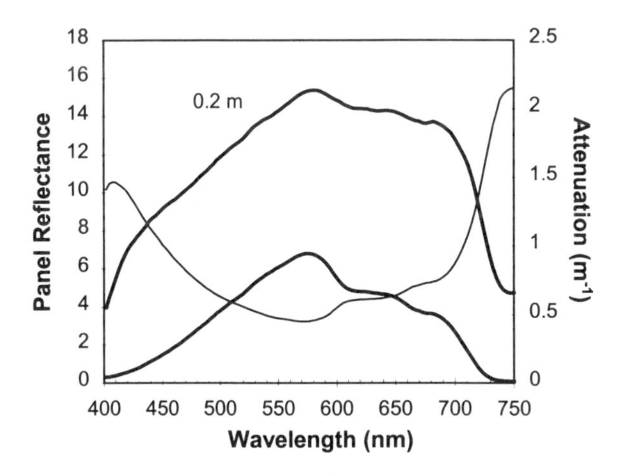

Figure 7. Series of downwelling spectra, measured in Stagnone Lagoon, Sicily (thick lines), depicting the changing nature of the spectral light field with depth, in this case indicating a narrowing of the downwelling spectrum into the middle wavelengths. Calculated spectral attenuation is also shown (thin line).

Subsurface reflectance *R*(0-) is the ratio of subsurface upwelling (ir)radiance and incident (ir)radiance and represents the most important parameter necessary for the development of algorithms using analytical approaches. Relations between *R*(0-) and water quality parameters have multi-temporal validity because *R*(0-) is relatively stable under variations in absolute irradiance (due to differences in Sun angle and atmospheric conditions) and states of the water surface. Reflectance has also been termed a so-called quasi-inherent optical property, as it is considered to be less dependent on the incident light field than other apparent optical properties.

9. Optical classifications of natural waters

A number of classification systems have been developed which attempt to categorise different water types on the basis of their optical characteristics. In remote sensing, the most frequently used classification is that of Gordon and Morel (1983) who differentiated waters into Case 1 waters, for which only phytoplankton and their derivative products predominate (oligotrophic to eutrophic), and Case 2 waters, in which significant contributions to optical properties may come from suspended sediments and dissolved yellow substances. For marine waters, it can be seen that Case 1 waters are more typical of open oceanic waters and Case 2 waters of coastal zones.

10. Modelling underwater optical processes

The total absorption $a(\lambda)$ is derived as the sum of all absorbing optical components:

$$a(\lambda) = a_w(\lambda) + a_{DYS}(\lambda) + a_{PH}(\lambda) + a_t(\lambda). \tag{5}$$

Similarly, the total scattering $b(\lambda)$ is regarded as the sum of scattering components:

$$b(\lambda) = b_w(\lambda) + b_{PH}(\lambda) + b_t(\lambda). \tag{6}$$

From a knowledge of these processes the nature of the underwater optical conditions can be studied. A number of bio-optical models have been developed which allow for greater understanding of the nature of underwater light processes; however, more importantly from the remote sensing perspective, they allow for the study of variations in optical water quality parameters to be investigated through the simulation of their effects on reflectance spectra. Such models range from radiative transfer models describing relations between inherent and apparent optical properties of the water (e.g. Smith and Baker 1978, Preisendorfer and Mobley 1984, Gordon et al., 1988, Sathyendranath et al. 1989, Lee et al. 1994) to Monte Carlo methods in which the fates of a large number of individual photons are separately determined based on the statistical probabilities of absorption and scattering for a given water body (e.g. Kirk 1984a, Morel and Gentili 1991, Gordon 1987, Sathe and Sathyendranath 1990, Malthus et al. 1997). The results of such approaches have led to the derivation of simple relations between apparent and inherent optical properties. For example, Morel and Prieur (1977) have shown that reflectance can be simply related to b_B/a:

$$R(0-) = R_I \, b_B / a \tag{7}$$

where R_I is a coefficient dependent on solar zenith angle, volume scattering function and hence on water type. For oceanic waters, Morel and Prieur (1977) defined $R_I = 0.33$. This was supported by Kirk (1991) who found that $R_I = 0.33$ for b_B values less than $0.25 m^{-1}$ and remains true for a wide range of scattering phase functions. On the other hand, more appropriate models for more turbid waters may take the general form

$$R(0-) = R_I \, b_B \quad (a + b_B) \tag{8}$$

or variations thereof. Dekker (1993) argued that it is not possible to predict a value for R_I that is widely applicable to a range of water bodies, but showed that the dependence of R_I can be circumvented by the use of algorithms employing spectral band ratios.

Similarly, for other quasi-inherent optical properties, Kirk (1984a) developed a model for the calculation of attenuation based on Monte Carlo studies:

$$K_d = \frac{1}{\mu_0}\left[a^2 + (0.425\mu_0 - 0.19)ab\right]^{\frac{1}{2}} \tag{9}$$

with coefficients included applicable to coastal waters.

The advantages of such approaches are that, if the spectral variations of individual optical components can be predicted (e.g. through the use of relations similar to those in Section 7), equations can be developed in spreadsheets for the simple study of the effects of variations in water column optical parameters on remotely- sensed reflectance and other quasi-inherent optical parameters.

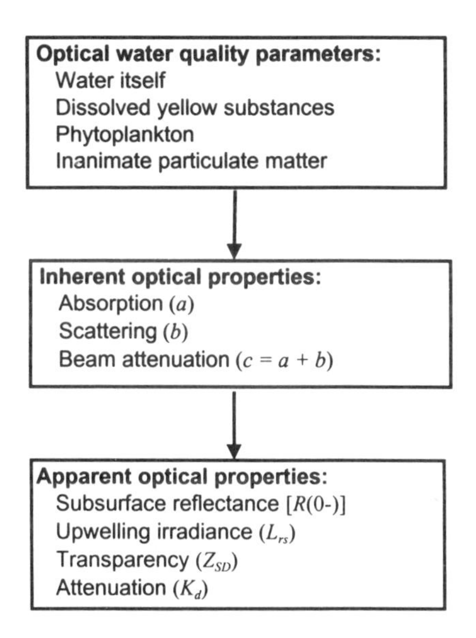

Figure 8. The relationship between water quality parameters and inherent and apparent optical parameters (after Davies-Colley et al. 1993).

11. Conclusions

The relation between the inherent optical properties, water quality parameters and the more measurable apparent optical properties of waters is shown in Figure 8. It is via their effects on the inherent optical properties of absorption and scattering that fluctuations in concentrations of optical water quality parameters affect subsurface reflectance and other apparent optical properties. As the inherent optical properties are well-defined and objective physical properties, they effectively form the bridge between the data we have (apparent optical properties) and the concentrations of the parameters we would wish to know (water quality). Thus, "it is the subsurface physical light field that must be directly compared to and interpreted in terms of the above-surface light field recorded at a remote platform" (Bukata *et al.* 1995). There is still much research required to further our understanding of the optical properties of coastal waters. In particular, there is a lack of knowledge of the range of spectral signatures from Case 2 waters and their variation in waters with different concentrations of particulate and dissolved substances. Similarly, we lack sufficient knowledge of the range in concentrations of optically important components, and their variations over time for many coastal areas.

The challenge to coastal remote sensing is to transform the knowledge of subsurface optical processes into the development of algorithms which provide multi-temporally meaningful information from remote sensing. However, the optical complexities of coastal zone waters means that robust algorithms will be harder to derive. While radiative transfer theory can become extremely complex, there are a sufficient number of well-established modelling approaches that can be used as an aid to such a goal.

References

Bagheri S and Dios R A, 1990, Chlorophyll a estimation in New Jersey's coastal waters using Thematic Mapper data. *International Journal of Remote Sensing*, **11**, pp 289-299

Bricaud A, Morel A and Prieur L, 1981, Absoption by dissolved organic matter of the sea (yellow substance) in the UV and visible domains. *Limnology and Oceanography*, **26**, pp 43-53

Bricaud A, and Stramski D, 1990, Spectral absorption coefficients of living phytoplankton and non-algal biogenous matter - a comparison between the Peru upwelling area and the Sargasso Sea. *Limnology and Oceanography*, **35**, pp 62-582

Bukata R P, Jerome J H, Kondratyev K Ya and Pozdnyakov D V, 1995, Optical properties and remote sensing of inland and coastal waters. CRC Press, Boca Raton

Davies-Colley R J, Vant W N, and Smith R G, 1993, Colour and clarity of optical water quality. Ellis-Horwood. Chichester

Dekker A G, 1993, Detection of optical water quality parameters for eutrophic waters by high resolution remote sensing. PhD thesis, Free University, Amsterdam

Estep L and Arnone R, 1994, Effect of whitecaps on determination of chlorophyll concentration from satellite data. *Remote Sensing of Environment*, **50**, pp 328-334

Gordon H R, 1987, Bio-optical model describing the distribution of radiance at the sea surface resulting from a point source embedded in the ocean. *Applied Optics*, **26** pp 4133-4148

Gordon H R and Morel A, 1983, Remote assessment of ocean colour for interpretation of satellite visible imagery. A review. Springer, New York

Gordon H R, Brown O B, Evans R H, Brown J W, Smith R C, Baker K S and Clark D W, 1988, A semianalytic radiance model of ocean color. *Journal of Geophysical Research-Atmospheres*, **93**, pp 10909-10924

Gordon H R and Wang M H, 1994, Retrieval of water-leaving radiance and aerosol optical-thickness over the oceans with seawifs - a preliminary algorithm. *Applied Optics*, **33**, pp 443-452

Gould R W and Arnone R A, 1997, Remote sensing estimates of inherent optical properties in a coastal environment. *Remote Sensing of Environment*, **61**, pp 290-301

Hakvoort J H M, 1994, Absorption of light by surface water. Published PhD. Thesis, Delft University of Technology, The Netherlands. Delft University Press. ISBN 90-407-1023-6

Jerlov N G, 1976, Marine optics. Elsevier Oceanography Series, 14

Kirk J T O, 1984, Dependence of relationship between inherent and apparent optical properties of water on solar altitude. *Limnology and Oceanography*, **29**, pp 350-356

Kirk J T O, 1989, The upwelling light stream in natural waters. *Limnology and Oceanography*, **34**, pp 1410-1425

Kirk J T O, 1991, Volume scattering functions, average cosines and the underwater light field. *Limnology and Oceanography*, **34**, pp 455-467

Kirk J T O, 1994a, Light and photosynthesis in aquatic ecosystems. Second Edition. Cambridge University Press

Kirk J T O, 1994b, Estimation of the absorption and scattering coefficients of natural waters by use of underwater irradiance measurements. *Applied Optics*, **33**, pp 3276-3278

Lee Z, Carder K L, Hawes S K, Steward R G, Peacock T G and Davis C O, 1994, Model for the interpretation of hyperspectral remote-sensing reflectance. *Applied Optics*, **33**, pp 5721-5732

Malthus T J, Ciraolo G, La Loggia G, Clark C D, Plummer S E, Calvo S, Tomasello A, 1997. Can biophysical properties of submersed macrophytes be determined by remote sensing? Proc Fourth International Conference on Remote Sensing for Marine and Coastal Environment, Orlando, Florida, 17 March 1997

Morel A, 1973, Diffusion de la lumiere par les eaux de mer. Resultats experimentaux et approche theorique. In: Optics of the sea, NATO, Neuilly-sur-Seine. pp. 3.1-1 to 3.1-76

Morel A, 1988, Optical modelling of the upper ocean in relation to its biogenous matter content: Case 1 waters. *Journal of Geophysical Research*, **93**, pp 10749-10768

Morel A and Gordon H R, 1980, Report on the working group on water colour, *Boundary Layer Meteorology*, **18**, pp 343-355

Morel A and Gentili B, 1991, Diffuse reflectance of oceanic waters - its dependence on sun angle as influenced by the molecular-scattering contribution. *Applied Optics*, **30**, pp 4427-4438

Morel A and Prieur L, 1977, Analysis of variations in ocean colour. *Limnology and Oceanography*, **22**, 709-722

Palmer K F and Williams D, 1974, Optical properties of water in the near infrared. *Journal of the Optical Society of America*, **64**, pp 1107-1110

Petzold T J, 1972, Volume scattering functions for selected oceanic waters. Scripps Institution of Oceanography, SIO Ref. 72-78, 79pp

Plass G N and Kattawar G W, 1972, Monte Carlo calculations of radiative transfer in the Earth's atmosphere-ocean system. I. Flux in the atmosphere and ocean. *Journal of Physical Oceanography*, **2**, pp 139-145

Preisendorfer R W, 1961, Application of radiative transfer theory to light meaurements in the sea. *Union Geod. Geophys. Inst. Monogr.*, **10**, pp 11-30

Preisendorfer R W, 1986, Secchi disk science: visual optics of natural waters. *Limnology and Oceanography*, **31**, pp 909-926

Preisendorfer R W and Mobley C D, 1984, Direct and inverse irradiance models in hydrologic optics. *Limnology and Oceanography*, **29**, pp 903-929

Prezelin B B, Boczar B A, 1986, Molecular bases of cell absorption and fluorescence in phytoplankton: Potential applications to studies in optical oceanography. *Progress in Phycological Research*, **4**, pp 349-464

Sathe P V, Sathyendranath S, 1990, Fortran programs for computation of optical-properties of the sea from radiation data collected by in situ spectrometers. *Computers and Geosciences*, **16**, pp 1085-1103

Sathyendranath S, Prieur L and Morel A, 1989, A three-component model of ocean colour and its application to remote sensing of phytoplankton pigments in coastal waters. *International Journal of Remote Sensing*, **10**, pp 1373-1394

Smith R C and Baker K S, 1978, The bio-optical state of ocean waters and remote sensing. *Limnology and Oceanography*, **23**, pp 247-259

Smith R C and Baker K S, 1981, Optical properties of the clearest natural waters. *Applied Optics*, **20**, 177-184

Topliss B J, Almos C L and Hill P R, 1990, Algorithms for remote sensing of high concentration, inorganic sediments. *International Journal of Remote Sensing*, **11**, pp 947-966

Topliss B J, Payzant L and Hurley P C F, 1988, Monitoring offshore water quality from space. Proceedings of IGARRS'88 Symposium, Edinburgh, Scotland 13-16 Sept. 1988. ESA SP-284

Tyler J E, 1968, The Secchi disk. *Limnology and Oceanography*, **3**, pp 1-6

Radiometric calibration of an optical spectrometer

Yan Gu

University of Dundee

1 Introduction to radiometric calibration

One of the critical issues of current Earth observation/remote sensing is the quality of the data. This has two aspects, according to Sakuma and Ono (1993). The first is the clear characterisation of the nature of the data in spectral, radiometric and spatial properties. The second is the compatibility among different data sets, which means that data sets taken by the same instrument at different times and different places should be compatible in a quantitative sense or, alternatively, data sets taken by different instruments at separate times should be capable of inter-comparison.

With the development of remote sensing technology, further Earth observation instruments will operate in orbit for longer periods. It is expected that accurate measurements taken during the mission will make it possible to derive geophysical parameters describing correctly both the long term and short term changes taking place from time to time and from place to place. To satisfy these requirements, a reliable measurement is necessary, which can discriminate geophysical parameter changes from instrument performance changes resulting from instrumental instability or degradation in the course of operation. Such measurements constitute an important aspect of instrument radiometric calibration.

An overview of many important image processing procedures is given in Figure 1 and an overview of instrument radiometric calibration procedures for space-borne instruments is given in Figure 2. The important steps in the latter are as follows.

1. Choose the appropriate primary standards for each spectral.
2. Use the primary standards to calibrate the secondary standard.
3. Calibrate the spectrometer with the secondary reference source.
4. Retrieve the observation data/on-board radiometric calibration data.
5. Measure the atmospheric parameters.
6. Ground radiometric observation.
7. Image radiometric correction.
8. Verify the calibration result using ground radiometric observations.

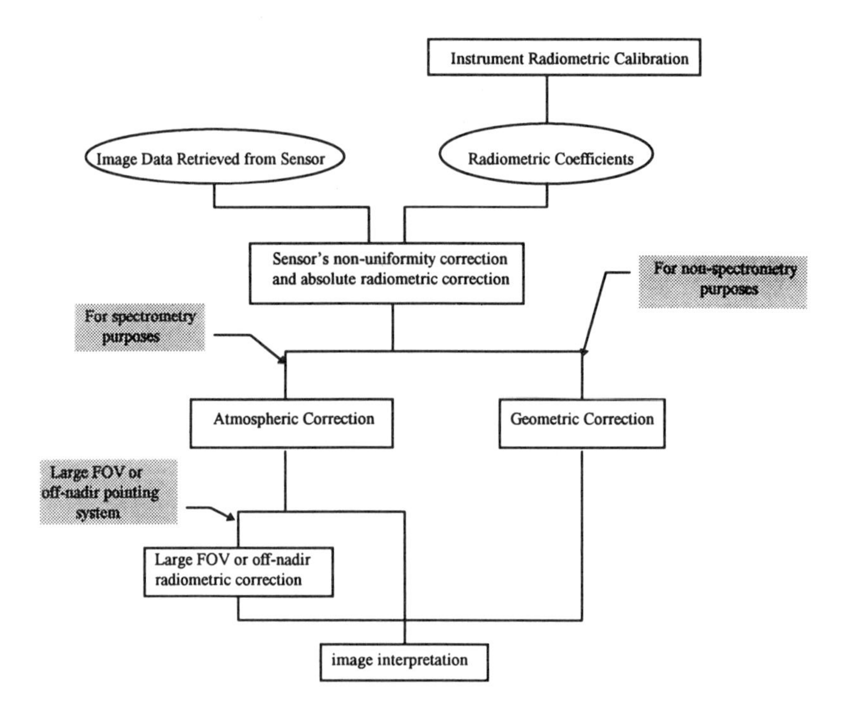

Figure 1 Image Processing Procedures

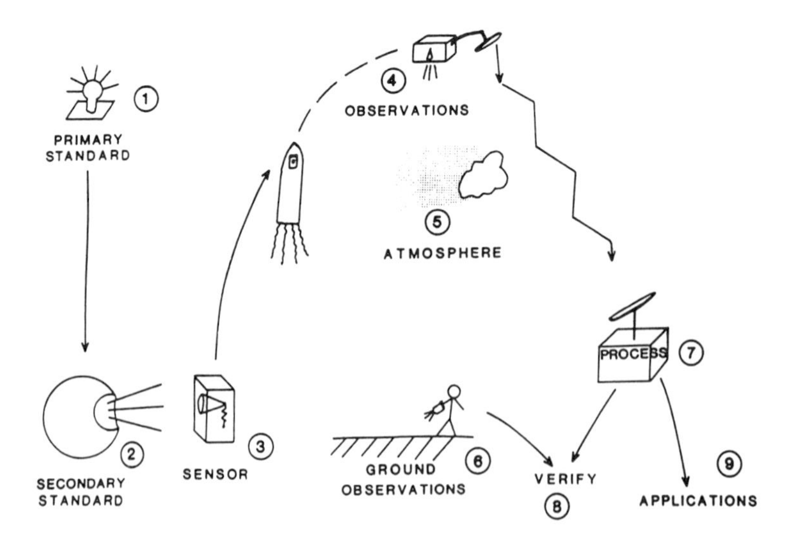

Figure 2 Steps involved in establishing the calibration of spaceborne spectrometers

The basic approach involved in radiometric calibration is to measure a system's radiometric performance against a known radiance target. We define a few important terms:

Spectral calibration accurately determines the spectrum of each channel or each band. The tasks involved include (Sun and Anderson 1993) (a) determining the centre wavelength of spectral passband, or its position on the detector, (b) determining the wavelength bandwidth of each passband and (c) checking the linearity of the spectral passband distribution.

Band-to-band radiometric calibration (or relative radiometric calibration) determines the multiplicative (gain) and additive (bias) factors to normalise the detector's response to an average or reference response. The tasks involved include (a) removing the effects of non-uniformity of response of detectors on the same band, or from different bands of the same instrument and (b) removing the effects of non-uniformity of response of detectors from different instruments.

Absolute radiometric calibration is a method that allows the digital data to be converted to radiance. This involves the retrieval of the solar-reflectance/spectral reflection or emission features of remote sensing targets (Thorne *et al.* 1997).

Radiometric calibration algorithm

As a pre-condition of a radiometric calibration algorithm we assume that the spectrometer has a linear response in its operation range; *i.e.* the radiance *Rad* satisfies $R_{min} < Rad < R_{max}$. The calibration point should be chosen in this range and near the centre of the region, so that the instrument has the largest dynamic range. The basic mathematical model for the raw digital reading *DN* is

$$DN(x,y,S) = K(ch)RC(x,y,S)\,[Rad(x,y,\lambda) + N(x,y,t,Rad)] \tag{1}$$

where:

- Rad = radiance on the entrance pupil of the spectrometer
- $K(ch)$ = channel gain factor
- RC = radiant sensitivity coefficient of the corresponding pixel
- N = the mathematical expected value of system equivalent noise at input end (including all random and fixed pattern noise)
- x,y,t = continuous co-ordinates on the sensor plane and time
- S = spectrometer settings
- λ = wavelength

The system equivalent noise may be decomposed as

$$N(x,y,t,Rad) = N_1(x,y) + N_2(x,y,t) + N_3(x,y,t,Rad) \tag{2}$$

where:

- $N_1(x,y)$ fixed pattern noise, such as DC, image smear (for CCD array), electronic offset (for digital sensor), etc. (Healey and Kondepundy 1994)

$N_2(x,y,t)$ = temporal noise, such as output noise, tape noise, frame grabber noise, most of device noise, etc.

$N_3(x,y,t,Rad)$ = radiance related noise, such as shot noise. This is the higher order non-linear factor. Since it is controlled by the manufacturer, and is reasonably small, it is negligible.

Substituting for $N(x,y,t,Rad)$ from Equation (2), we obtain

$$DN(x,y,S) = K(ch)\, RC(x,y,S)\, [Rad(x,y,\lambda) + N_1(x,y) + N_2(x,y,t)] \quad (3)$$

The calibration task can be simplified to that of removing N and retrieving RC by acquiring the $DN(x,y,S)$ from a known $Rad(x,y,\lambda)$ reference source. If we oversample the same reference source n times, from Equation. (3) we get

$$\sum_{i=1}^{n} DN_i(x,y,S) = nK(ch)RC(x,y,S)\, [Rad(x,y,\lambda) + N_1(x,y) + \sum_{i=1}^{n}(N_2(x,y,t))_i\,] \quad (4)$$

According to their distribution, the mathematical expected value of $N_2(x,y,t)$ is zero. So,

$$\sum_{i=1}^{n} DN_i(x,y,S) = n\, K(ch)\, RC(x,y,S)\, [Rad(x,y,\lambda) + N_1(x,y)] \quad (5)$$

$$K(ch)RC(x,y,S) = \frac{1}{n} \frac{\sum_{i=1}^{n} DN_i(x,y,S)}{Rad(x,y,\lambda) + N_1(x,y)} \quad (6)$$

Equation (6) is the original formula used in the data process of radiometric calibration.

The task of geometric calibration is to minimise the spatial sampling error in the instantaneous regional flux measurement, so as to meet the spatial resolution and geolocation requirements. The spatial sampling error usually comes from optical abbreviation, system design (FOV size, data rate, detector time response and scan period), the platform attitude variation, and spatial co-registration among the fields of view of different channels or different spectrometers. The co-registration among different channels can be established by cross-spatial calibration to the same scene at the same time.

It was recently noted that the calibration of optical abbreviation can be undertaken in two ways, according to M A Folkman (personal communication, 1997): (a) by measuring the system's Modulation Transfer Function (MTF) and (b) by measuring the Point Spread Function of the system in the overall FOV.

2. Pre-launch (ground) radiometric calibration

2.1 Introduction

The pre-launch radiometric calibration of a space-borne instrument is the most fundamental and important calibration. It allows the system to be tested to ensure that it operates properly and within the error budget before being integrated into a system for launch into space.

The system's pre-launch radiometric calibration activity was summarised into three steps by Ono and Sakuma (1996): (1) calibrate the spectrometer against a standard ground calibration reference, (2) calibrate the on-board calibration source reference and (3) recalibrate the spectrometer against the on-board calibration source. We consider these processes in turn.

2.2 Spectral calibration

The requirement for spectral calibration derives from the need to determine accurately the channels' bandwidths or the wavelength positions of the diagnostic spectral features. Although this calibration is fairly independent, its accuracy is closely related to the accuracy of the radiometric calibration. The spectral calibration procedures involve the following steps:

1. Illuminate the system with a calibrated (standard) monochromatic source. Possible standard sources include:

    - tungsten filament lamps with standard spectral lines
    - laser beams
    - sunlight with filters (on-board spectral calibration).

2. Acquire the images of the spectral references.

3. Process the data. The procedures involved include:

    - filter temporal noises
    - acquire the spectrum curves (see Figure 3)
    - find the peak values
    - calculate half peak width and central wavelength
    - verify the channel bandwidth or the spectral position on the detector array
    - retrieve the linearity of the spectral passband distribution by calculating several spectral positions across the passband and applying the least-square-fit regression.
    -

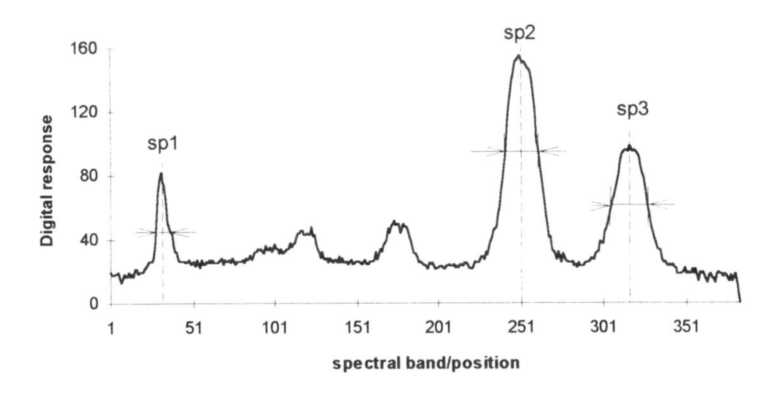

Figure 3. Spectral features of spectral reference source

2.3 Pre-launch calibration system

The pre-launch calibration system is usually composed of a primary standard source, a secondary standard source, a standard spectrometer, a comparison spectrometer and auxiliary facilities. The function of the calibration system is to transfer the standard radiance from the primary standard to the final spectrometer.

For an ideal standard radiance source, its spectral distribution should be similar to the spectral features of a real target and the magnitude of the radiance should be of the same level. In the visible and near infrared region, it should be similar to the solar spectral radiance. Unfortunately, such a standard is not available except for the Sun itself and the choice is limited within the practical available sources.

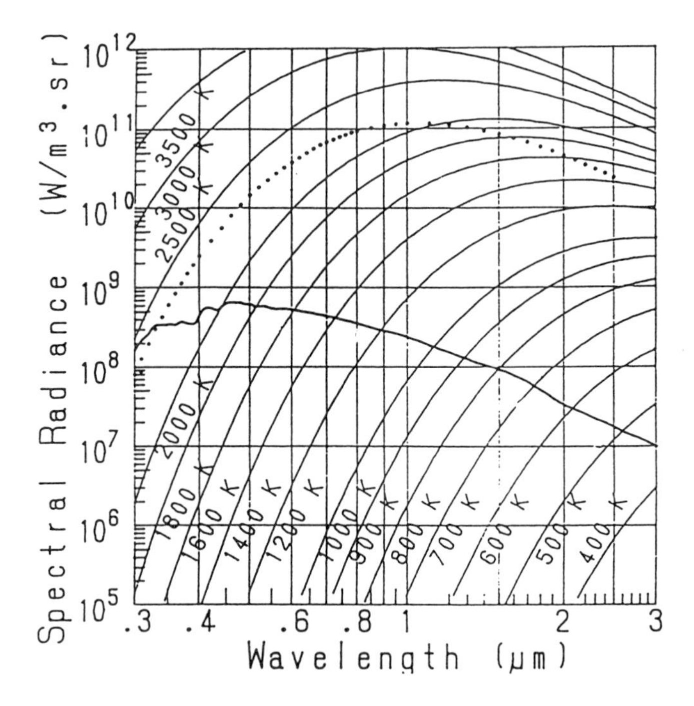

Figure 4. Spectral radiance of different sources, after Ono and Sakuma (1996). (Thick solid line: spectral radiance of an ideal diffuse reflector placed normal to the Sun outside the atmosphere. Dotted line: spectral radiance of a spectral radiance standard lamp. Thin solid lines: spectral radiances of blackbodies at various temperatures)

From Figure 4 it may be seen that the spectral radiance standard lamp (maintained by NIST, the National Institute of Standards and Technology) provides higher levels of radiation than the reflected solar radiation does. The difference is of two to four orders of magnitude depending on the spectral region. Its level is too high to be used as a standard source. The blackbody's radiation provides levels almost identical with the high level input radiance required for calibration in the visible and near infrared region, if an appropriate black-body temperature can be chosen.

It is reasonable at present to take limited fixed-point blackbodies as standard sources whose spectra can be combined according to their radiance distribution and magnitude to cover the required spectral range, and whose temperatures are precisely given by the International Temperature Scale of 1990 (ITS-90). Examples are: tin (505.08°K), lead (600.61°K), zinc (692.677°K), silver (1235.02°K) and copper (1357.77°K). These are used as primary standard sources for the calibration of secondary standard sources, see Figure 5.

As secondary standards there are (a) spectral radiance lamp and variable temperature intermediate temperature blackbody furnaces. According to Sakuma and Ono (1996), the radiation characteristic of variable-temperature blackbody furnaces can be expressed as

$$L(\lambda) = \varepsilon_C L_B(\lambda, T_{\text{eff}}) \tag{7}$$

where:

$L(\lambda)$ = radiance of the variable temperature black-body furnace

$L_B(\lambda, T_{\text{eff}})$ = Planck's expression for black-body spectral radiance

T_{eff} = effective temperature

ε_C = cavity emissivity

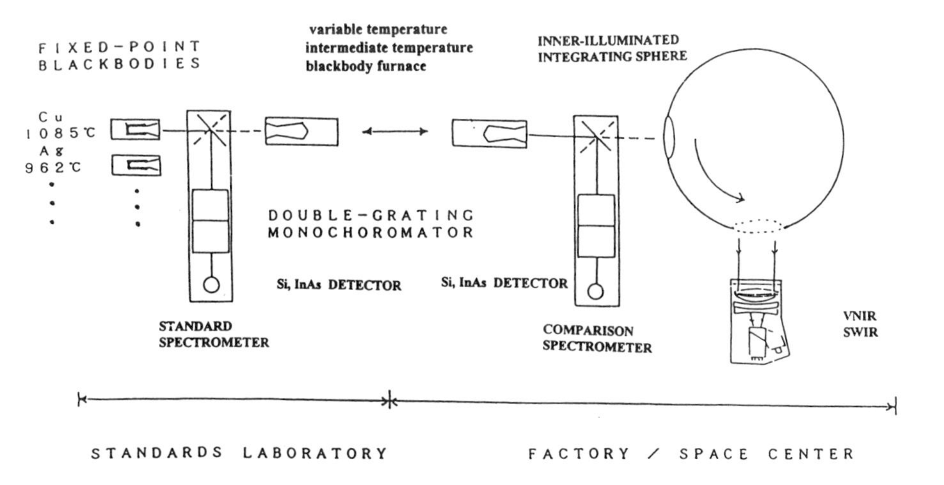

Figure 5. Schematic of Pre-launch Calibration System, after Sakuma and Ono (1991)

We now turn our attention to the integrating sphere. Its structure is illustrated by a certain number of "equal intensity" halogen lamps which are placed symmetrically around the exit port inside the sphere, supplied with a stabilised direct current power and cooled by fans. The inner wall of the sphere is covered with barium sulphate which can provide near-Lambertian reflectance. The requirements of an integrating sphere include: (a) that the aperture diameter is large enough to fill the spectrometer aperture, (b) homogeneous radiance over the source aperture, (c) long-term temperature stability in the required spectral range and (d) accurately calibrated spectral radiance.

The requirements for the pre-launch calibration of a spectrometer come from the application purpose as well as the operation principle of the system. Generally, we demand

(a) calibration accuracy of absolute responsivity, (b) offsets of individual bands and (c) the stray-light rejection.

The data processing procedure includes the steps.

- Acquire frames of image DNs of a known radiance reference source.
- Eliminate the temporal noise influence.
- Subtract the dark-current.
- Compensate the image frame smear influence.
- Calculate the *RC* and *K(ch)*.
- Calibration evolution.
- Accuracy analysis.

For the calibration of an on-board calibrator, the integrating sphere in Figure 5 is replaced by the on-board calibrator. The radiances from the furnace and the on-board calibrator are compared by the same comparison spectrometer, then an accurate radiance distribution data set is built up.

To study the evolution of the spectrometer calibration against an on-board calibrator, the calibrated on-board calibrator is used as the reference source for the spectrometer, employing the procedure described above, to calculate the radiance output of the calibrator. The result is used to establish the on-board radiometric coefficients table. In this way, the transfer of the radiometric calibration from the ground to the on-board mission is completed.

3. In-flight radiometric calibration

3.1 The necessity of in-flight calibration

There are several factors which make in-flight calibration necessary:

- Vibration and high G influence during the launching period will cause space and position changes between optical parts, which will eventually influence the optical performance of the spectrometer system.
- Outgassing and contamination influence - the low pressure in the vacuum space environment will make gas inside the system, especially inside the reference lamp, release gradually so that the filament temperature is higher and the lamp is brighter than it is on the ground. The contamination is due to the propellant waste felt on the optical surface which might change the optical transmission characteristic.
- Long term degradation of optical components, detectors, electronics etc. - the long time accumulation of high energy particles and γ-rays on the system will degrade the performance of optical components, detectors and electronic parts.
- Environmental changes (temperature, vacuum).
- Temporal noise influence is the problem of electronic and magnetic compatibility, which means that the operation of other on-board systems (especially the switch on and switch off operation) will influence the operation of the spectrometer.

All these factors make the in-flight radiometric calibration necessary; there are three approaches: (a) using an on-board radiometric calibrator, (b) using ground reference and (c) using the Sun or the Moon as a reference. These are considered in the following sections.

3.2 On-board radiometric calibration

An on-board radiometric unit introduces a well-defined reference beam to the spectrometer at the front end. The method applied simulates that of the pre-flight calibration. To obtain good calibration accuracy, the system should have the following features (Thorne *et al.* 1997).

- The reference beam should be wide enough to illuminate the full field of view (FOV) and full aperture. It should also go through the whole optical system so that the degradation caused by any elements can be detected. Because of technical problems, however, this requirement cannot be satisfied perfectly.
- The calibrator system should be as simple and reliable as possible, in order to minimise the degradation and contamination influences of the calibrator system.
- The reference source should be more stable than the spectrometer system.
- Detectors should be used to monitor the radiometric changes of the calibrator.
- A redundant calibrator system is necessary.

The reference sources which can be used include: (a) halogen filament lamps, (b) a solar illuminated diffuser, (c) moonlight and (d) blackbodies for thermal bands. The stability of the lunar reflectance (change of the order of one part per million per year) allows an excellent check of long-term drift of the spectrometer

Because of the uniformity of the on-board reference source, which can reach most detectors, the on-board calibrator is the best source to eliminate the non-uniformity among detectors of the same band. It can also be used to monitor the long term changes of the individual detector if the change introduced by the calibrator is known or negligible. Reference to the pre-flight, absolute radiometric calibration of each sensor allows the on-board calibration system to provide an absolute calibration.

3.3 Ground-reference in-flight calibration

The on-board radiometric calibration method cannot be relied upon to provide accurate absolute calibration even though its precision may be very high (Thorne *et al.* 1997); this is because it is very difficult to eliminate systematic errors caused by the long term degradation of system components. The most reliable way to confirm the absolute accuracy of a calibration is to compare the result with that of another totally independent calibration approach with the same estimated accuracy and precision. If the results of the two calibrations agree to within the uncertainties estimated for them, then there is a reasonable possibility that systematic errors are not significantly influencing the result.

To ensure highly consistent and accurate radiometric calibration over a long period, a reliable in-flight calibration method, besides on-board calibration, has to be found for all sensors. Reference to the radiance from ground scenes with well-characterised ground-atmospheric properties meets the above need. This is the so-called ground-reference method or vicarious method.

The basic approach of ground-reference methods is to use ground-measured radiance at certain test sites to predict the radiance at the top of the Earth-atmosphere system, and then to compare them with the radiance of the same sites acquired by the overpassing on-board spectrometer at the same time. The result can provide degradation information about the spectrometer system and can be used to estimate the measuring accuracy.

The requirements for a test site include (Slater *et al.* 1987): (a) the location is inside the spectrometer's FOV, (b) a high elevation far from a city where there is a low probability of clouds, low aerosol loading and low water vapour influence, (c) the surface is flat and extended, large enough to fill more than one pixel of the spectrometer's FOV and (d) the surface reflectance is high and uniform, being close to a Lambertian reflector.

TMs on-board Landsat-5 and HRVs on-board SPOT-2 use White Sands Missile Range in New Mexico, Rogers Dry Lake in California and Maricopa Agricultural Centre in Arizona as ground-reference test sites.

The reflectance-based method has been in use for many years, primarily for the calibration of TMs and the HRVs in the solar-reflective range. The steps involved are as follows (Slater *et al.* 1996) (see Figure 6).

1. Choose the calibration site.
2. Obtain the average nadir reflectance of the site for each spectral band of interest at the time the satellite sensor is acquiring an image of the area.
3. Retrieve the reflectance factor data by computing the above result with those recorded when viewing a calibrated near-Lambertian reference panel.
4. Get atmospheric information by monitoring the atmosphere during the calibration.
5. Calculate the Rayleigh, aerosol and ozone optical depth and size distribution.
6. Input the above data into a Gauss-Seidel iteration radiative transfer code.
7. Compute the top-of-the atmosphere radiance in the sensor's spectral bands.
8. Average the site DNs from satellite image data to filter the noise.
9. Calibrate the spectrometer.

This method has an estimated accuracy of 4.9%.

The irradiance-based method is a modification of the above method. It does not depend on the accuracy of the size distribution measurement of the aerosol particles; instead, the additional measurements of diffuse and global downwelling irradiances are made. The ratios of these measurements are corrected for the diffuse component, then extrapolated or interpolated to determine the values for the satellite and solar zenith angles at the instant of image acquisition. Then the same procedures mentioned above are adopted. This method has an estimated accuracy of 3.9% (Slater *et al.* 1996).

The radiance-based method uses an airborne, well-calibrated radiometer to measure the radiance instead of estimating the surface reflectance. Since only a minor correction of the measured radiance is needed to account for the intervening atmosphere between the aircraft and satellite, this is the most direct and, potentially, the most accurate of the vicarious calibration methods. The procedure involved is as follows (see Figure 7).

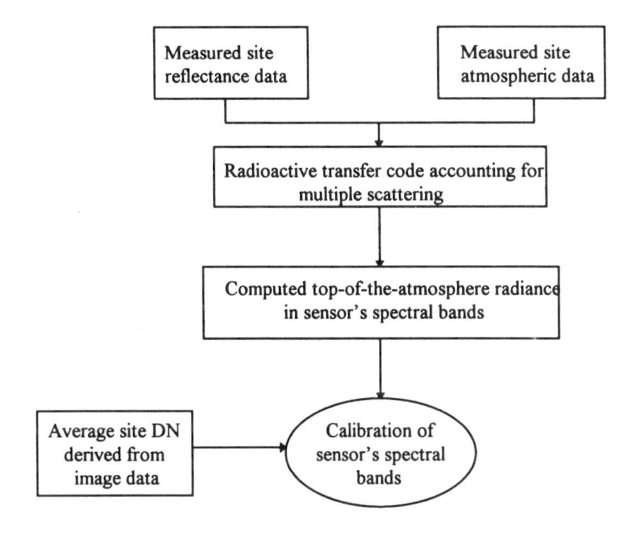

Figure 6. Reflectance-based calibration

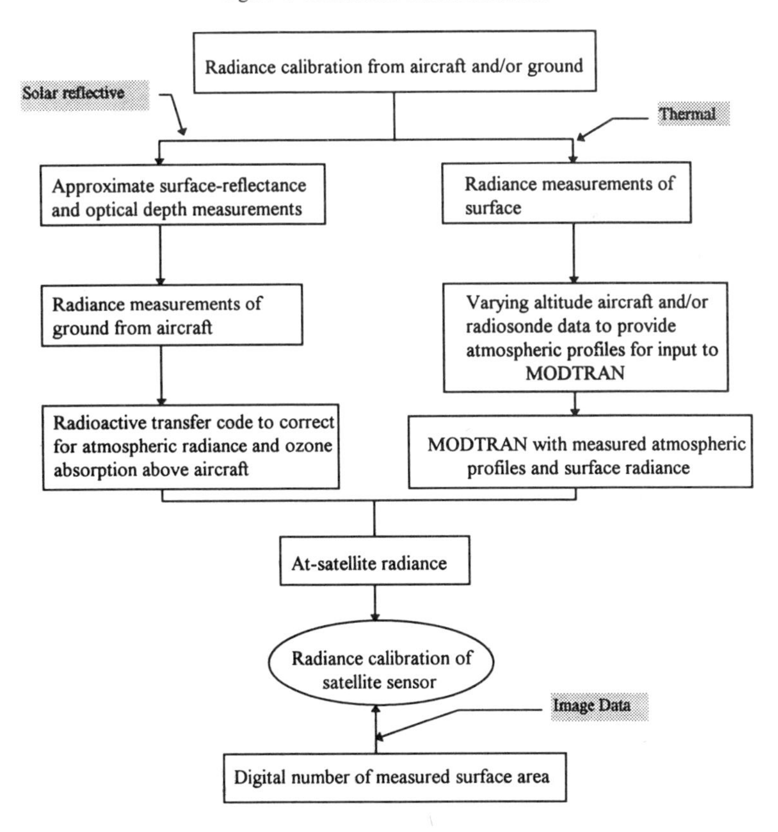

Figure 7. Radiance-based calibration

1. Measure upwelling radiance from the test site using a well-calibrated radiometer from an aircraft at an altitude above most of the aerosols and water vapour.

2. Acquire the atmospheric parameters.

3. Input the parameters measured above into the atmospheric radioactive transfer code, the result providing the correc90tion for the atmospheric path from the aircraft altitude to the satellite.

4. Compute the top-of-the atmosphere radiance in the sensor's spectral bands.

5. Average the site DNs from the satellite image data to filter the noise.

6. Calibrate the spectrometer.

This method has an estimated accuracy of 2.8%.

3.4 Comparison of on-board and ground-reference in-flight calibration

On-board in-flight calibration

- Calibration can be performed with high temporal frequency. It is obvious, then, that the on-board data are better at showing temporal variability of sensor calibration. Besides regular radiometric calibration during each cycle, some spectrometers record the calibrator's signal on each frame.
- Best method for non-uniformity calibration. (Since pixels can be calibrated by the on-board calibrator).
- Excellent for short period absolute radiometric calibration. In a short time period, (months after the launch), the degradation of the on-board calibrator is negligible. The radiance given by the calibrator can be used for absolute calibration. Pre-flight radiometric calibration coefficients can be applied jointly with on-board calibration. It does not test full optical path and full aperture, so some contamination and degradation may not be monitored, especially the first optical window where the contamination and degradation are serious.
- The calibrator's degradation is very difficult to detect accurately. As a long term absolute radiometric calibration source, the systematic error is difficult to eliminate. Even when a detector is used to monitor the calibrator system, there is still no way to measure the degradation of this detector.
- Making use of a full-aperture full-path solar diffuser provides an additional way to determine the radiometric calibration of the data. But the solar radiance distribution is to be considered, and intensity daily variation has to be taken into account. Accuracy of the result is questionable.

Ground-reference in-flight calibration

- Full-aperture full-path calibration. Any changes in the system can be monitored. The calibration is done while the system is sensing, so the system's calibration and operation share the same mode. This means that any factors involved in the sensing period are taken into account during the calibration.
- Very labour intensive. A lot of manpower is involved in the field work. An aircraft or helicopter and many other kinds of instruments are needed in this research: it is expensive and time-consuming.
- Calibration is limited by the test site condition, overpass time, weather, etc. Perfect weather is necessary. If the weather is poor (rain, mist or strong wind), or there are clouds over the test site when the spaceborne spectrometer is passing over, the calibration becomes impossible.
- For a high spatial resolution spectrometer, only limited pixels in each band can be calibrated. Only the pixels that are sensing the test site can be calibrated. So the result can be provided to check the degradation of these pixels, which can be supposed as systematic error.
- The measuring accuracy depends on the accuracy of each measurement, especially the measurement of weather parameters. Increasing the measuring accuracy of surface bi-directional reflectance and improving human knowledge about aerosol scattering phase function might therefore be the way to improve the accuracy.
- Difficult to use for relative calibration purposes.

3.5 In-flight cross-calibration concept

The objective of in-flight cross-calibration is to compare spatially and spectrally uniform sites with two or more spectrometers in orbit, that cover part or all of the same spectral region. By comparison, the extent to which calibration differences between well-characterised sensors providing the same data products at different scales will be known (Slater *et al.* 1996).

In-flight cross-calibration between sensors can be conducted in several different situations: (a) The same GIFOVs and spectral bands, and both image the same scene simultaneously. This is the situation when two same spectrometers operate on the same platform, such as two HRVs on the SPOT payload. (b) Similar spectral bands that cover part or all of the same spectral range and still image the same scene simultaneously but with different IFOVs. For example different spectrometers on-board the EOS platform. (c) Similar spectral bands and IFOVs but not imaging the same scene simultaneously. This is the most common situation, with different spectrometers on different platforms, such as EOS, ASTER, Landsat TM and SPOT HRV.

Cross-calibration results are self-verifiable by comparing spatially and spectrally over the same test site with two or more spectrometers. If agreement is found between two spectrometers, but these two disagree with those from the third one, this will indicate that a change has possibly taken place in the calibration of the third spectrometer. Other methods will have to be found to check the source of changes. The precision of the method will be determined by statistically analysing the results of several hundred cross-comparisons.

4. Case study: radiometric calibration of EOS ASTER system

4.1 Instrument description and radiometric performance

The ASTER (Advanced Spaceborne Thermal Emission and Reflection Radiometer) is a multispectral optical imager of high spatial resolution for Earth remote sensing from space. The instrument is provided by the Japanese Ministry of International Trade and Industry (MITI) and is planned to be flown on NASA's Earth Observing System (EOS) AM-1 spacecraft in a polar orbit in 1998 (Ono and Sakuma, 1996).

The Mission task is to acquire high spatial resolution imaging of land surfaces, inland waters, ice and clouds in 14 bands for the use of solid Earth, climate and hydrology processes, biogeochemical dynamics and Earth system history, etc. The main parameters of the instrument may be split into two categories. First there are those that are independent of the wavelength of the observations:

Orbit:	705km Sun-Synchronous
IFOV:	4.9°
Swath:	60km
Off-nadir capability:	±24° (swath 314km)
Platform	S-AM-1 (Earth Observation System)

The wavelength bands and their properties are listed below. The abbreviations are :Visible and Near Infra Red (VNIR): Short Wave Infra Red (SWIR): Thermal Infra Red (TIR).

	VNIR		SWIR		TIR	
Band and	1.	0.56 ± 0.08	4.	1.65 ± 0.10	10.	8.30 ± 0.35
wavelengths (μm)	2.	0.66 ± 0.06	5.	2.165 ± 0.04	11.	8.65 ± 0.35
	3N.	0.81 ± 0.10	6.	2.205 ± 0.04	12.	9.10 ± 0.35
	3B.	0.81 ± 0.10	7.	2.260 ± 0.05	13.	10.60 ± 0.70
			8.	2.33 ± 0.07	14.	11.30 ± 0.70
			9.	2.395 ± 0.07		
GIFOW resolution (m)	15		30		90	
Scanning Mode	Push-broom, scanning mirror (cross-track pointing				Whisk-broom	
Data Format	8-bit		8-bit		12-bit	
Detector	5000 linear CCD for each band		2048 linear CCD for each band		10-element linear detector array for each band	

4.2 Pre-flight calibration/characterisation methodology

The spectral characterisation concept (see Figure 8) involves the following:

- centre wavelength
- bandwidth
- band edge response - wavelength difference between 10% and 80% of peak response
- flatness of in-band responses - the responsivity between two wavelengths corresponding to 80% peak responsivity should be better than 80%
- suppression of out-of-band responses - the integrated contribution from the region below 10% peak should be less than 3% of the integrated in-band response.

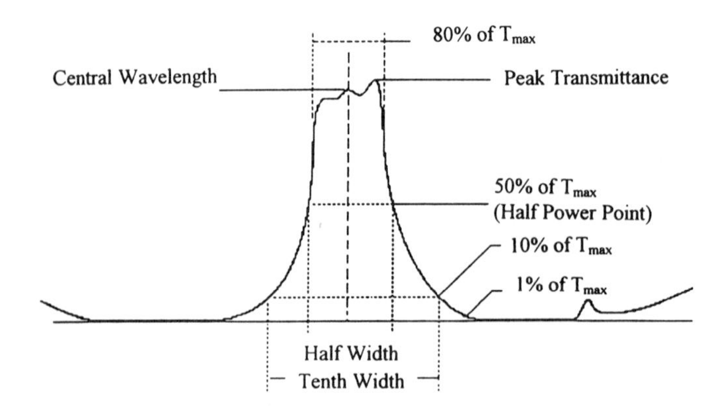

Figure 8. Schematic of spectral definitions

The radiometric calibration concept involves the radiometer, the internal on-board calibration and the transfer of the ground calibration result to the instrument.

4.3 In-orbit radiometric calibration methodology

(a) The on-board calibration concept for the VNIR and SWIR involves:

- Reference source - high stability halogen lamps.
- Calibration approach - the basic approach is the two-point calibration to determine the offset and the reponsivity. High level input is given by the on-board calibrator, while a zero-level input is given by the night side of the Earth.
- Reliability measures - to minimise the influence of degradation and contamination over a long period, each radiometer is equipped with two identical on-board calibration units, and for each unit, two photodetectors are used to monitor the lamp radiance changes. These measures make it possible to cross-check the performance of each other periodically.

(b) The onboard calibration for the TIR involves:

- Reference source - variable temperature blackbody.
- Calibration approach - the basic approach is also the two-point calibration. High level input is provided by the blackbody at temperature 340K, and low level input is provided by the blackbody at temperature 270K. Interpolation and extrapolation have to be used to calculate the temperature scale.
- Reliability measures - single blackbody radiator with multiple thermometer for cross-checking each other.

Ground-reference based calibration for the VNIR and SWIR involves the reflectance-based method or the radiance-based method. For the TIR, the radiance-based method is used to establish the thermal uniformity of the targets. The calibration involves the use of water, land and cloud top targets. Water targets will play an important part in the thermal radiometer calibration, for the following reasons.

- The emissivity of water is known.
- The temperature of large areas is uniform and can be determined with a minimum number of measurements.
- The temperature changes only slowly.

The sunlight reflected from the moon can also be used to provide a check on the responsivity of VNIR and SWIR bands. The moon is an object of 440 pixels on VNIR and 220 pixels on SWIR bands. The lunar radiance as a function of phase angle and libation is being determined by ground-based telescopes; the absolute radiometric accuracy is better than 2%.

4.4 Instruments cross-calibration

The pre-flight cross-calibration is illustrated in Figure 9.The steps are as follows.

1 TIS is characterised by comparing with SBB on SR.

2 RR radiometers record the radiance from TIS.

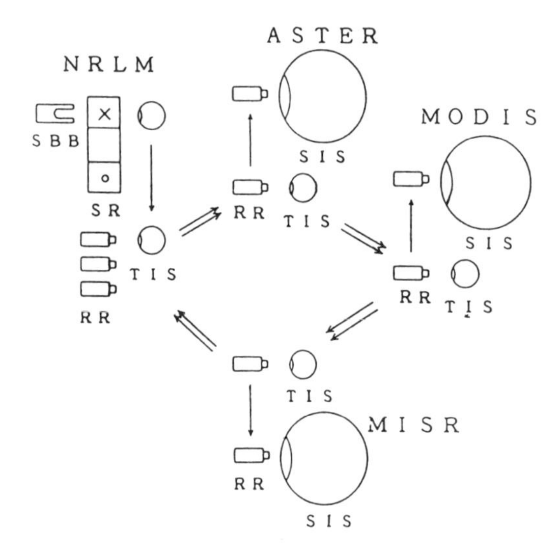

Figure 9. Round-robin cross-calibration of EOS AM-1 spherical integrating sources, after Ono and Sakuma 1996 (RR: Transfer Radiometer, TIS: Transfer Integrating Sphere, SIS: Spherical Integrating Source, SR: Spectral Radiometer, SBB: Standard Blackbody

3 RR radiometer and TIS are transported to the manufacturer of each instrument.

4 the stability of RR is checked against the TIS.

5 RR radiometers record the radiance from SIS.

6 SISs are used for each instrument.

In-orbit cross-calibration involves the following steps.

- Cross-sensor calibration within the platform (ASTER TIR and MODIS).
- Cross-platform calibration among sensors (ASTER with SPOT and Landsat).
- Target-related calibration with an aircraft spectrometer.

5. Conclusion

From the above description, it can be seen that the radiometric calibration of an optical spectrometer consists of two major parts: pre-flight calibration and in-flight calibration. The in-flight calibration consists of on-board calibration, ground-reference calibration and cross-calibration between different spectrometers.

The pre-flight calibration has higher accuracy and its result can be transferred into the in-flight radiometric calibration; it therefore plays very important roles in the calibration.

The on-board calibration and ground-reference radiometric calibration provide means to monitor the sensor's behaviour with relatively high accuracy over both short-term and long-term periods. The most serious limitation to the on-board radiometric calibration is obviously caused by the inaccessibility of calibration scientists. Consequently, the accuracy of on-board calibration is considerably lower than that of the pre-flight calibration and is difficult to verify.

The combination of different on-board calibration results provides the possibility to increase the accuracy and minimise the systematic error. However, multiple calibration inputs introduce the problem of how to combine these results in the most reliable manner, and human knowledge about contamination and degradation in orbit is inadequate to meet the calibration requirements. Research on radiometric calibration is therefore far from finished. Further efforts are needed to satisfy the increasing demand for high quality geophysical information.

Bibliography

Ahmad S P and B L Markham, 1992, Radiometric Calibration of a Polarization-Sensitive Sensor, Journal of Geophysical Research, **97**(D17), 18815 - 18827

Biggar S F, M C Dinguirard, and D I Gellman, 1991, Radiometric Calibration of SPOT 2 HRV - A Comparison of Three Methods, SPIE, **1493**, 155 - 162

Biggar S F, P N Slater, D I Gellman, 1994, Uncertainties in the In-flight Calibration of Sensors with Reference to Measured Ground Sites in the 0.4-1.1μm Range, Remote Sensing of Environment, **48**, 245 - 252

Che N, B G Grant and D E Flittner, 1991, Results of Calibrations of the NOAA_!! AVHRR made by Reference to Calibrated SPOT imagery at White Sands N.M., SPIE, **1493**, 182 - 194

Fraedirch D S, 1991, Methods in Calibration and Error Analysis for Infrared Imaging Radiometers, Optical Engineering, **30**, 1764 -1770

Gao B C, 1993, An Operational Method for Estimationg Signal to Noise Ratios from Data Acquired with Imaging Spectrometers, Remote Sensing of Environment, **43**, 23 - 33

Grosser Z A and J B Collins, 1991, Determination of the Wavelength Positioning Accuracy of a Sequential Scanning ICP Spectrometer, Applied Spectroscopy, **45**, 993 - 998

Guenther B, W Barnes and E Knight, 1996, MODIS Calibration: A Brief Review of the Strategy for the At-Launch Calibration Approach, Journal of Atmospheric and Oceanic Technology, **13**, 274 - 285

Jackson R D and M S Moran, 1987, Field Calibration of Reference Reflectance Panels, Remote Sensing of Environment, **22**, 145 - 158

Jackson R D, P M Teillet and P N Slater, 1990, Bidirectional Measurements of Surface Reflectance for View Angle Corrections of Oblique Imagery, Remote Sensing of Environment, **32**, 189 - 202

Manalo N D and G L Smith, 1991, Spatial Sampling Errors for a Satellite-borne Scanning Radiometer, SPIE, **1493**, 281 - 291

Markham B L, S P Ahmad and R D Jackson, 1991, Radiometric Calibration of an Airborne Multispectral Scanner, SPIE, **1493**, 207 - 214

Moran M S, R D Jackson, G F Hart, P N Slater, 1990, Obtaining Surface Reflectance Factors from Atmospheric and View Angle Corrected SPOT-1 HRV Data, Remote Sensing of Environment, **32**, 203-214

Sakuma F, B C Johnson and S F Biggar, 1996, EOS AM-1 Preflight Radiometric Measurement Comparison Using the Advanced Spaceborne Thermal Emission and Reflection Radiometer(ASTER) Visible/Near-infrared Integrating Sphere, SPIE, **2820**, 184 - 196

Slater P N and S F Biggar, 1996, Suggestions for Radiometric Calibration Coefficient Generation, Journal of Atmospheric and Oceanic Technology, **13**, 376 - 382

Slater P N, S F Biggar and J M Palmer, Unified Approach to Pre- and In-flight Satellite-sensor Absolute Radiometric Calibration, SPIE, **2583**, 130 - 141

Suzuki N, U Narimatsu, F Sakuma and A Ono, 1991, Large Integrating Sphere of Prelaunch Calibration System For Japanese Earth Resources Satellite Optical Sensors, SPIE, **1493**, 48 - 57

References

Healey G E, and R Kondepudy, 1994, Radiometric CCD Camera Calibration and Noise Estimation, IEEE Transactions on Pattern Analysis and Machine Intelligence, **16**, 267 - 276

Ono A and F Sakuma, April 1996, Preflight and In-Flight Calibration Plan for ASTER, Journal of Atmosperic and Oceanic Technology, **13**, 321 - 335

Sakuma F and A Ono, 1991, Prelaunch Calibration System for Optical Sensors of Japanese Earth Resources Satellite, SPIE, **1493**, pp 37 - 47

Sakuma F and A Ono, 1993, Radiometric Calibration of the EOS ASTER Instrument, Metrologia, **30**, 231 - 241

Slater P N, S F Biggar and R G Holm, 1987, Reflectance- and Radiance-Based Methods for the In-Flight Absolute Calibration of Multispectral Sensors, Remote Sensing of Environment, **22**, 11 - 37

Slater P N, S F Biggar and K J Thome, 1996, Vicarious Radiometric Calibrations of EOS Sensors, Journal of Atmospheric and Oceanic Technology, **13**, 349 - 359

Sun X and J M Anderson, 1993, A Spatially Variable Light-Frequency-Selective Component-Based, Airborne Pushbroom Imaging Spectrometer for the Water Environment, Photogrammetric Engineering & Remote Sensing, **59**, 399 - 406

Thorne K, B Markham, J Baker, P Slater and S Biggar, 1997, Radiometric Calibration of Landsat, Photogrammetric Engineering & Remote Sensing, **63**, 853 -858

Developments in radar and satellite rainfall measurement

Mohan L Nirala and Arthur Cracknell

University of Dundee

1 Introduction

This chapter serves as a brief review of some of the major steps in the monitoring of rainfall from space-based systems and from ground-based radar during the past few decades and prospects for the future. It considers the development of satellite and radar rainfall estimation, active and passive microwave instruments, past and present sensors and the Precipitation Intercomparison Projects/Programmes with their aims/objectives, areas of analysis and period of the studies.

Precipitation is an important environmental parameter which affects the hydrology of land surfaces, coastal processes, climate and global heat circulation. The understanding of rainfall distribution and its intensity can improve the protection of the environment and the knowledge of geophysical processes of land, ocean and atmosphere.

Precipitation measurements are typically made at ground-based stations using rain gauges which provide accurate estimates of precipitation at a point over a fixed area. However, the local conditions, winds, topography etc. affect the reliability of the estimates of the spatial and temporal patterns of precipitation.

Quantitative estimation of rainfall can be derived from remote sensing observations involving electromagnetic radiation emitted by or reflected from precipitating particles or implied from precursors of precipitation processes such as clouds. In principle, such measurements can be active, in which the observed features are actively illuminated by a rain radar and the reflected energy measured, or they can be passive in which the natural radiative or reflective properties of the precipitating particles are observed.

Indirect methods of estimating precipitation from satellite observations are typically based on observation of cloudiness. While many algorithms have been developed over the past 30 years, few have offered the temporal continuity and spatial coverage necessary for routine evaluation of the space mean precipitation (Arkin and Ardanuy 1989). Virtually all indirect estimation techniques use visible and/or infrared (IR) window observations of the brightness or temperature of the tops of the clouds.

Direct precipitation estimation algorithms, based on observations of precipitating particles, for instance by ground-based radars, have an evident advantage over indirect methods in that they are more physically related to the intensity of rainfall at the surface and less subject to systematic errors due to inhomogeneities in the cloud-rainfall relationship. Active measurements have yet to be made from satellite, but the Tropical Rainfall Measurement Mission (TRMM), which is a joint United States - Japan venture, will include a rain radar. But passive measurements from space, however, have been available for about ten years. The Special Sensor Microwave/Imager (SSM/I) passive multichannel microwave radiometer has been operating on Defense Meteorological Satellite Program (DMSP) satellites since June 1987. This instrument observes microwave radiation at four frequencies (19, 22, 37 and 85.5GHz) with vertically and horizontally polarized radiation observed separately at all but 22GHz. Observations over a given site are made twice daily, with spatial resolutions ranging from roughly 15km for 85.5GHz to 50km for 19GHz over a swath 1400km in width. The swaths are not contiguous over the tropics and much of the mid-latitudes, but the orbit of the satellite preccsses at a rate sufficient to give complete global coverage (with several observations at each point) in about 6 days. The SSM/I series is expected to continue in operation; two instruments are planned to be continuously in orbit within a few years.

2 Satellite precipitation estimation techniques

2.1 An overview

The atmosphere has several nearly transparent windows in the microwave spectrum that enable a microwave radiometer to observe the scattering and absorption of radiation by hydrometeors whose dimensions are comparable to those of the wavelength of radiation. The non-precipitating cloud droplets only have a minor effect on the transmission of such radiation, so that microwave radiometers respond mainly to the effect of precipitation rather than to the surrounding non-precipitation cloud field. A more detailed discussion of absorption by various constituents of the atmosphere is given by Meneghini and Kozu (1990).

The measured microwave energy is mainly emitted as blackbody radiation from the Earth's surface and by the hydrometeors. The measured microwave radiance is commonly presented in units of brightness temperature of an equivalent blackbody, T_B. The temperature of the Earth's surface and the temperature profile of the atmosphere are needed to derive those radiances. Water vapour also absorbs a small amount of radiation in windows, and its vertical distribution must be considered in any radiative transfer model. The microwave radiances emerging from a precipitating cloud depends most significantly upon the amount and the vertical distribution of the precipitating liquid and ice hydrometeors. Raindrops mainly interact with microwave radiation by absorption, although scattering assumes increasing significance as the frequency increases or as the rainfall rate increases, thereby increasing the population of large drops. As the raindrop density increases, the radiance emerging from liquid water drops increases significantly over the ocean, whereas little change occurs over land. Some reduction in the measured radiance occurs in heavy rainfall cases because the drops become larger and they scatter more radiation back to the surface. Ice hydrometeors exist near the cloud tops where they reflect radiation from the lower parts of the cloud back to the surface. The radiance is thus reduced as the ice hydrometeors concentration increases both over land and over water. The size of the hydrometeors is conveniently represented by the Marshall-Palmer size distribution that is characterised by single parameter, the "rainfall rate" R (Meneghini and Kozu, 1990).

Once radiances measured as brightness temperature (T_B) at various frequencies and at two orthogonal polarisation's can be defined on a grid with uniform resolution, the rainfall rate distribution may be retrieved from those brightness temperatures. The vertical distribution of precipitation cannot simply be characterised in terms of one variable such as the surface rainfall rate; the vertical structure must be allowed to vary in the retrieval. The surface rainfall rate, cloud liquid water, temperature and fraction of the footprint covered by rain were varied randomly to generate a large set of possible cloud models.

2.2 VIS and IR techniques

The network of geostationary meteorological satellites, including GOES, Meteosat, *etc.*, which provide half-hourly coverage of that part of the Earth viewed, is the primary source of precipitation estimates. For those areas where geostationary satellite data are not available, however, polar-orbiter data are used. Other data sources used in estimating precipitation include NMC (National Meteorological Centre) analyses and forecasts, conventional atmospheric soundings, radar data, satellite-derived soundings, and rainfall climatology. Satellite-derived precipitation estimation techniques ranging in uses from the mesoscale to the global scale are discussed by Barrett and Martin (1981).

There are two basic types of VIS and IR precipitation estimation techniques: cloud history and indexing. VIS and IR methodologies infer precipitation amounts for specific cloud features. Cloud history works best where geostationary satellite data are available. The frequent views from geostationary satellites allow the life cycle to be followed and precipitation estimates to be computed for each stage of the cloud development. In contrast, cloud indexing is the principal method used when only polar-orbiter satellite data are available; only two pictures a day can be obtained from a single polar-orbiting satellite. Cloud indexing involves characterising a cloud by an index number according to its appearance in imagery and then using a look-up table or regression equation to estimate the precipitation from the cloud. Both the cloud history and cloud indexing methods have procedures for modifying the estimates for different climates and environments. Estimates of rainfall are a function of the following.

- Dominant organisation of the synoptic weather.
- Proportion of sky that is cloud covered.
- Intensity of rain, which is influenced by the types of clouds present, cloud top temperature/cloud growth, overshooting tops, mergers, etc.

2.3 Passive microwave techniques

The primary advantage of microwave measurements is their ability to probe through clouds, rain being the major source of attenuation for the window frequencies below 50GHz. Furthermore, over low-emissivity sea surfaces the brightness temperature measurements in clearer areas are high-lighted against the more emissive, warmer measurements in precipitating regions. The large contrast (>50°K) in brightness temperature between the rain and its surroundings has stimulated much interest in applying microwave radiometry over oceans for determining rainfall. Satellite and aircraft instruments have been used to obtain rain data for tropical storms (Jones *et al.* 1981). The global distribution of rainfall over the oceans was also attempted from satellite measurements (Rao *et al.* 1976).

2.4 Single frequency technique

In most situations it is sufficient to relate rain intensity directly to increase in brightness temperature over oceans. Effects such as those of sea surface winds on emissivity or the contributions due to cloud and water vapour absorption are of second order and can usually be neglected.

The greater variation in the emissivity of land presents problems because of the similarity in brightness temperature of rain and land. Also, decreases in surface emissivity due to wet ground and snow cover can produce false precipitation signatures if not accounted for (Rodgers and Sidalingaiah 1983). Only in the case of heavy thunderstorms can the rain measurements be much lower (>50°K) than the surrounding brightness temperatures. The enhancement for the case of storms is principally due to the scattering of upwelling radiation by large ice particles at the tops of rain layers. This scattering effect increases with increasing frequency, and was first noted during observation of convective storms over the United States, using scanning multi-channel microwave radiometers (SMMRs) (Spencer *et al.* 1983). Brightness temperature as low as 163°K were obtained at the centre of an intense storm for the highest frequency (37GHz) channel.

In the case of stratiform rain the contribution due to scattering is small. To identify rain one must use the smaller difference between the thermal emission from precipitation drops and that from the land background. Dual frequency and/or polarisation techniques are being developed to enhance the precipitation signature over land by minimising the effect of surface emissivity on the microwave measurements. These approaches use statistical correlations between emissivities at different frequencies or between the two polarisations.

2.5 Frequency screening

For many surfaces the emissivity has very little dependence on frequency increase. This is true for wet and dry soil, vegetation, open water, melting snow, lake, ice, and new sea ice. By contrast, the brightness temperatures of rain cells decrease with increasing frequency i.e. the temperature due to scattering and thermal emission. It is therefore possible to isolate the precipitation signature by measuring the differences between the brightness temperatures at two frequencies (i.e. ν_1 and ν_2): $T_B(\nu_1) - T_B(\nu_2)$, where $\nu_2 > \nu_1$ and displaying only the possible values. However, surfaces such as dry snow have a frequency response similar to that of rain and would not be screened out. This problem can be alleviated by viewing sequential measurements and associating the rapid changes with precipitation. Also, the brightness temperature sum at two frequencies $T_B(\nu_1) + T_B(\nu_2)$ can be used to make the final discrimination (i.e. stratiform rain has higher brightness temperatures than dry snow).

2.6 Polarisation algorithm

Frequency screening uses the correlation between emissivities at different frequencies to enhance the precipitation signature. Alternatively, one can employ polarisation information to minimise the effect of emissivity components which are bound by the curves representing a specular and diffuse surface. To minimise the brightness temperature variations due to surface variations, it is sufficient to combine the vertical T_V and horizontal T_H polarisation measurements according to the linear transformation $(T_V - BT_H)/(1 - B)$. The parameter B physically represents the linear slope of vertically polarized to horizontal polarized emissivity and can be estimated from plots of T_V against T_H for land and water surfaces (Weinman and

Guetter 1977). It is found that B has a value near 0.52 at 37GHz. By using the radiative transfer equation, it can be shown that the transformed measurements approximately represent the brightness temperature of a unity emissivity surface (Grody 1984). As such the weighted measurements show a large contrast between the atmospheric emission (and scattering) due to precipitation and the higher background emission.

3. Precipitation estimation using radar

Probert-Jones presented the radar equation for precipitation in 1962. His results applied strictly to a region of homogeneous reflectivity which completely fills the beam. However, the beam may not be homogeneously filled and the precipitation reflectivity is characterised by both vertical and horizontal gradients, which are convolved with the antenna radiation pattern. These effects increase with range as the beam rises above the Earth and samples progressively larger volumes. The problems of inferring precipitation rate at the surface from such distorted reflectivity measurements is further exacerbated by the many factors which alter the particle size distribution between the radar pulse volume and the surface near the gauge. Such effects include growth, evaporation, melting, coalescence and break-up, and wind shear sorting. It is no surprise then that the radar measurement of rainfall has been marked by such great difficulties for so many years.

Severe discrepancies between the observed reflectivities and those that would be observed by an ideal radar with an infinitely narrow beam often occur due to beam spreading and partial beam filling. These distortions increase with beamwidth and reflectivity gradients. The beam spreading smears the real radar reflectivitity Z field, and thus lowers the peak reflectivities. Rogers (1971) calculated that a negative bias of up to 7dB occurs in regions of large reflectivity gradients such as 20dB.km^{-1}, due to the effects of signal averaging. At the same time, beam spreading and side lobes expand the precipitation echoes beyond the actual precipitation areas. Austin (1987) suggested that these added echo areas may cause radar overestimates of the true rainfall in situations with large reflectivity gradients. Others have shown quantitatively that partial beam filling, caused by large horizontal reflectivity gradients (also about 20dB.km^{-1}), can result in gross overestimates of convective rainfall.

The net result is that it is virtually impossible to relate measured or effective radar reflectivity Z_{eff} to rain rate R in a deterministic manner unless the radar pulse volume is small and very close to the rain gauge. Good agreement between gauge measurements of drop size distribution (DSD) and calculated rain rate and reflectivities at a height of 200m above the gauge, as measured by a vertically pointing radar, were found (Joss and Waldvogel 1967). However, at larger distances only a small fraction of the variability between radar and gauge measurements can be explained by changes in the DSD.

Over the past two decades Professor Robert Houze and his group at the University of Washington have developed an algorithm that distinguishes between convective and stratiform precipitation (Steiner *et al.* 1995). Radar reflectivity patterns associated with active convection form well-defined cores with vertical extents of several kilometres or more, often extending well above the level of the 176°K isotherm. Convective cells are identified by a combination of the local peakedness and intensity of the radar reflectivity field at an altitude of 2km above ground level within the 2km x 2km gridded reflectivity field. For each grid point identified as a convective centre, all surrounding grid points within an intensity-dependent radius around the centre are also classified as convective. Grid points not classified as convective are classified as stratiform. This algorithm provides an overlay mask for the radar-derived rain fields to classify the areas of convective or stratiform rain and consequently their respective contributions to latent heating. The University of Washington algorithm also includes a program to produce contoured frequency (of reflectivity) by altitude diagrams.

Two approaches for rainfall estimation are being evaluated. One is being developed by Professor Daniel Rosenfield and his group at the Hebrew University in Jerusalem, Israel (Rosenfield *et al.* 1994). This approach employs the Probability Matching Method, which uses probability distributions of radar reflectivity Z and rain gauge G rain rates to determine the appropriate conversion of observed Z values into rain rates. Once a suitable database of Z and G measurements has been obtained, (about 200mm of rainfall for all gauges for an associated rain type) an objective classification scheme is used to determine the rain type and hence the appropriate $Z_{eff} - R$ relation to use. The objective classification scheme uses 3-dimensional windows to determine the rain type via parameterisations of the radial reflectivity gradient, cloud depth and vertical reflectivity profiles.

Another group, headed by Professor Witold Krajewski at the University of Iowa, has developed a different method of rainfall estimation. In their approach, the radar reflectivity is converted into liquid water content in atmospheric columns in polar coordinates (Krajewski and Smith, 1991). The height of the radar echoes taken into consideration is limited to the level of the 176°K isotherm appropriate for a given situation. The liquid water content estimates are vertically averaged and converted to rainfall accumulation on the ground using a power-law transformation and an assumed accumulation interval.

4 The global precipitation climatology project

Both the success of the indirect estimation techniques and the prospect for continuing direct estimates from operational microwave radiometers led the World Climate Research Programme (WCRP) to institute a Global Precipitation Climatology Project (GPCP) to provide global rainfall analysis for the various WCRP research programmes (World Climate Research Programme, 1986). Over the past few years a number of intercomparison projects have been sponsored by the Global Precipitation Climatology Project i.e. Algorithm Intercomparison Programme (AIP-1/-2/-3). One expression of this has been the selection and implementation of a simple geostationary based infrared technique for quasi-global rainfall monitoring on a monthly basis for 2.5° latitude/longitude square, namely the GOES Precipitation Climatology Centre in Offenbach, Germany, to develop the best possible rainfall climate data bases from the gauge records. A third important activity of the GPCP has been the inauguration of an Algorithm Intercomparison Programme to evaluate alternative (satellite rainfall) estimation techniques, and determine if improvement can be made upon the present algorithms and to evaluate new precipitation estimation methods.

4.1 Algorithm intercomparison programme

The Algorithm Intercomparison Programme series of projects (AIP-1, AIP-2 and AIP-3) was planned to evaluate the precipitation merits of the techniques and their improvement. Each project focused on a small region and limited time-frame, with satellite data (visible, infrared and passive microwave data) and surface validation data.

AIP-1 was conducted over Japan and the surrounding ocean for two periods in 1989, *viz*. 1–30 June and 15 July – 15 August, based on GMS visible and infrared imagery, SSM/I imagery, radar and rain gauge data, and numerical forecast outputs from the Japan Meteorological Agency, incorporating both frontal and convective precipitation regimes, typical of mid-latitude summertime precipitation. Results have been presented in an "Atlas of Products" (Lee *et al.* 1991).

AIP-2 was conducted over Western Europe during February–April 1991, investigating winter/spring precipitation over land areas. This area proved particularly challenging due to the known discrepancies between cloud-top temperatures or microwave radiances and the rainfall intensities. For the GPCP-AIP-2 dataset, a statistical analysis of the NOAA AVHRR data in five channels, DMSP-SSM/I dataset at 19, 22, 37GHz (The 85GHz channel data in both polarisations were not available because of a problem with the instrument), Meteosat data on an hourly basis from the three channel (infrared (10.5–12.5μm), visible (0.5–0.9μm) and water vapour (5.7–7.1μm)) were collected for the period 1 February to 9 April 1991, covering the area between 10° W and 10° E at 42° N and between 14° W and 14° E at 55° N shown in Figure 1.

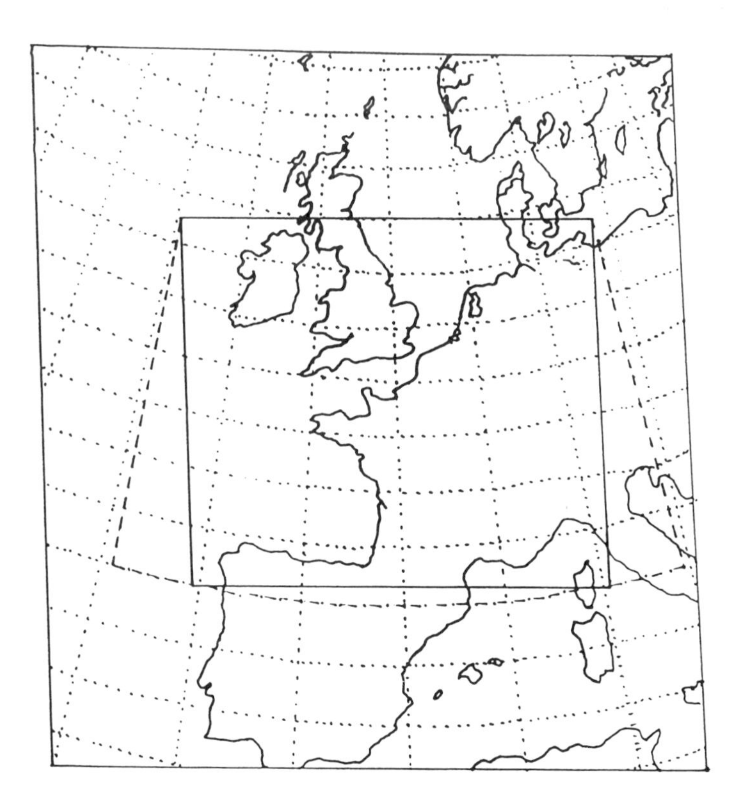

Figure 1 The GPCP-AIP/2 area is bounded by the solid line. The dashed line contains the area analysed (after Liberti 1992).

Statistical analysis was carried out by the Meteorological Office in the United Kingdom (Liberti 1992) on the data set in order to :

1. test the quality of the data,
2. study the statistical properties of the data,
3. select interesting cases from the point of view of estimating precipitation.

AIP-3 was conducted over the TOGA-COARE (Tropical Ocean-Global Atmospheric Coupled Ocean-Atmosphere Response Experiment) region of the Western Equatorial Pacific, covering the period from November 1992 to February 1993. This area was oceanic in nature, and employed two ships with Doppler radars for three periods, each 20 days long, to help validate satellite estimates for a mainly oceanic area more poorly provided with rain gauges than either of the two areas investigated in the previous two AIP studies.

5. Precipitation intercomparison project

The Precipitation Intercomparison Project was devised as part of the NASA WetNet project, aimed at the improvement of our understanding of, and retrieval of, hydrological parameters using passive microwave data. The main instrument to be used for this project was the DMSP-SSM/I series of imagers. Three intercomparison projects were devised to investigate the relative merits of passive microwave rainfall estimation techniques.

PIP-1: The first Precipitation Intercomparison Project would be distinctive, specifically in that it would be the first such project to adopt a global viewpoint. It would have the following principal objectives. It was organised by the University of Bristol UK (Barrett *et al.* 1994) and investigated global precipitation retrieval for the period from August to November 1987. Participants were asked to retrieve estimates of monthly global precipitation at a resolution of 2.5° by 2.5°, with an optional resolution of 0.5° by 0.5°. In addition, instantaneous rain rates were sought for latitudinal and longitudinal cross-sections. The four primary objectives of PIP-1 were as follows.

1. To demonstrate, test and intercompare and, where possible, validate techniques for global monthly rainfall monitoring by SSM/I.
2. To "validate" SSM/I rainfall monitoring techniques over land, ocean and other surfaces (e.g. ice, snow) through comparisons with other datasets.
3. To gain insights into rainfall algorithms and radiative processes through studies of their differences in terms of the specific assumptions, strengths and weaknesses of each.
4. To confirm to the broader global weather and climate modelling communities the relevance of the WetNet inter-disciplinary and multi-laboratory approach to data distribution, analysis, interpretation and application.

PIP-2: This study was coordinated by Florida State University (Smith 1997). This study investigated instantaneous rainfall retrievals for 28 (mostly multi-overpass) case studies, based upon contributors' area(s) of interest and the availability of surface validation data. Special areas of interest were the rain/no rain delineation and satellite radiance to rain rate conversions. However, rather than calculate monthly rainfall products, PIP-2 intercompared instantaneous rainfall products for about 120 cases ranging from hurricanes to mid-latitude cold season depressions. The importance of validation data was stressed in this project.

PIP-3: The third Precipitation Intercomparison Project followed the success of the previous efforts, but put an increased emphasis on the evaluation of the satellite-based global monthly precipitation fields that are already available through various sources, in addition to the products generated by algorithm developers. The period of study was the whole of 1992, with the addition of January and July of 1991 and 1993 (to study inter-annual variation), and August 1987 (to allow comparison with the PIP-1 results).

The main objective of PIP-3 was to determine the utility of the current quasi-standard global/monthly precipitation products to the climate modelling/diagnostic community and the potential improvement expected with the latest satellite algorithms. PIP-3 was designed to produce an evaluation of the current products and facilitate the exchange of information on future directions.

SSM/I antenna temperature data from the F-8 satellite for August 1987, from the F-10 and F-11 satellites for all of 1992, and from the F-10 satellite for January and July 1991 and 1993 were provided to the participants.

6. The tropical rainfall monitoring mission (TRMM)

This is a joint venture between NASA and NASDA (Japan) (Simpson *et al.* 1988). The suite of sensors is designed to measure tropical precipitation between 35°N and 35°S. The satellite will be launched by NASDA in November 1997, with a designed lifetime of three years. Sensors on board the satellite include: Precipitation Radar (PR), TRMM Microwave Imager (TMI), Visible and Infrared Scanner (VIRS), Lightning Imaging Sensor (LIS), Cloud and Earth's Radiant Energy System (CERES). The TRMM orbit will be circular, non-Sun-synchronous at an altitude of 350km and with an inclination of 35° to the equator.

The TRMM Microwave Imager (TMI) uses the heritage of the Special Sensor Microwave Imager (SSM/I), and the Visible/Infrared Sensor (VIRS) uses five channels in common with operational visible/ infrared radiometers currently in use. In addition to these two facility instruments provided by NASA for the rain package, a Precipitation Radar has been developed and supplied by NASDA, in collaboration with the Communications Research Laboratory of Japan. The Precipitation Radar and the VIRS operate in a cross-track scanning mode, with swath widths of approximately 220km and 700km respectively. The TMI, whose channels are dual polarised, has a conical scanning mode with an angle of incidence at the surface of 52°.

The TRMM goal is to measure tropical rainfall and its variations much more accurately than before, particularly over the tropical oceans. A major objective is to obtain a 3-year data set of monthly estimated rainfall rates averaged over grid boxes of about 5° by 5° on a side, with a sampling error of about 10% over oceans and somewhat more over land. Such data are essential to achieve progress in our knowledge of the overall Earth system and its hydrological cycle. Evaporation of water vapour into the atmosphere and its condensation to produce rainfall are at the heart of the Earth's habitability for man and other species. In the atmosphere, the release of latent heat by precipitation is the primary source of the energy that drives the wind circulations. In the ocean, rainfall is a source of fresh water which influences the density of water (buoyancy) and surface fluxes in a coupled ocean-atmosphere system. Over land, rainfall is the source of soil moisture, which has a strong influence on atmospheric circulation either by evaporation or by evapo-transpiration from plants.

The observational results from the 1992–1993 field programme TOGA–COARE have already helped greatly with the testing of models and algorithms relevant to TRMM. Instruments similar to all three of the TRMM rain package instruments were flown on NASA aircraft and calibrated C-band Doppler radars were aboard two research ships. This data set will continue to be useful to TRMM and other Earth Science programs for the foreseeable future. The main use for the TRMM products are climate diagnostic studies, better understanding of rain processes and their improved formulation in Global Circulation Models (GCMs) including those linking atmosphere and ocean, and improvements to Climate Models. Other important objectives include model impact studies leading to improved long-range weather and climate prediction, better knowledge of the global hydrological cycle as required by GEWEX and improved assessment and understanding of global change.

It is expected that the sea experiment will be complemented with oceanographic measurements while hydrological measurements will likely complement the land field campaign. The main focus of these campaigns will be to (a) validate the TRMM observations using aircraft instrumentation, (b) validate the rainfall algorithms and quantify the error models, (c) validate convective/stratiform separation by means of kinematic and microphysical measurements, (d) validate latent heating retrievals by more direct measurements of vertical gradients of heating and (e) provide heat and moisture budget constraints on precipitation estimates.

At least four types of modelling research are relevant to TRMM, namely: (1) the use of models to retrieve rain-related information from TRMM, (2) the use of TRMM data to initialise and test forecast models from mesoscale to global, (3) the use of multiple nested models for studies of large-scale impacts of rain processes and improved parameterisation schemes and (4) the use of statistical models for sampling and other research. So far, the only means of latent heat retrievals is by algorithms derived from a cloud-resolving model; proposals of other ways to obtain and test the latent heat release over a range of space and time scales are welcome, as are research plans to analyse various types of precipitating cloud systems in the tropics and subtropics. The component parts of the TRMM are described below:

1. **Visible Infrared Scanner (VIRS)** is a 5-channel, cross-track scanning radiometer operating at 0.63, 1.6, 3.75, 10.8 and 12μm, which will provide information on the integrated column precipitation content, cloud liquid water, cloud ice, rain intensity and rainfall types (e.g. stratiform or convective).

2. **Precipitation Radar (PR)** is the first of its kind in space. It is an electrically scanning radar operating at 13.8GHz, and it will measure the 3-dimensional rainfall distribution over both land and ocean and define the layer depth of the precipitation.

3. **Lightning Imaging Sensor (LIS)** is an optical sensor operating at 0.7774μm which will observe the distribution and variability of lightning over Earth.

4. **Clouds and Earth's Radiant Energy System (CERES)** will measure emitted and reflected radiative energy in 0.3–50μm from Earth surface and atmosphere.

The table below lists the advantages and limitations of the microwave radiometer, radar and the visible infrared scanner and how each contributes to obtain the best possible data retrieval. It is envisioned that all the three types of instruments will be necessary in order to derive the optimum retrieval algorithm.

	Advantages	Limitations
Microwave Radiometers (6-9GHz)	Quantitative measure of rain Wide swath	Less quantitative over land for low rainfall Moderate spatial resolution
Radar (14GHz)	Quantitative measure of rain Better spatial resolution Vertical profile of rain Can provide layer thickness Works well over land	Narrow swath Largely untested in space
VIS / IR Radiometer	Best spatial resolution Distinguish between convective and stratiform precipitation Transfer standard to geosynchronous and to polar orbiters	Less quantitative measure Obscuration by cirrus shields

7. ATMOS-A1

This is another cooperative effort of NASA and NASDA (Japan) to develop precipitation measurement between 55° latitude north and south for launch after 2000. This mission will be conducted within the framework provided by the NASDA ATMOS-A1 satellite. This initiative is in response to pressure from several international committees that have addressed the needs of the Global Energy and Water Cycle Experiment, GEWEX, to improve our understanding of the Earth's climate and possible changes of the climate.

The main new technical development proposed for ATMOS-A1 is a dual frequency precipitation profile radar, designed as a precipitation radar-2 (PR2) operating at 14 and 35GHz to reveal the vertical structure of heavy rain, light rain and snow. The PR2 will be the first dual-frequency precipitation radar flown in space, utilising two frequencies to improve precipitation measurements (see, for example, Fujita 1983, Koju and Nakamura 1991). The PR2 is the centrepiece instrument of the mission because it provides information on the horizontal and vertical structure of precipitation over both ocean and land. This is especially significant because passive microwave radiometers are able to measure only the most intense rainfall events on land. Taken in conjunction with the ATMOS-A1 Microwave Imager, AMI, the PR2 provides information regarding the behaviour of frozen hydrometeors in maritime storms. The PR2 thus provides the unique information that is needed to yield heating rate profiles required by various atmospheric models.

Because fewer and less intense snow storms will be encountered in mid-latitudes than in the tropics, it becomes necessary to enhance the radar sensitivity to low reflectivities found in light rain and snow. Such measurements are vital to ensure that suitable precipitation statistics will be obtained at mid-latitudes. These considerations not only require the PR2 to consist of a 14GHz component, as was done on TRMM, but to also have an additional 35GHz component. The 35GHz radar component of PR2 is expected to operate in a nadir staring mode. It is hoped that the sensitivity of this radar can be maximised by employing whatever antenna is compatible with the constraints imposed by the spacecraft to achieve that sensitivity.

The objectives of the GEWEX programme to which ATMOS-A1 can contribute are (1) to model the global hydrological cycle and its impact on the atmosphere and oceans, (2) to develop the ability to predict variations of global and regional hydrological processes and water resources and their response to the environmental changes and (3) to foster the development of observing techniques and data management and assimilation systems. Figure 2 shows the additional latitudinal coverage provided by ATMOS-A1 compared to TRMM. ATMOS-A1 will provide measurements of the mid-latitude peaks in precipitation .

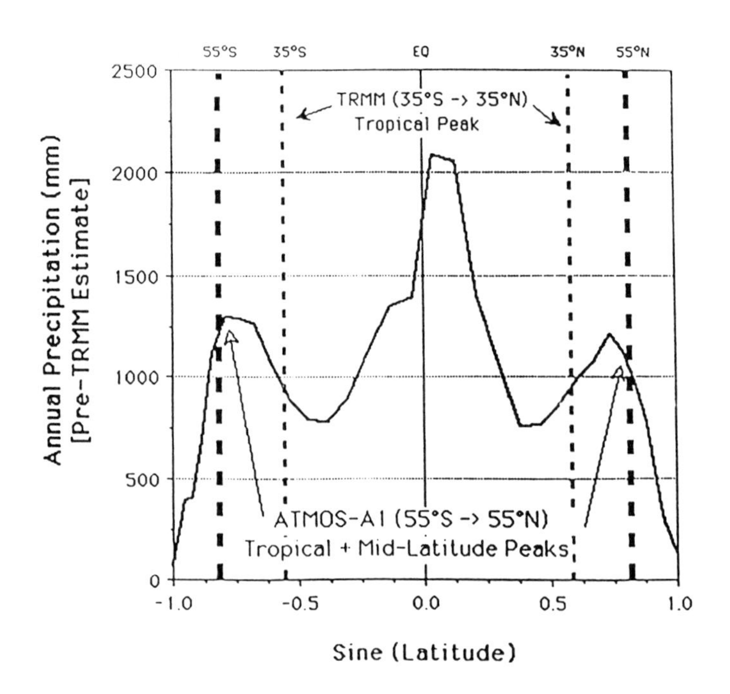

Figure 2 Comparison of the pre-TRMM estimates of the zonal precipitation in swath observed from TRMM and ATMOS-A1. The distribution (Jaegar 1967) is presented as a function of the sine of the latitude angle to weight the results in proportion to the zonal area (TRMM GSFC NASA).

Beside the PR2, the instrument ensemble should include an ATMOS-A1 Microwave Imager (AMI) radiometer with frequencies from 10 to 86 or 150GHz to observe rain, snow and the vertically integrated water vapour amount. In addition, the PR2 and AMI on ATMOS-A1 will provide the most extensive measurement of snowfall rates ever obtained from spacecraft, enabling relations between microphysical processes in clouds and mesoscale meteorological phenomena to be established.

8. Conclusion

Results of these intercomparisons indicated that for monthly rainfall totals infrared techniques do well in the tropics. This is primarily due to the clear delineation of rain-areas with an associated infrared cloud-top temperature and the better sampling rate obtainable from geostationary satellites. In mid latitudes passive microwave techniques proved to be as good as the infrared despite their poorer sampling rate. For instantaneous estimations, the passive microwave techniques perform significantly better due to the direct nature of the radiance to rain-rate conversions. The importance of the validation data were discussed in all the intercomparisons. The combination of visible and infrared techniques is stressed. This unites the benefit of two techniques with more direct detection of rainfall and better temporal resolution.

The understanding of the processes involved in microwave radiometry has improved significantly over the past few years. Spaceborne radiometric measurements have been used to map precipitation on a global scale. Unfortunately, light stratiform rain cannot be adequately measured over land by the currently available radiometer systems. However, this limitation will be overcome by future spaceborne radar measurements.

It is expected from TRMM (Simpson *et al.* 1988) (1) to calibrate the visible infrared radar estimation techniques that are presently in use from geosynchronous meteorological satellites, (2) to measure oceanic precipitation directly to validate models of the oceanic thermohaline circulations, (3) to establish biomass-precipitation relations through diagnostic studies and examine the droughts and anomalous rainfall effect on the biomass and (4) to improve understanding of climate variations and improve the models in order that advances can be made in the prediction of climate change.

References

Arkin P A and Ardanuay P E, 1989, Estimating climate scale precipitation from space: A review, Journal of Climate, **3**, 1229-1238

Austin P, 1987, Relationship between measured radar reflectivity and surface rainfall, Monthly Weather Review, **115**, 1053-1070

Barrett E C and Martin D W, 1981, The use of satellite data in Rainfall Monitoring, Academic Press, New York,

Barrett E C, Adler R F, Arpe K, Bauer P, Berg W, Chang A, Ferrado R, Ferriday J, Goodman S, Hong J, Janowiak J, Kidd C, Kniveton D, Morrissey M, Olson W, Peety G, Rudoff B, Shibata A, Smith E and Spencer R, 1994, The first WetNet Precipitation Intercomparison Project (PIP-1): Interpretation of results, Remote sensing reviews, **11**, 303-373

Fujita M, 1983, An algorithm for estimating rainfall rate by dual frequency radar, Radio Science, **18,** 697-708

Grody N C, 1984, Precipitation monitoring over land from satellite by microwave radiometry, Proceedings, International Geoscience and Remote Sensing Symposium (IGARSS 1984), Strasbourg, France, 27-30 August, European Space Agency SP-215, 417-423

Jaeger L, 1967, Manatskarten des Niederschlags fur die Ganze Erde Berichte des Deutschen Wetterdienstes, Germany

Jones W, Black P G, Delnore V E and Swift C T, 1981, Airborne microwave remote sensing measurements of Hurricane Allen, Science, **214**, 274-280

Joss J and Waldvogel A, 1967: Ein spektrograph fur Niederschlagostrophen mit automatisher Auswertung, Pur Applied Geophysics, **68**, 240-246

Krajewski W F, and Smith J A, 1991, On the estimation of climatological Z-R relationships, Journal of Applied Meteorology, **30**, 1436-1445

Koju T, Nakamura K, 1991, Rainfall parameter estimation from dual radar measurements combining reflectivity profile and path-integrated attenuation. Journal of Atmosphere Oceanic Technology, **8**, 259-270

Lee T H, Janowiak J E and Arkin P A, 1991, Atlas of products from the Algorithms Intercomparison Project 1: Japan and Surrounding Oceanic Regions, June-August 1989, University Corporation for Atmosphere Research, Washington D C

Liberti G L, 1992, Short range forecasting reseach, Global Precipitation Climatology Project, Algorithm Intercomparison Project, Meteorological Office, United Kingdom, Report 1-7, pp 1

Meneghini R and Kozu T, 1990, Spaceborne Weather Radar, Artech House, Boston, Massachusetts, pp 197

Probert-Jones J R, 1962, The radar equation in meteorology, Quart. Journal of Royal Meteorological Society, **8**, 485-495

Rao M S V, Abbott W V and Theon J S, 1976, Satellite-derived global oceanic rainfall atlas (1973 and 1974), NASA SP-410, Washington D.C. (NTIS #N77-19709)

Rogers R R, 1971, The effect of variable target reflectivity on weather radar measurements, Quarterly Journal Royal Meteorological Society, **97**, 154-167

Rodgers E and Siddalingaiah H, 1983, The utilization of Nimbus-7 SMMR measurements over land, Journal of Climate and Applied Meteorology, **22**, 1753-1763

Rosenfield D, Wolff D B and Amitai E, 1994, The window probability matching method for rainfall measurements with radar, Journal of Applied Meteorology, **33**, 682-693.

Simpson J, Adler R F and North G R, 1988, A Proposed Tropical Rainfall Measuring Mission (TRMM) Satellite, Bulletin of American Meteorological Society, **69**, 278 - 295

Smith E, 1997, Results of the Second Precipitation Intercomparison Project (PIP-2), accepted by Journal of Atmospheric Sciences

Spencer R W, Olson W S, Rongzhang W, Martin D, Weinman J A and Santek D A, 1983, Heavy thunderstorm observed over and by the Nimbus-7 scanning multichannel microwave radiometers, Journal of Climate and Applied Meteorology, **22**, 1041-1046

Steiner M, Houze R A and Yuter S E, 1995, Climatological characterization of three-dimensional storm structure from operational radar and rain gauge data, Journal of Applied Meteorology, **34,** in press

Weinman J A and Guetter P J, 1977, Determination of rainfall distributions from microwave radiation measured by the Nimbus-6 ESMR, Journal of Applied Meteorology, **16**, 637-442

Mixing of sediments in estuaries

John McManus

University of St Andrews

1. Introduction

Estuaries are geologically ephemeral systems within which sediments derived from many sources accumulate. In his early review of the sources of sediments in estuaries Guilcher (1967) identified three principal regions from which sediment could be derived, namely from the catchment of the river system, from offshore marine areas, or from the cliffs, marshes and tidal flats of the estuaries themselves. In suitable locations it is possible for the identities of the mineral assemblages present to provide indication of the sediment sources, and their systematic changes to indicate directions of migration. However, in many sites the relatively recent glaciations have ensured that both river and offshore areas are characterised by essentially the same deposits, so that in high latitude areas this technique is often inappropriate.

Postma (1967) showed that the migration of sediment within an estuary is entirely dependent upon the nature of the dynamic processes active in the water body. While the regular and highly predictable tidal flows may dominate the marine segments, the largely unpredictable fluvial discharge of freshwater, controlled mainly by the convectional or cyclonic-orographic rainfall is of prime importance in the upper reaches. Within the estuary itself the erosion of banks or cliffs is often controlled by the activity of wind-generated waves. Currents and wave action may combine, leading to erosion of earlier deposits and enhanced transport to areas of relative quiescence. Gravitational circulation induced by density differences between the saline marine waters and the lighter freshwater may also carry fine sediments into the upper parts of estuaries.

In this chapter examples of the sources and process acting will be commonly drawn from the Tay estuary of eastern Scotland (Figure 1), but the sources and processes will be discussed in a context of sediment from estuaries world-wide. The nature of the dynamics and how these interact with mobile and deposited sediments are examined, and controls of the turbidity maximum zone and sedimentation from it onto marshy margins are addressed. Finally the methods of identifying the derivations of the sediments are examined, with reference to clay minerals, heavy minerals, and the use of natural and radioactive tracers.

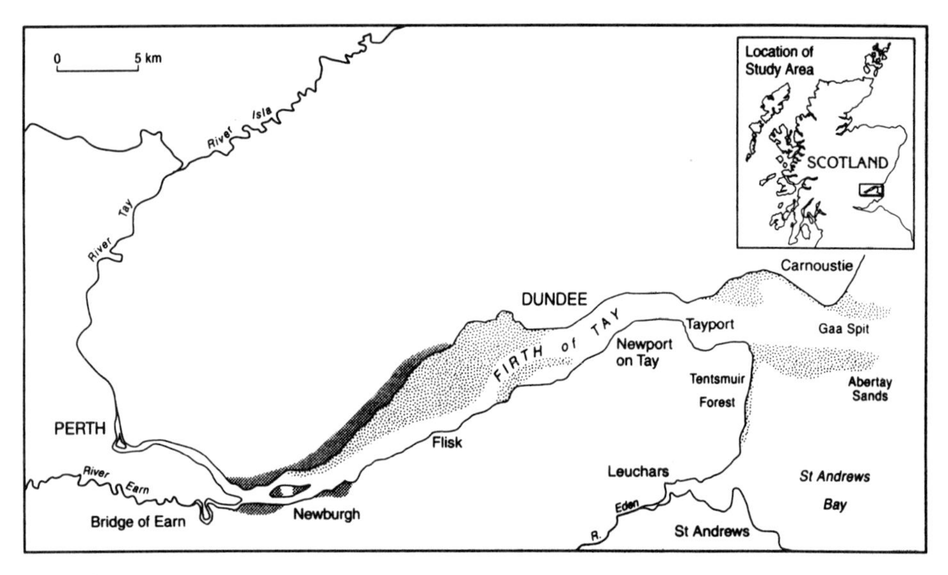

Figure 1 The Tay estuary, showing channels, tidal flats (light stipple) and reed beds (heavy stipple).

2. Dynamics of the river system

In most parts of the world the quantity of water discharged into the head of an estuary is largely decided by the precipitation reaching the catchment. The hydrograph of water discharge against time reveals that the flows carried by a small river are likely to vary rapidly through time. The river discharges increase rapidly as the run-off becomes concentrated into the channels during the early part of the storm, but later in the rainfall event, after peaks of flow have passed, the water levels fall more slowly until a base level of flow is re-attained as the final waters are drained from the land surface. In larger catchments the flow variability is decreased as the cumulative effects of rainfall arriving at adjacent catchments sequentially is reflected in a gradual increase in discharge and a more prolonged decrease as the storm passes out of the system.

When rainfall runs off over the ground surface with sufficient speed it entrains exposed particles of sediment, the more intense the rain, the longer and steeper the slopes, the wetter the soils before the storm, and the more bare earth exposed to attack the greater the quantities of material moved (Zingg 1940, Wischmeier *et al.* 1958, Morgan 1979, 1986). The quantity of water flowing over the land surface may be greatly increased during the annual period of snow-melt.

In any river system the bulk of rock debris is carried as fine particles in suspension, the coarser fragments move as bed load or wash load, at least partly in contact with the stream bed. Other material is transported in solution within the water column. The response of sediment transport to increased flows is as rapid as the variation of the water discharge (Walling and Webb 1992). In a large catchment the variability of sediment yield from each sub-catchment is reflected in changes in the nature and quantities of material arriving at the head of the estuary throughout the duration of a single storm. However, not all material removed from the land surface in the upland parts of the catchments arrives at the estuary head in a single phase of transport, for a portion of the load normally becomes deposited in the lower reaches of the river system and is available for transport early during later storms (Walling 1983).

During periods of high flows very substantial quantities of sediment are carried into the estuary waters, principally in suspension, for bed and wash loads rarely exceed 10% of the total solids load carried. In many rivers the bulk of the annual load of sediment is carried within a few days, and during a single catastrophic flood of a few days duration the load transported may exceed that carried during the preceding decade. The classic example of this was recorded by Meade and Parker (1985) on the Eel River of California, where during three days of storm flows the load was estimated to be equivalent to that of the seven previous years. In the Tay estuary the 100 year floods of 1993 showed discharges of $2200m^3s^{-1}$ (Anderson and Black 1993), more than ten times the long term mean daily flow of $198m^3s^{-1}$ (McManus 1968). Exceptionally high concentrations of suspended sediments were detected throughout the length of the estuary and beyond, into the North Sea coastal waters. Similar conditions were recorded by Mill (1885). During average years the bulk of the load reaches the Tay estuary within 15–20 days (Al-Jabbari *et al.* 1980), and similarly the majority of the loads entering the Forth estuary do so within 12–15 days (Asaad and McManus 1986).

The rivers responding to rainfall and snow melt provide water and sediment to the estuaries in varying and largely unpredictable time scales. The Tay catchment experiences on average 20 cyclonic rainfall events annually, but these are irregularly distributed throughout the year. Few events occur between April and July, but outside these months their arrival is unpredictable, so that sediment is carried to the estuary mainly in the winter months, but in bursts of activity spaced at intervals varying from a few days to a few weeks. In monsoon climates all of the sediment is carried within a few weeks of heavy river discharge, with virtually no solids transport during the rest of the year.

3. Tides and tidal currents

The second principal energy source for sediment transport in estuaries is to be found with the penetration of the tidal wave, which enters from and drains back into the sea. Tidal rise and fall is controlled largely by gravitational forces exerted by the Sun and Moon which determine the timing of high water and low water at any site, and they also determine the range of water level variation to be experienced at that site. The tidal wave established in a sea basin circulates around an amphidromic point, at which no variation of water level occurs. As one progresses away from the amphidrome the tidal range increases. In the northern North Sea the amphidrome is near Stavanger, so that tidal water level varies little along this part of the Norwegian coast (Doodson and Warburg 1941). However, the Scottish coast is remote from the amphidrome and tidal ranges of 4m in the Ythan estuary, north of Aberdeen, rise to 5m in the Tay, and 6m in the Forth estuaries (McManus 1992). Ranges of 10m are known in the Severn estuary, 13m in the Mersey, and 18m in the Bay of Fundy.

A tidal wave, which normally has a period of 12.6 hours enters the estuary mouth with a symmetrical sinusoidal form. As it penetrates into the shallowing and progressively confined waters the wave becomes distorted, and rather like a much shorter period breaker on a shoreline, its height increases landwards, the rising limb becomes shortened as the flood tide is compressed in time whereas the ebb tidal phase increases in length. At the head of the Tay estuary the rising tide at Perth lasts for 3.5 hours and the falling tide for about 9 hours. In the upper reaches of some estuaries the wave creates the surficial feature of a bore travelling landwards. It is in this sector that the truly tidal waters meet the effluent river discharges and there is temporary ponding of the freshwater.

As tidal waters flow into and out of the estuary they create currents which normally peak in strength towards mid-tide and decrease towards High and Low Water. The strengths of the currents vary greatly with tidal range and local bathymetry. In the Tay estuary current speeds of $2.5ms^{-1}$ are detected both widely in the lower reaches and for short periods in the channels of the upper reaches. Such currents have the ability to transport sediments both landwards during the flood tide and seawards during the ebbing tide. When taken as an average across the estuary in the middle and lower reaches, where the tide is symmetrical in nature, the motion of sediment in transport is roughly in equilibrium. In the upper reaches the landward currents are strong for only a short period of the tidal cycle and there is a net seaward transport of material during normal and high river flows. The imbalance is reduced during low summer flows, and permits more landward motion of the fine sediments in suspension and their deposition during high water slack conditions.

Within the estuary waves may be generated by winds blowing across the surface of the waters. These waves, often exceeding 1m in height in the lower reaches of the Tay during storms, create small circulation patterns of currents which are able to lift deposited fine sediments into suspension, thereby eroding the surface of the tidal flats. While the strength of the wind is an important factor in the degree of erosion which takes place, in an estuarine context the direction from which the wind is blowing is of fundamental importance (Weir and McManus 1987). If it is blowing across the estuary there is little opportunity for large waves to develop within the restricted fetch, but if the winds blow along the estuary they may generate large waves capable of destroying shoreline structures as well as entraining deposited sediments. In the Tay estuary winds from the seaward end commonly produce waves which erode the surface of the tidal flats in the middle and upper estuary. Winds from the southwest blow along the estuary and may generate large waves in the deep water zones near the mouth, but are less effective in generating waves in the shallow waters of the upper estuary, and have no effect once the water has drained from the tidal flats.

The nature of the gravitational circulation induced by the meeting and intermixing of the light freshwaters with the denser saline marine waters led Pritchard (1955) to identify four principal types of estuaries, referred to as Types A, B, C and D. The salt wedge, stratified, partially mixed and mixed estuarine systems are recognised world-wide. In salt wedge and stratified flows there is advection of salt waters from depth into the upper part of the water column, so that additional landward flow is generated in the near bed waters to compensate. These net non-tidal flows are slow, but have the ability to aid in the landward transport of fine suspensions in the main channels. In the more mixed water bodies Pritchard suggested that Coriolis effects might permit inward flooding waters to become diverted towards one shore or the other, according to the hemisphere in which the estuary is situated. Thus an eastward facing estuary in the northern hemisphere, such as the Tay, might experience waters of higher salinity towards the northern shore on the flood tide. A westward facing estuary such as the Mersey would show greater salinities on the southern side on the rising tide, for currents are deflected to their right in the northern hemisphere. Ideally this phenomenon is best detected in wide estuaries, but such systematic variation has been recorded in the middle reaches of the Tay (McManus and Wakefield 1982).

The recognition by Simpson and Nunes (1981) that distinct boundaries may be detected between water masses of differing salinities enabled them to study the salt wedge migration in the Conwy estuary in North Wales. Subsequently frontal systems have been identified in many estuaries, not always trending along the face of the salt wedge, but much more commonly trending along the length of the estuary (Anderson 1989). These also separate water masses of differing salinity, temperature and suspension concentrations. Thus the rising tidal waters are no longer to be viewed as rather homogeneous masses penetrating the estuary, but are now seen to consist of discrete slices of water moving past each other along shear surfaces whose position is often defined by submerged or emergent bathymetric features. Waters passing a headland are often differentiated by salinity and temperature from those sheltered behind it. The intersection of the plane of shearing separating the water masses with the water surface itself is commonly represented by foam bands along which circulation eddies a few metres in diameter are present. In terms of sediment transport each longitudinal slice of the water mass will have its own flow speed, and will carry both suspension and bed load at rates differing from those of its neighbours. The researcher attempting to address the problem of sediment source has a multiplicity of water masses to consider. However, the general behaviour of neighbouring bodies of water is similar and it is still possible to examine the generality of water motion due to tides as outlined above.

4. Wave activity

While the tidal motion is essentially that of a long period wave, the water surface of coastal and estuarine waters is characterised by the presence of short period waves. Generated by the friction of winds blowing across the surface, these waves are most powerful when created by strong winds blowing across long distances of open sea (fetch) for long periods of time. Single gusts have little effect in forming waves whereas protracted storms may create surprisingly large waves in coastal waters.

Although waves crashing on the cliffs of the open coast provide evidence of the power of the waves, for the estuarine sedimentologist it is the interaction between the wave and the sea bed which is of crucial importance. The water within waves follows regular orbital paths which produce upward motion capable of lifting particles off the bed and into suspension. This state may prevail for a short time before sand particles settle back to the bed. Where waves disturb the bed sediments any along-shore currents, however slow, are able to induce the particles to migrate in the direction of current. As waves approach the coast the essentially circular motion of the water becomes progressively deformed as the base of the wave slows while the crest progresses shoreward unimpeded. The wave form distorts, and becomes higher, forming breakers, and in the breaker zone large quantities of sediment are thrown into suspension, to be carried shorewards, seawards or along-shore, according to the conditions prevailing.

When waves approach the shoreline at any angle other than directly up the beach face the wave front becomes refracted through contact with the bed occurring earlier at one point on the wave front than another. As a result the wave crest moves progressively along the beach face as the wave breaks. The uprushing water transports sediment obliquely across the beach face and the backwash returns the water directly downslope, so that entrained particles migrate along the coast. Waves approaching from different directions often show contracting refraction patterns. In the Tay mouth area Sarrikostis and McManus (1987) demonstrated that wind from almost any onshore direction out of the North Sea created waves which refracted in such a way as to sweep sediments southwards along the coast north of the estuary and northwards along the coast south of the estuary (Figure 2). In consequence the Gaa and Abertay Sand banks line the sides of the estuary channel in the entrance reaches for distances of 2 and 6km respectively.

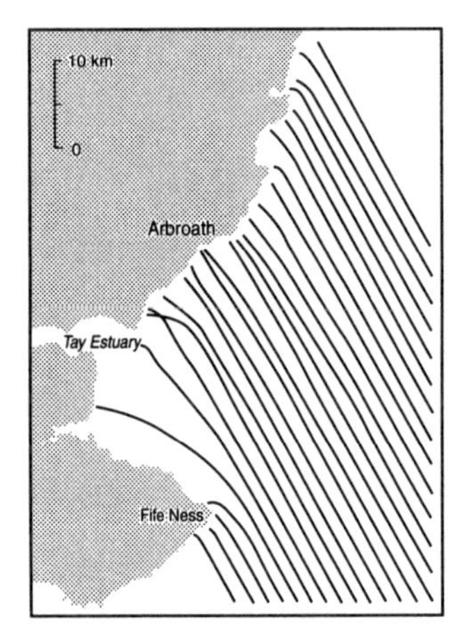

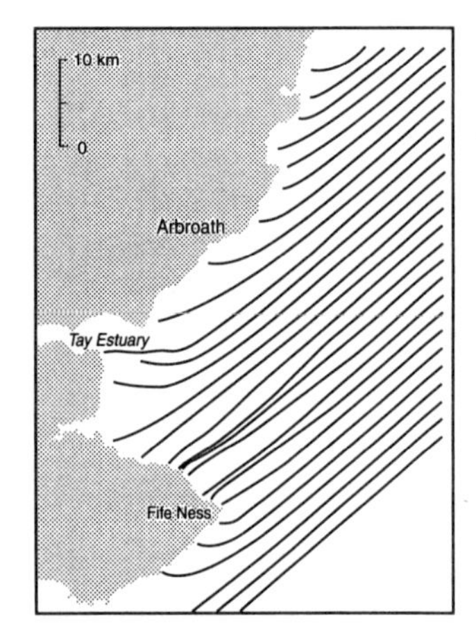

Figure 2 Wave refraction in the North Sea coasts at the entrance to the Tay estuary showing approach of 8.0 second period wave of 0.60m significant height from (a) the south east, (b) the north east.

Within the confines of estuaries winds may create water surface waves, which, under protracted storm conditions may attain heights of more than one metre. However, these are unusual, because they require a combination of high wind fetch along the length of the estuary and continuous high winds. At low water wind–created waves are confined to the channels, but as the tide rises the wave activity is introduced across the tidal flat areas, leading to wave–induced erosion and re-suspension of deposited fine material into the water column. When wind–blown waves travel in the direction of the tidal currents they enhance flow rates, but when they have opposed directions of motion the waves steepen, create broken water, and dissipate their energy into the water column, again stimulating local erosion.

5. Sediment movement, deposition and source of fine particles

Fine material in suspension, coarse material on the bed and intermediate size sandy materials, mostly on the bed, are the main sediments of estuaries. They react differently to current flows and to wave activity. All particles on the bed respond to water dynamics by becoming mobile once some critical threshold of flow speed or bed shear stress has been reached. Movement of pebbles requires greater energy expenditure than entrainment of sand, but, due to their cohesive nature the very fine silts and clays are more difficult to dislodge from the bed, but they may become entrained through the action of small relatively high frequency waves acting across submerged tidal flats (Anderson 1972). Indeed, within most estuaries the principal wave activity is characterised by high frequency low amplitude waves, which expend a great deal of energy on the bed and banks, achieving slow but almost continuous change in contrast to the lower frequency, higher amplitude waves accompanying storm events and often associated with catastrophic changes to shorelines.

Silts and clays are carried into the estuary buoyed up in the turbulent flows. In many cases they arrive in the form of aggregates rather than as discrete primary particles. As they are carried into the partly saline waters the suppression of interparticle ionic repulsion is overcome in the presence of the electrolytes in the waters. Aggregation or flocculation progresses rapidly in the upper parts of the estuary, and this leads to the creation of a zone in which the marginal banks may become deeply coated in mud deposits.

The lightly bound aggregates may become re-suspended by tidal currents or wave activity, and as a result the waters of the upper estuary normally contain much higher concentrations of suspended matter than either the waters to landward or seaward. This zone of enhanced suspended sediment concentration is commonly known as the turbidity maximum zone of the estuary. In the Tay estuary the turbidity maximum rarely shows concentrations above $400mgl^{-1}$ whereas the waters to landward and seaward show concentrations of $5–10\ mgl^{-1}$. In other estuaries the greatest concentrations rise to over $5000mgl^{-1}$. The turbidity maximum and its behaviour have formed the focus of many studies, notably those of Schubel (1968), Mehta and Partheniades, (1975) Migniot (1977), Kranck (1979), Allen *et al.* (1980), Officer (1981), Faas (1981), Kirby and Parker (1983), Burt (1986), Mehta (1986), Nichols (1986), and Sills and Elder (1986) to which the reader is referred for further information.

The sediment in the turbidity maximum zone is largely land–derived or resuspended from internal sources. Dobereiner and McManus (1983) demonstrated that the concentrations varied in distribution according to the freshwater flows entering the estuary. During low discharges the greatest concentrations were well up estuary and formed a single clear peak, whereas during high flows they were recognised much further seaward and often showed the presence of multiple peaks. Buller (1975) suggested that multiple peaks resulted from creek-focused run-off from emergent tidal flats. During the winter months of high river discharge and wave action the turbidity maximum migrates to the middle reaches and leads to deposition in that sector of the estuary. During the summer months of low flow and light winds it is detected in the upper and riverine reaches, where silt deposition hinders netting for salmon during exceptionally dry years.

The variation of suspension concentrations in waters over the tidal flats at high water was shown by Weir and McManus (1987) to be very sensitive not only to wind strength, but also to wind direction (Figure 3). As these waters are carried over the marginal tidal flats they pass onto extensive Phragmites reed beds and salt marshes. Here the silts may settle directly onto the surface of the marsh or into the broken stems of the reeds. As the depth of water increases over the open reed stems so the quantity of silt entrapped increases, so that spring tide sedimentation is far greater than that on neap tides (Alizai and McManus 1980). Furthermore, as wind strength increases and suspension concentrations increase, so the stem-entrapped quantities on each tide increase (Figure 4). The marsh sediment surface rises and falls through the year in response to the sedimentation and erosion, with up to 2cm variation and showing repeated cyclic patterns of change. However, such dynamic marsh surfaces are not ubiquitous.

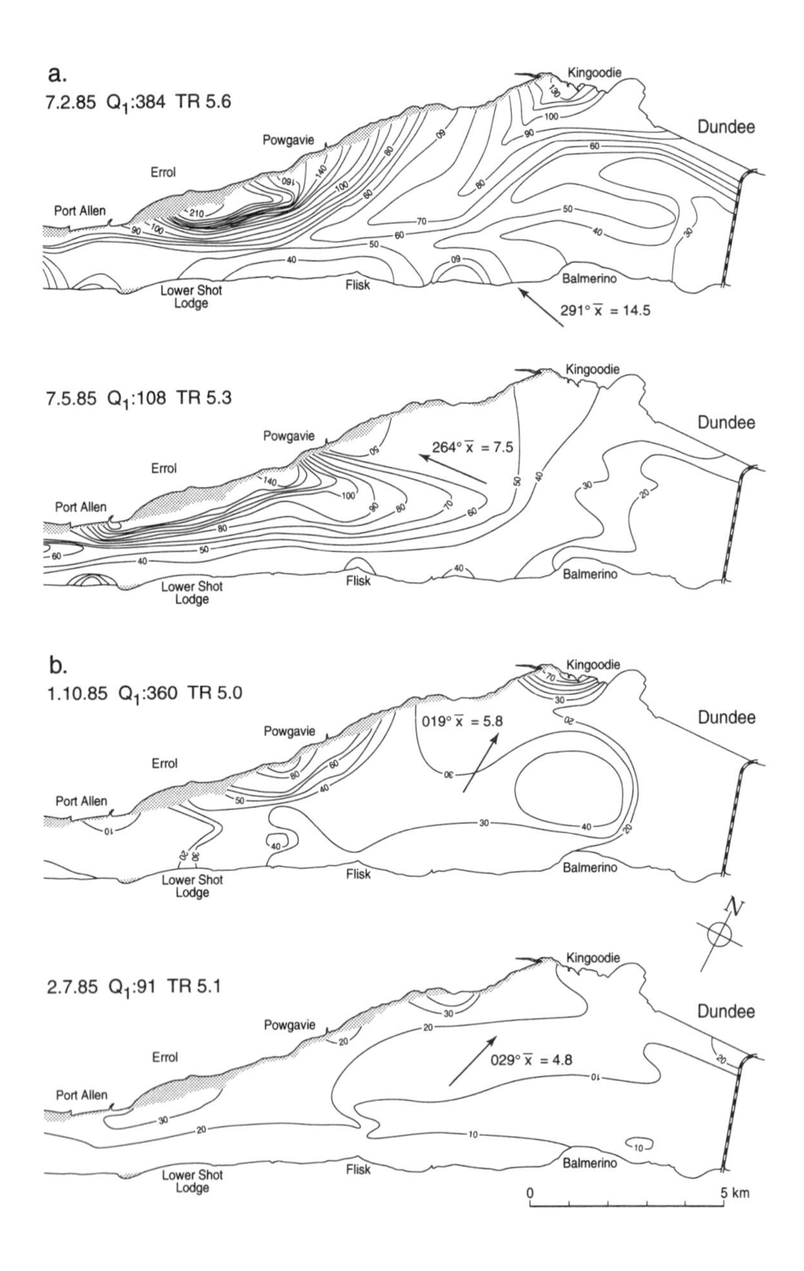

Figure 3 Suspended sediment concentrations in waters above the tidal flats at high tide under relatively calm conditions and under direct wind action.

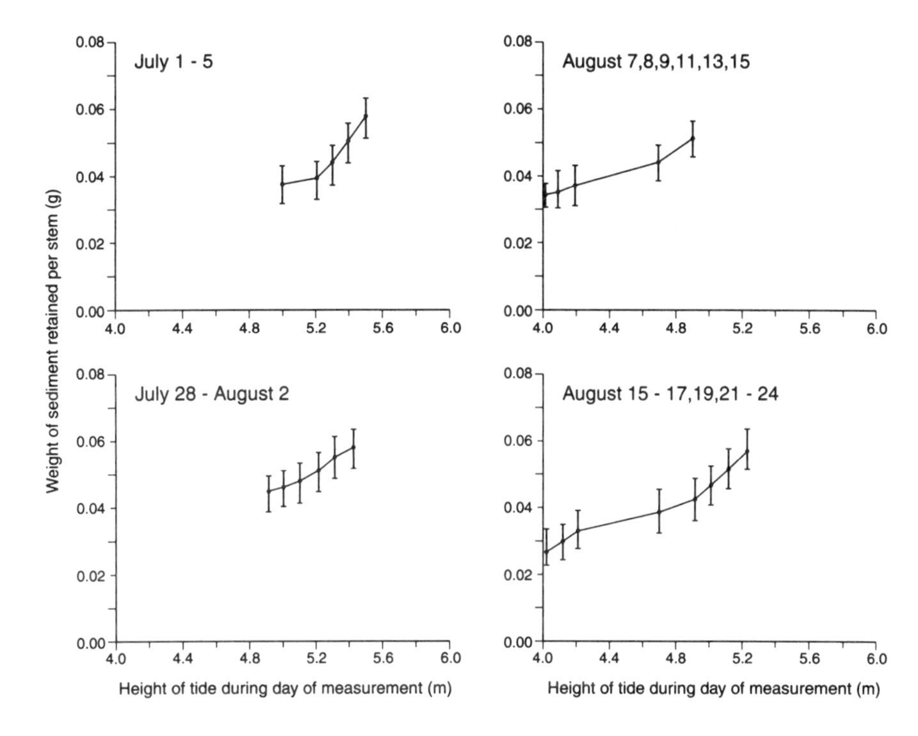

Figure 4 Pattern of instem sedimentation during single tides on the northern shore of the Tay estuary. The low gradient curve was measured during calm conditions, the steep gradient during windy conditions.

In many estuarine marsh systems where the vegetation lacks open stems of broken reeds the accumulation of sediment is principally directly onto the plants and inter-plant surfaces. Where plant communities are closely spaced the deposited sediment is protected against re-entrainment and sedimentation is a near-continuous process. At such sites continuity may be detected by stratigraphic records of the caesium content of the deposits which parallel the variations levels known to have occurred as a result of the atomic weapons testing programmes of the mid-twentieth century. Short term events such as the Chernobyl explosion have left their mark in radioactivity detected in agricultural fields in many parts of Britain, and doubtless further afield (Walling 1990). It is to be expected that in suitable sites this marker will be detected in estuarine marsh deposits. It serves to remind us that particles in estuaries may also be derived from air-fall, whether from man made events or natural activities such as fine air-borne particles blown from hot or cold deserts.

The materials entrapped are readily examined by X-ray diffraction techniques which reveal the proportions of various minerals present. In addition to the common quartz, feldspar and micas the families of clay minerals are readily recognised. Within the Tay estuary the changes in proportions of these minerals are slight, with a decrease of chlorite towards the sea and a compensating increase in the importance of illite. However, in some estuaries, as along the US coastlines of the Gulf of Mexico, the clay mineral families present show significant variations in many locations, and help to point to the sources of the suspended mineral matter, pointing not only to landward but also to seaward sources of the materials.

Much of the deposited fine material is of organic origin, whether originating from eroded soils, leaf debris of catchment plants, phytoplankton from the rivers, lochs, estuary or coastal waters. Analysis of the isotopic composition of the oxygen, carbon sulphur and nitrogen in the organic matter recorded from all of the local environments revealed that more than 80% of the suspended matter in the upper Tay estuary is fluvially derived (Thornton and McManus 1992).

6. Sediment movement, deposition and source of coarse particles

Rivers entering an estuary carry 5–10% of their sediment load as sand and gravel, which migrate along thc bed, only the finer sands rising into suspension when exposed to strong tidal currents or wave action. Once deposited into a sheltered position the pebbles may become entrained rarely, during times of peak tidal currents a few times each month, or during periods of high river discharge. The longer a pebble remains undisturbed the more difficult it becomes to dislodge from the bed for it increasingly becomes used as a holdfast for algal material which produces a coating on the surface and serves to bind it to its neighbours. Eventually the pebble may also attract coatings of iron and manganese oxides, but even then it is possible for it to become disturbed when subjected to sufficiently severe dynamic conditions.

The sands, however, remain mobile and migrate along the estuarine channels, temporarily resting and creating bedforms such as dunes and ripples on mid-channel banks. Where the sands are washed onto the outer parts of the tidal flats they may take up more permanent residence, in deposits which become progressively finer across the tidal flats away from the channels, reflecting a gradient of decreasing water dynamism. As Oomkens and Terwindt (1960) demonstrated in the Haringvliet, much of this material may become reworked as creeks migrate across the tidal flats.

At the seaward end of the Tay estuary the wave refraction pattern ensures that sediment is carried into the entrance, where tidal currents may then transport them landwards. Similar patterns of refraction and landward motion of sediment are recorded at the mouth of the Gironde, the Loire and Seine, most of the estuaries of the Appalachian coastal plains and many of the Australian and New Zealand coastal inlets (Meade 1972, Woodroffe 1996, Healy *et al.* 1996). In detail it has been shown, largely on the basis of grain size variation, that the motion of the material into the Tay takes place principally along the beaches through continued wave action, the progression becoming increasingly influenced by tidal currents (Al Mansi 1990). While it is possible to postulate that the tidal currents sweep the marine sands into the estuary it is important to be able to prove that there genuinely is a landward transport of material.

Two lines of approach, each yielding support to the contention have been used. The introduction of tracer sediment stained with fluorescent dyes is precluded by the sheer quantity which would be required to be deployed for any probability of tracer being recovered among the volumes in transit. However, there are natural tracers in the form of the blue coloured shells of the common mussel, Mytilus, which lives in extensive banks in the lower reaches and in decreasing abundance as far west as the railway bridge off Dundee, which marks its landward limit of growth. The channel floor sediments contain fragments and entire shells of Mytilus for up to 6km west of the railway bridge, providing an unequivocal indication of landward motion of the sand sized particles (Al-Dabbas and

McManus 1987). Similar patterns of landward movement of coarse shells occurs on coastal inlets on the western side of Florida (R A Davis Jr, personal communication).

In a second line of approach, Mishra (1969) used the heavy mineral fraction, which is commonly concentrated by wave action on beaches and tidally exposed estuarine sand banks, to show that three discrete populations of sediment could be recognised in the Tay estuary region. Using the nature of the garnet and hornblende grains he demonstrated that one fluvial population could be distinguished and two marine populations, one from the coast north and the other from coasts south of the Tay estuary entrance, could be distinguished also. The three populations become intermixed in sand banks between the bridges off Dundee, and there were traces of the marine materials on tidal flats west of the rail bridge.

In tracer experiments in the Firth of Forth, Smith and Parsons (1965) and Smith *et al.* (1965) used radioactive tracers to examine the movement of dredged spoil taken from Rosyth docks and released into the waters of the Forth in a series of experiments. Having released the tracers into the water column or on the sea bed at high water of spring tide, they tracked the plume of material migrating back up estuary for more than three months, demonstrating that some of the dredged spoil returned to the dock within a month of removal and release. The release point was towards the northern shore in what today we would regard as a flood dominant channel. In the second set of experiments, Smith *et al.* recorded that deposition of sediment nearer the southern shore into an ebb-dominant channel led to no return of spoils to the dock, all radioactivity being detected to the east of the release point. This latter site has since been used for dumping of the dock spoils and there is no sign of build-up on the channel floor receiving the spoils.

7. Conclusions

There is strong evidence that sediments derived from catchment sources and carried into the estuaries become intermixed with materials derived from the coastlands on either side of the estuary mouth. Within the estuary the particles are moved principally by tidal currents, in water masses which maintain separate identities for parts of their excursion into the estuary. Individual flood- or ebb-dominant channels permit passage of sediment to seaward and landward, but there is always a tidal residual of transport. In the Tay, as in most other temperate climate estuaries, the normal residual permits a landward drift of bed material. The presence of a turbidity maximum zone encourages deposition of fine sediments in the upper reaches of the estuary, on the inner parts of the tidal flats and marshes. Since estuaries world-wide are seen as sites of net sediment accumulation, their long-term survival in a geological sense is very much in doubt. Perhaps this is one natural system whose preservation will depend on the widely predicted global rise of sea-level, and continued observation and recording in relation to features which will survive will be needed in years to come.

References

Al-Dabbas M A M and McManus J, 1987, Shell fragments as indicators of sediment transport in the Tay Estuary, Proceedings of the Royal Society of Edinburgh, **92B**, pp 335–344

Al-Mansi A M A, 1990, Wave refraction patterns and sediment transport in Monifieth Bay, Tay Estuary, Scotland, Marine Geology, **91**, 299–312

Alizai S A K and McManus J, 1980, The significance of reed beds on siltation in the Tay Estuary, Proceedings of the Royal Society of Edinburgh, **78B**, pp 1–14

Al-Jabbari M H, McManus F, and Al-Ansari N A, 1980, Sediment and solute discharge into the Tay Estuary from the river system, Proceedings of the Royal Society of Edinburgh, **78B**, pp 15–32

Allen G P, Salmon J C, Bassoulet P, Du Penhoat Y and De Grandpre C, 1980, Effects of tides on mixing and suspended sediment transport in macrotidal estuaries, Sedimentary Geology, **26**, 69–90

Anderson F E, 1972, Resuspension of estuarine sediments by small amplitude waves, Journal of Sedimentary Petrology, **42**, 602–607

Anderson J and Black A, 1993, Tay flooding: act of God or climate change? Circulation, British Hydrological Society, **38**, 1–4

Asaad N M, and McManus J, 1986, Fluvial sediment transport in the Forth drainage basin, Journal of Water Resources, **5**, 19–39

Buller A T, 1975, Sediments of the Tay Estuary II Formation of ephemeral zones of high suspended sediment concentration, Proceedings of the Royal Society of Edinburgh, **75B**, pp 65–89

Burt T N, 1986, Field settling velocities of estuarine muds, In A J Mehta, Estuarine Cohesive Sediment Dynamics, Springer Verlag, New York, pp 126–150

Doodson A T and Warburg H D, 1941, Admiralty Manual of tides, H M S O London

Faas R W, 1981, Rheological characteristics of Rappahannock Estuary Muds, SE Virginia, USA In Holocene Marine Sedimentation in the North Sea Basin, International Association of Sedimentologists, Publication **5**, pp 505–515

Guilcher A, 1967, Origin of sediments in estuaries, In G H Lauff (Ed.) Estuaries, American Society for the Advancement of Science, **83**, pp 149–157

Healy T R, Cole Rand de Lange W, 1996, Geomprphology and ecology of New Zealand shallow estuaries and shorelines, In K F Nordstrom and C T Roman (Eds.) Estuarine Shores, J.Wiley, Chichester,115–144

Kranck K, 1979, Dynamics and Distribution of Suspended Particulate Matter in the St Lawrence Estuary, Naturaliste Canadienne, **106**, 163–173

Kirby R and Parker W R, 1983, Distribution and behavior of fine sediment in the Severn Estuary and Inner Bristol Channel, UK Canadian Journal of Fisheries and Aquatic Science, **40**, 83–95

McManus J, 1968, Hydrology of the Tay basin, In S J Jones (Ed.) Dundee and District, British Association for the Advancement of Science, Winter, Dundee, pp 107–124

McManus J, 1992, A hydrographic framework for marine conservation in Scotland, Proceedings of the Royal Society of Edinburgh **100B**, pp 3–26

McManus J and Wakefield P, 1982, Lateral transfer of water across the Middle Reaches of the Tay Estuary, In J McManus (Ed.) Sedimentological, Hydrological and Biological Papers, Tay Estuary Research Centre Report, 25 –38

Meade R H, 1972, Transport and deposition of sediments in estuaries, Geological Society of America, Mewmoir, **133**, 91–120

Meade R H and Parker R S, 1985, Sediment in rivers of the United States, In National Water Summary, 1984. U.S.Geological Survey Water-Supply Paper 2275, pp 49–60

Metha A J (Ed.), 1986, Estuarine Cohesive Sediment Dynamics, Springer Verlag, New York, 468pp

Mehta S A J and Partheniades E, 1975, An investigation of the Depositional Properties of Flocculated Fine Sediments Journal of Hydraulics Research, **13**, 361–381

Migniot C, 1977, Effect of currents, waves and wind on Sediment. La Houille Blanche **32**, 9–47

Mill H R, 1885, On the salinity of the estuary of the Tay and of St Andrews Bay, Proceedings of the Royal Society of Edinburgh, **13**, pp 347–350

Mishra S K, 1969, Heavy mineral studies of the Firth of Tay Region, Journal of the Geological Society of India, **5**, 37–49

Morgan R P S, 1979, Topics in Applied Geography, Soil Erosion, Longman, London, 113pp

Morgan R P S, 1986, Soil Erosion and Conservation, Wiley, New York, 298pp

Nichols M M, 1986, Effects of fine sediment resuspension in estuaries, In A.J.Mehta (Ed.) Estuarine Cohesive Sediment Dynamics, Springer Verlag, New York, pp 5–42

Officer C B, 1981, Physical dynamics of estuarine suspended sediments, Marine Geology, **40**, 1–14

Oomkens E and Terwindt J H, 1960, Inshore estuarine sediments in the Haringvliet (Netherlands), Geologie en Mijnbouw, **39**, 701–710

Postma H, 1967, Sediment transport and sedimentation in the estuarine environment, In G.H.Lauff (Ed.) Estuaries, American Association for the Advancement of Science, **83**, 159–179.

Pritchard D W, 1955, Estuarine circulation patterns, Proceedings of the American Society of Civil Engineers, **81**, 717–729

Sarrikostis E and McManus J, 1987, Potential longshore transports on the coasts north and south of the Tay Estuary, Proceedings of the Royal Society of Edinburgh, **92B,** pp 297–310

Schubel J R, 1968, Suspended sediment of Northern Chesapeake Bay, Technical Report 35, Chesapeake Bay Institute, Johns Hopkins University

Sills G C and Elder D McG, 1986, The transition from sediment suspension to settling bed. In A J Mehta (Ed.) Estuarine Cohesive Sediment Dynamics, Springer Verlag, New York, pp 192–205.

Simpson J H and Nunes R A, 1981, The tidal intrusion front: an estuarine convergence zone, Estuary, Coastal and Shelf Science, **13**, 257–266

Smith D B and Parsons T V, 1965, Silt movement investigation in the Oxcars spoil ground, using radioactive tracers, 1961 and 1964. UK AEA Unclassified Report AERE-R4980. H.M.Stationery Office, London, 41pp

Smith D B, Parsons T V, and Cloet R L, 1965, An investigation using radioactive tracers into silt movement in an ebb channel, Firth of Forth, 1965, UKAEA Unclassified Report AERE-%5080, H.M.Stationery Office, London, 11pp

Thornton S F and McManus J, 1994, Applications of Organic Carbon and Nitrogen stable isotope and C/N Ratios as source indicators of Organic matter provenance in Estuarine systems: Evidence from the Tay Estuary, Scotland. Estuary, Coastal and Shelf Science, **38**, 219–233

Walling D E, 1983, The sediment delivery problem, Journal of Hydrology, **65**, 209–237

Walling D E and Webb B W, 1992, Water quality I Physical Characteristics. In P Calow and G E Petts (Eds.), The river handbook, Blackwell, Oxford. pp 48–72

Walling D E, 1990, Linking the field to the river: Sediment delivery from Agricultural Land, In J Boardman, I D L Foster and J A Dearing (Eds), Soil Erosion on Agricultural Land, J Wiley, Chichester, pp 129–152

Weir D J and McManus J, 1987, The role of wind in generating turbidity maxima in the Tay Estuary, Continental Shelf Research, 7, 1315–1318

Wischmeier W H, Smith D D, and Uhland R E, 1958, Evaluation of factors in the soil-loss equation, Agricultural Engineering, **39**, 458–462

Woodroffe C, 1996, Late Quaternary infill of macrotidal estuaries in northern Australia, In K F Nordstrom and C T Roman (Eds.) Estuarine Shores, J.Wiley Chichester, 89–114

Zingg A W, 1940, Degree and length of land slope as it affects soil loss in runoff, Agricultural Engineering, **21**, 59–64

Computer modelling and Earth observation: providing data for coastal engineering

Keiran Millard

HR Wallingford Ltd[1]*. Wallingford, UK*

1. Introduction

1.1 Coastal engineering

Fundamental to coastal zone management (CZM) is shoreline management. While CZM covers a large boundary stretching from inland regions to the boundary of the Exclusive Economic Zone (extending to 200 nautical miles offshore, or to the boundary of a neighbouring EEZ), shoreline management is concerned with the crucial interface between the terrestrial and marine environments. As such, an effective shoreline management plan (SMP) is an integral component of a wider CZM plan. Coastal engineering is concerned with providing the technology to manage the land sea divide necessary for SMP. It chiefly embraces the topics outlined below in Table 1. The threat posed by global warming and resulting sea level rise suggests that the coastal engineer is going to be very occupied throughout the next few decades and beyond.

Topics	Subjects
Hydrodynamics	Analysing the behaviour of waves, tides and currents for the circulation and transport of water
Morphodynamics	Analysing the mechanisms for the movement of sediments; this includes both in shoreline environments such as beaches, cliffs and saltmarshes, but also on the seabed
Hard engineering	Design of 'hard' sea defences such as revetments, seawalls and breakwaters
Soft engineering	Design of 'soft' sea defences such as beach re-nourishment and managed retreat schemes

Table 1 Topics embraced by coastal engineering

To undertake their work, coastal engineers need a comprehensive understanding of the environment they are operating in, i.e. the project site. This is typically a small region such as a harbour, a marina or an area of beach in front of a town or hotel; accordingly the immediate study area is usually only a few kilometres square, e.g. 3km longshore and 2km offshore.

[1] HR Wallingford is a high technology research and consultancy organisation specialising in environmental and civil engineering hydraulics.

Example data requirements for predicting shoreline change are specified in Table 2. Quite often good site specific data of this type are difficult to obtain or are simply not available and coastal engineers have to make approximations using what data is available or use models synthesised variables, e.g. synthesis of a wave climate from wind data.

Data Requirement	Data Uses
Historical maps	Establishing historical large-scale shoreline change (e.g. cliff top retreat or beach erosion/accretion)
Remotely-sensed images (aircraft)	Detailed measurements of shoreline change
Tidal levels	Interpretation of shoreline
Tidal currents	Primary input into tidal flow models
Seabed surveys	Primary input to hydraulic models
Wind data	Primary input to wave generation models
Wave measurements	Calibration (sometimes replacement) for wave numerical models
Sediment sampling	Analysis and input to models of beach response
Geotechnical sampling and analysis	Input to models of cliff recession/collapse
Cliff monitoring	Calibration data for cliff retreat models
Beach surveys	Calibration data for beach response models

Table 2 Example data requirements for predicting shoreline change

1.2 Computer modelling in coastal engineering

Forecasting is an integral component of coastal engineering. Given the nature of the environment, time scales and costs involved, coastal engineers do not have the luxury of testing their engineering designs before accepting a final solution. It is simply not practical to construct a trial breakwater and then monitor its performance over ten years to decide if it is the best solution! Modelling is essential to compress both cost and time scales.

In coastal engineering both physical and computer modelling are used. Physical modelling involves construction of a reduced scale representation of the environment under investigation in a wave basin and forcing that environment with mechanically generated waves. (Note that a reduced scale does not simply mean a literal dimension reduction: scaling is non-linear such that the physics of the miniature remain consistent with the full size environment). Computer modelling involves constructing a mathematical simulation of the processes under investigation. Such computer models are only as accurate as our present understanding of coastal processes. Accordingly while computer models are quicker and less expensive to run, physical models continue to be necessary where complex hydrodynamic or morphodynamic systems need to be understood.

Computer models vary in complexity from one-dimensional models to complex two-dimensional and fully three-dimensional models. Example of models used by HR Wallingford are shown in Table 3. In all cases, however, it needs to be remembered that a model is only presenting one possible answer to a problem and as such the answer given by a model should never be treated blindly as the correct answer. Model outputs need to be assessed in the light of expert opinion and other analysis to determine whether the results supplied by the model are reasonable.

Model	Description	Application
PISCES	Coastal area morphodynamic modelling framework	Determination of sediment (mud and sand) transport under wave and tidal conditions and associated seabed evolution.
COSMOS-2D	Interactive simulation of waves, currents and sediment transport	Wave transformation due to refraction, shoaling, friction and breaking. Longshore tidal and wave driven currents. Vertical distribution of cross-shore currents. Cross-shore and longshore sediment transport.
BEACHPLAN	One-line model for long term beach plan shape evolution	Long term (decadal) evolution of beaches under the action of breaking waves. Impact of structures (groynes, breakwaters etc.) on beach evolution.
HINDWAVE	Numerical simulation of wave climates based on wind data	Simulating wave climate and basic wave climate analysis.
TELEMAC	3D modelling system for hydrodynamics, sediment transport and water quality	Tidal, wind and wave driven hydrodynamics, tracer transport, wetting and drying, suspended sediment and bedload, water quality and dispersal of pollutants, including thermal pollution.

Table 3 Examples of models used by HR Wallingford

1.3 Satellite remote sensing

Satellite remote sensing has the potential to provide a large amount of wide area synoptic data to coastal engineers. Coastal engineering data requirements typically falls into two categories; data for design and data for operation. Design requires large quantities of data, usually in the form of time series for several parameters, of the type specified in Table 2. This data is required on a one-off basis while the design is being undertaken. Operation requires smaller quantities of data, but on a regular basis. One example would be providing data on the performance of a beach (e.g. aerial surveys) that was renourished as the outcome of the design phase.

Coastal engineers have been using remote sensing techniques such as airborne photography for a number of years; until recently remote sensing from satellites has not been particularly useful however, since the parameter set available did not match what was required. This is unfortunate because the over-selling of satellite remote sensing to coastal engineers while the technology was in this evolving phase has resulted (not surprisingly) in a scepticism of data available from Earth observation sources. Although existing data from these early missions is still largely all that is available, the improvement in sensor technology on future missions and the improved understanding of the requirements of coastal engineers by the remote sensing community does offer greater potential for the use of Earth observation data in coastal engineering in the future.

Accordingly this chapter examines the important relations that exist between coastal engineering, computer modelling and satellite remote sensing. The overall aim is to show the role satellite remote sensing can play in the field of coastal engineering by providing valuable environmental data. Three main topic areas are covered in this chapter. An example is given of a beach management project, to illustrate the data requirements for a typical coastal engineering project, together with an illustration of the role of computer modelling. Extending from this, consideration is given to the wider issue of the performance of satellite remote sensing in providing coastal engineers with the information they require. Finally we shall discuss a recently completed project that uses remote sensing as a basis to provide data on the transport of suspended sediment in coastal waters.

2. Shingle beach design and management

2.1 The shingle beach at Seaford

It is now widely accepted that a properly managed beach can provide an effective form of sea defence that is in many respects superior to traditional hard engineering solutions. Developing these schemes is not straightforward and requires a detailed analysis of the various interacting processes that affect beach morphology. This inevitably involves the use of mathematical models to simulate coastal processes, and predict the response of a beach several years into the future. It is only by using such approaches that conclusions can be drawn as to whether a technique such as beach renourishment represents a suitable coastal defence option at a given site.

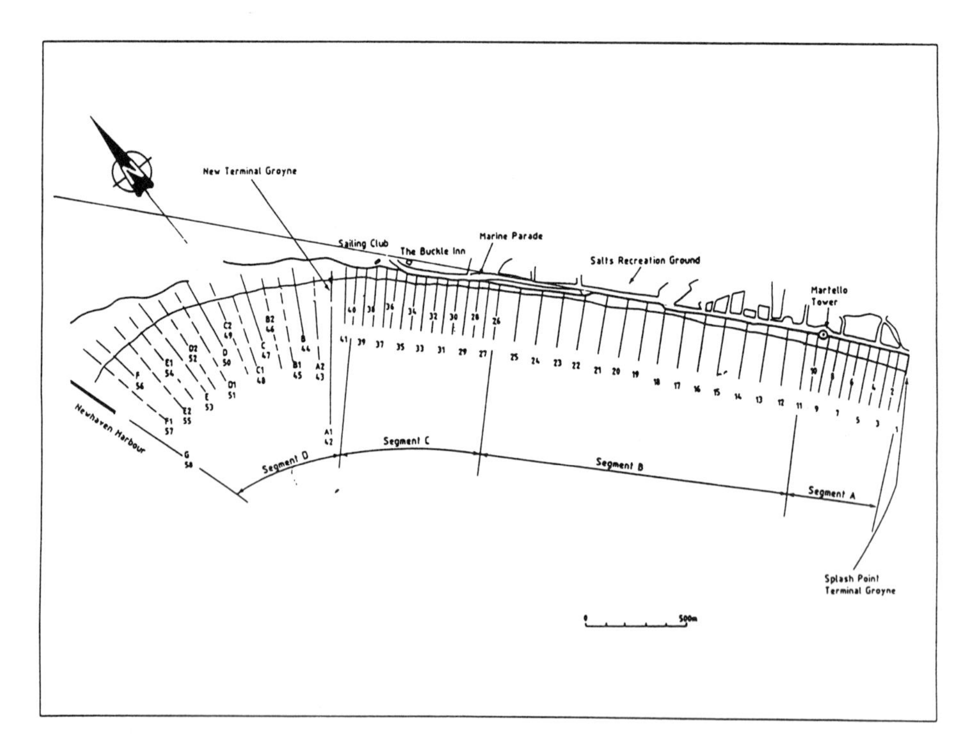

Figure 1 The beach at Seaford

A map of the Seaford shoreline area is shown in Figure 1. Like much of the Sussex coastline, the seafront between Newhaven and Seaford is composed of a shingle beach lying on a wave-cut chalk platform. Throughout recorded history, the shingle ridge which formed the main sea defence for the low lying hinterland has been subject to breaching in severe storms. River training works on the Ouse enabled Newhaven to develop as a port from the 17th century onwards and the completion of the long west breakwater in 1890, have accelerated this development. As a result, from about 1830 onwards remedial measures have been necessary so that by about 1912 the full length of the shingle beach between Newhaven and Seaford was backed by a near-vertical sea wall, and was groyned at approximately 80m intervals. From that time the levels of the shingle beach dropped substantially, and a constant battle was waged to maintain the sea wall and the groyne system.

In April 1981 responsibility for protecting this length of coastline passed from the Newhaven and Seaford Sea Defence Commissioners to the Southern Water Authority. The Authority initiated a major review of the existing defences and outlined possible courses for future action. It was concluded that artificial renourishment of the beach with shingle was the cheapest alternative to provide an adequate standard of sea defence, and environmentally had most to offer. However the report recognised that with current knowledge the littoral transport in Seaford Bay was not fully understood, and any such scheme should proceed only after a full investigation to establish a reasonably stable beach profile shape, and to verify rates of littoral transport. These two items have a direct influence on the capital and maintenance costs respectively of such a scheme.

To investigate the design of a suitable beach nourishment scheme and the littoral drift regime within the bay, HR Wallingford completed a series of physical and mathematical modelling studies supplemented with site data collection. With this package of studies it was expected to be possible to predict with some confidence the behaviour of the proposed nourishment scheme, and also any alterations to the scheme which might be necessary to ensure its success. The main recommendations arising from these studies were to renourish the beach to the profile determined by the study, provide a large terminal groyne at Splash Point to trap sediment transported by the dominant eastward drift, and thereafter recycle the beach material periodically.

Beach renourishment was carried out between April 1987 and October 1987 and since then periodic re-cycling of beach material has been carried out. In addition HR Wallingford has conducted regular studies to assess the performance of the renourished beach. The latest review was conducted in 1995. The map in Figure 1 shows the main features of the area, together with the transects used to measure the beach profiles (1-58) and the four segments of combined profiles (A-D) used to give a more meaningful analysis of beach volume changes.

2.2 Renourishment, monitoring and recycling

Beach renourishment at Seaford took place between April and October 1987 and involved pumping 1,450,000m^3 of shingle from an offshore site over a beach length of 2,500m. This was landscaped to form a profile with a 1:7 slope and a crest width around 25m. The crest was to be 6.5m above ordnance datum; this compares to a MHWS tide level of +3.1m. The final beach was completed in spring, 1988.

Since renourishment, the beach has been regularly surveyed using a combination of topographic, hydrographic and aerial techniques. Hydrographic and topographic surveying effort has been gradually decreased over the years following renourishment and at present there are 58 transects. These were surveyed approximately every three months between October 1987 and June 1991, with a hydrographic survey taking place on every other survey. Aerial surveys have taken place every spring since 1988, and continue as part of the then NRA Southern Regions overflight of the south coast.

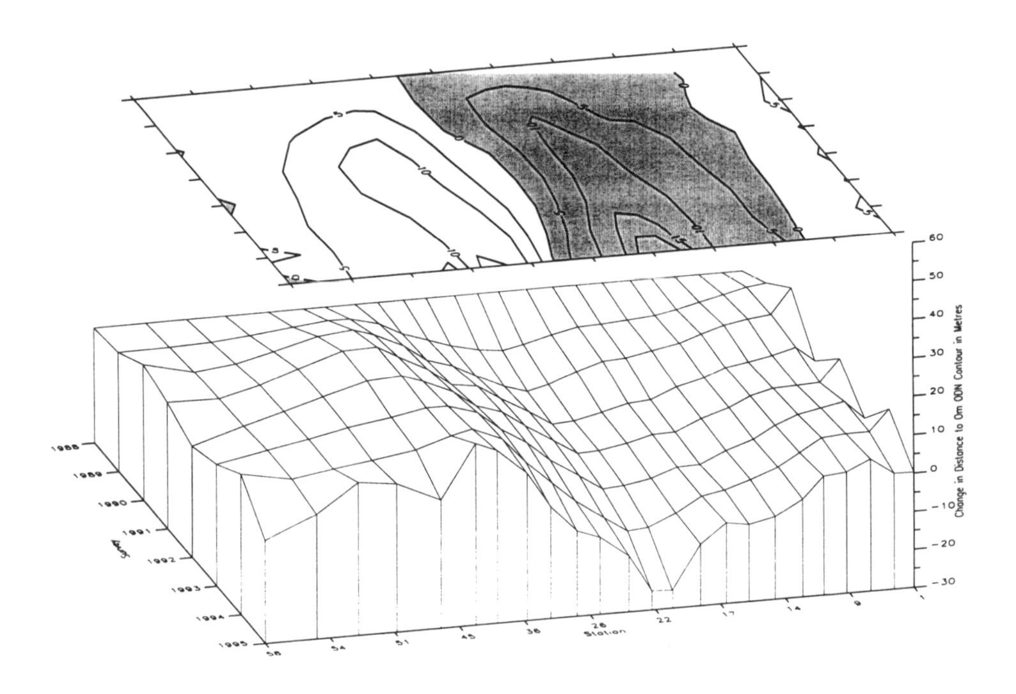

Figure 2 Aerial surveys of Seaford beach 1988 - 1995

The coarse temporal resolution of the aerial surveys means that they are not able to represent seasonal changes in beach levels, and also they can only provide data above the low-water mark. They do however, provide useful information on the general performance of the beach. A comparison was made between data from the two survey methods and a close agreement was found. Figure 2 shows the change in distance to the 0m beach contour at Seaford, based on the NRA survey data.

In order to maintain the renourished profile, periodic recycling of beach material is required. The amount of recycling needs to match the drift of material along the beach; the original study predicted a predominant eastward drift of shingle and this was to be recycled from segment A of the beach to segments B and C. Figure 3 shows how the natural changes in beach volume caused by longshore drift match the recycling programme. From this it can be seen that at the eastern end of the beach, the recycling is closely matched to longshore drift with the removal of around $70{,}000m^3yr^{-1}$ resulting in a stable volume in this segment. However along the central part of the beach, particularly in front of the Salts recreation ground (Segment B), losses are still occurring, but these are matched by gains in volume at the western end of the beach in Segment D. In fact these losses almost exactly match the gains, with an overall small net increase in total beach material of around $5{,}000m^3yr^{-1}$. In consequence, the recycling operation needs to be modified only slightly with extra material, say $20{,}000m^3yr^{-1}$ being moved from segment D to segment B to provide a stable beach.

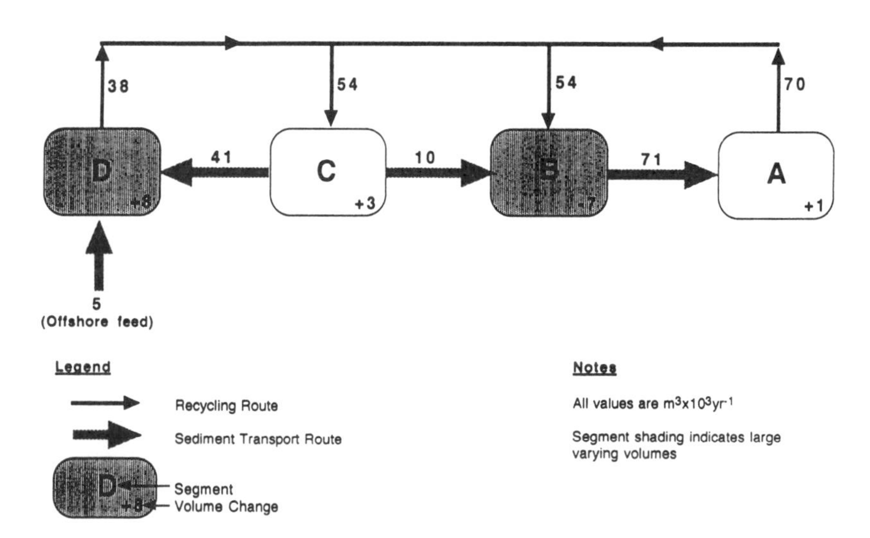

Figure 3 Shingle recycling operations at Seaford

2.3 Beach modelling (theory)

Modelling of the beach evolution involved three operations, incorporated into two models. First, using the HINDWAVE model an offshore wave climate is hindcast from wind data and calibrated against wave-rider buoy measurements. This is then refracted inshore and the longshore transport due to this wave climate calculated by the BEACHPLAN model. The sediment transport model is essentially a finite difference solution of the following equation which expresses the continuity of the volume of sediment moving along the shoreline:

$$\frac{\delta Q}{\delta x} + \frac{\delta A}{\delta t} = 0 \tag{1}$$

where

Q	is the volume rate of longshore sediment transport ($m^3 s^{-1}$)
x	is the distance along the shore (m)
A	is the beach cross-sectional area (m^2)
t	is the time (s).

The width of the beach at a given height can be obtained from the beach cross-sectional area. This model is accordingly of a type known as 'one line', as the beach position is given by the location of a single contour, e.g. the high water line. The sediment transport model is based on the widely used CERC (US Army, Coastal Engineering Research Centre) formulae for sediment transport which is dominated by breaking waves:

$$Q = \frac{K_1 E_B (nC)_B \sin(2\alpha_B)}{Y_S} \tag{2}$$

where

K_l is a non-dimensional calibration coefficient

E is the wave energy density = $0.125\rho g H^2$

H is the wave height (m)

ρ is the water density (kgm^{-3})

g is the acceleration due to gravity ($9.81ms^{-2}$)

Y_S is the submerged weight of the beach material

nC is the group velocity of the waves (ms^{-1})

α is the angle between the wave crests and local depth contours

B is a subscript that denotes the breaking wave condition

This simple model represents a good compromise between versatility and accuracy and has been widely used on a large number of engineering studies throughout the world. Equation (2) was originally developed for the transport of sand where a value for K_l of around 0.35 is used; however the model can be effectively used for shingle by reducing K_l by a factor of around 20. Tests at the University of Southampton using tracer pebbles suggested that a value of around 0.02 is an appropriate starting value for shingle.

2.4 Beach modelling (results)

For the modelling of Seaford, a baseline survey of August 1988 was chosen as this represented the beach in its most complete state following final beachscaping earlier in the year and allowing the beach time to settle. Wave data was synthesised from the Shoreham wind station from August 1988 to June 1991. Hence the model could be run using the actual wave climate present at the time of the bathymetric/topographic surveys. The model was also set to give an output at dates corresponding to these surveys and carry-out recycling between the actual dates it took place.

Figure 4 shows the results of the modelling to predict the drift rate along the main frontage of Seaford. These results show that to obtain a correct drift rate of $70{,}000m^3yr^{-1}$, a calibration factor of K_l=0.04 is required, double the typical value of 0.02. With the absence of any suitable calibration data, K_l=0.02 was used in the original study until survey data revealed that it was under-predicting the drift rate. It is also interesting to compare these results for the drift rate with those from the wave climate for the original study. This was synthesised from ten years of wind data from the Dungeness wind station. With this data drift rates are under-predicted, indicating that the 1970s was a particularly quiet decade from a wave activity perspective. In the light of these findings, future studies may require more than ten years of data to establish a long-term wave climate.

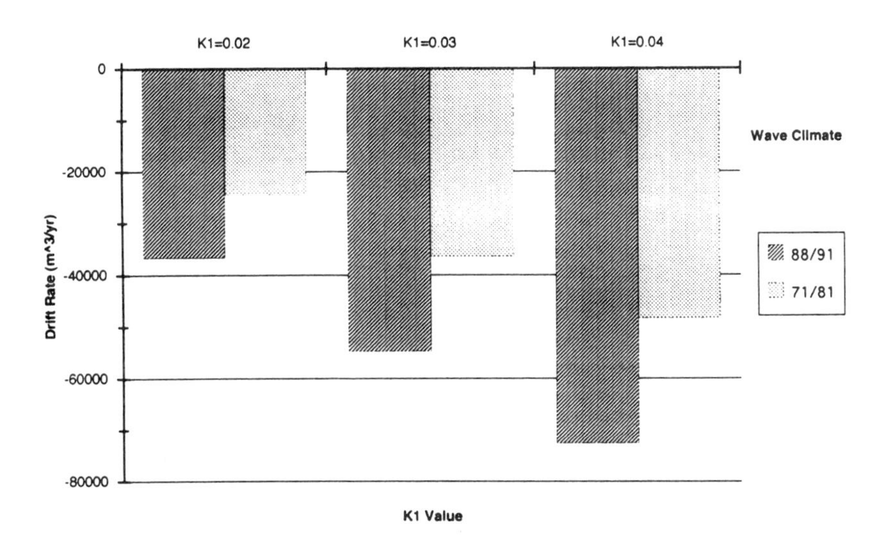

Figure 4 Modelled drift rates at Seaford

2.5 Analysis of beach modelling

In 1983 a series of mathematical modelling studies was carried out to design a beach renourishment and subsequent recycling scheme for Seaford. Since the completion of the beach renourishment in 1987, regular recycling has maintained an adequate beach width, confirmed by eight years of survey data. The performance of the completed beach as a viable method of sea defence was proved in the great storm of winter 1989/90. The success of the beach management scheme is due to a partnership between modelling and feedback from the environment. Complex systems such as coastal processes are difficult to model, and engineering judgement is always required to decide on the best mathematical representation of certain model components, since there will always be site specific criteria that cannot be accounted for in a general purpose model.

Since the original studies in 1984, the models used have been refined continually and the latest generation contains new refraction and diffraction algorithms. A new bed load transport algorithm for shingle has also been developed at HR Wallingford for incorporation in the BEACHPLAN model. The drift rate calculated from this model agreed well with the CERC drift rate with K_1=0.04. One-line models such as BEACHPLAN will continue to be used into the future, primarily because they are easy to use and computationally less demanding. However simple models rely on site specific calibration to account for factors not represented in the model, but once calibrated, their accuracy is equivalent to that of more sophisticated models. The skill lies in selecting which is the most appropriate modelling technique to use. Incorporating extra factors into models is always a trade-off between cost and the benefit that will be realised by this extra sophistication.

3. Satellite remote sensing and coastal engineering

3.1 Data evaluation

Data from Earth observation sensors will only be used by coastal engineers if it can provide useful information. This means that, no mater how sophisticated the techniques employed to process the images, unless this processing leads to data that can be used by coastal engineers then it will be of no more than academic interest. In other words, and in general terms, the data has to be valuable. Data is only valuable when it can be used to provide *information* and information enables us to *manage the environment*. As information demands are not wholly static, data requirements will also vary. Understanding why Earth observation data is not generally valuable to user groups such as coastal engineers is currently attracting a lot of interest as it is potentially a major barrier to the development of more commercially funded satellite missions.

Value Parameter	Description	EO Positive	EO Negative
Contribution	What impact the data have on solving the problem. Fundamentally, data on tidal currents will have a greater impact on solving a beach erosion problem than sea surface temperature measurements; similarily inshore waves data will typically make a greater contribution than tidal currents data.	Provides good baseline data and synoptic overview	Limited set of variables
Quality	This involves the accuracy of the data, the precision of the measurements and the reliability of the measurements. Associated with this is the concept of meta-data which enables assessments about data quality to be made.	Often comprehensive meta-data and traceable QC procedure	Poor spatial and temporal resolution. Accuracy of derived physical parameters is debatable
Location	This is particularly important as most coastal engineering projects are site specific. Time is important also as data may only be required after a given event has occurred, e.g. construction of a harbour jetty.	Often the only available data	Location (temporal and spatial) often does not match project site
Usability	How easy it is to process the data. This is particularly the case if the data is supplied in a digital form that requires specialist extraction algorithms, which may be explained in a foreign language.	Potential for direct input to GIS	Specialist algorithms required to extract physical parameters. Data files often large and integration in-situ time series data not straightforward
Deliverability	How easy it is to obtain the data. This particularly includes the delivery time for the data.	On-going work on pan-European data catalogues.	Not easy to locate specialist processors and data providers.

Table 4 Data value criteria

The ENVALDAT proje0ct (Customer Valuation of Environmental Data) examines the price organisations are prepared to pay for data and the amount for which that data can be sold. The project, led by HR Wallingford, runs until April 1998 and seeks to elicit the methodologies presently used for data valuation. ENVALDAT suggests that five parameters chiefly determine the value of data. These criteria, shown in Table 4 in order of importance, should really be referred to as describing 'potential value' as a data set may not exist, or it may exist but with its whereabouts not known. The table also shows the positive and negative aspects of Earth observation (EO) data with respect to the value criteria.

3.2 Data from satellites

This section gives four examples of data that is required for coastal engineering and gives a summary of the arguments why EO data is not ideally suited at present.

Hydrometric data

In the example of hydrometric data, the focus will be on the provision of wave data. Obtaining wave height data from satellites using altimeters such as ERS-1, GEOSAT and Topex-Poseidon is now well established and companies in the UK (Satellite Observing Systems) and Norway (OceanOR) have established large data archives and associated derived statistics that are sold on a commercial basis. While this data has found use in the offshore and shipping industries, it has limited benefit to coastal engineers. First, coastal engineers require three components to their wave data: height, period and direction; secondly they require the data in the form of a time series, say sampled at 1-3 hour intervals, with associated statistical measurements of height, period and direction.; thirdly the measurements should be at an inshore point (an area less than 10m square), say 1-2km from the shore. (Note that time series data are generally preferred to drive numerical models: statistical data can easily be derived from time series data, whereas it is not as straight forward to derive time series data from statistics).

The repeat cycles of satellites means that collecting time series data at three-hourly intervals for the same location is not possible and hence only statistical information can be derived, usually within an area typically several kilometres square. In addition only height information can be derived from altimeters, with direction data available separately from SAR techniques, and the two do not always correspond at the temporal scales required for engineers. Period information cannot presently be derived from satellite data with any great accuracy. Lastly, altimeters are only accurate when their entire footprint is over the sea and this in reality limits them to at best around 15km from the shoreline.

Most coastal engineers use models to simulate wave data and/or deploy devices such as wave rider buoys. Wave rider buoys are placed directly in the nearshore environment and are used to measure directly wave height period and direction. Such deployment is expensive, costing of the order of £20,000 per annum per buoy in total, and such models are often used to generate wave climates from wind data. The UK Meteorological Office (UKMO) operates a global wind-wave model and HR Wallingford purchases data directly from this and processes it to provide wave data for coastal engineers.

Topographic Data

Coastal engineers require accurate topographic information to monitor such parameters as the recession rates of beaches and cliff tops and also for mapping ground elevation in areas at risk from flooding. This typically requires a spatial resolution of around 1m with a vertical resolution of around 1cm. Surveys of beaches and cliffs should be conducted annually, but every three or four months is ideal. This kind of information is usually not provided by national cartographic agencies such as the Ordnance Survey as it is not required by the vast majority of their customers.

Air photography has been used with great success for this kind of survey work, and the UK Environment Agency has conducted trials using both airborne SAR and airborne LIDAR for mapping beaches and flood plains with excellent results. (LIDAR, 'Light Induced Direction and Ranging'. is an active sensor based on a scanning laser techniques). Satellite data has not been suitable for this kind of work in the past, due primarily to low spatial resolution; the best resolution available to date is 10m with Landsat. However there are exciting possibilities with the new range of high resolution optical data that will become available in 1998. Earlybird, Quickbird and Carterra offer spatial resolutions of 3m, 1m and 1m respectively.

Nearshore bathymetric data

There is a general shortage of good bathymetric data for coastal engineering. The prime reason for this is that coastal engineers require bathymetry data in the nearshore region, say from the MHWS level to the 10m depth contour. This region contains an element of both land and sea, but in the UK at least, it is not routinely surveyed by either the Ordnance Survey or the Hydrographic Office, except, in the latter case, it forms part of the approach to a harbour etc. For a typical project, coastal engineers would ideally require surveys once or twice a year over a five year period, or longer, to understand how the bathymetry had changed over time.

Traditionally, bathymetric data has been collected using ship-based sonar techniques, but these are difficult to use in shallow water due to the action of wave motion on the survey vessel and are also costly. Accordingly Earth observation techniques are being examined closely as a way of providing accurate data on the nearshore region at a reasonable cost. Optical techniques have been used in clear waters such as coral seas where the bathymetry is visible, e.g. by analysis of Landsat imagery. These techniques are claimed to be able to map bathymetry to 20m, but this technique does not lend itself to the more turbid waters typically found around the UK coastline.

SAR techniques have been used in two methods. The first method developed at the University of Reading produces diagrams of inter-tidal elevation (Mason *et al.*, 1996). This uses SAR imagery to detect the land/sea interface and this is coupled with an accurate tidal model to derive the elevation at this point. Repeating this procedure for several images enables the derivation of an elevation map. The drawback with this technique is that it can only be used as low as the MLWS mark and the accuracy is less certain on steep beaches where a large change in tide is required to cause a significant shift in the shoreline.

The second approach uses SAR imagery coupled with hydrodynamic models and requires a sparse set of in-situ depth measurements and favourable meteorological and hydrodynamic conditions (winds 3–5ms^{-1} and tidal currents about 0.5ms^{-1}). Essentially the SAR imagery detects variations in the surface wave spectrum. Variations in the surface wave spectrum are caused by modulations in the surface flow velocity as a result of interactions between tidal flow and the bottom topography. Initial trials of using this technique claim depth accuracies of ±30cm.

Aircraft based LIDAR, as described for topographic mapping, also has applications for bathymetric mapping. The technique employed is the same as for land mapping, except that a different wavelength for the laser has to be used. Using this technique, depths up to three times the secchi depth can be mapped and accuracies of ±30cm are claimed.

Morphodynamic data

There is a great lack of available morphodynamic data and consequently morphodynamic understanding. The main reason for this is the huge costs incurred in collecting this type of data using traditional ship-based methods. The increased understanding in the need for strategic planning of coastal defence based around natural sediment cells has stimulated the need for greater understanding of morphodynamic processes. The concept of a shoreline management plan (SMP) advocates a cascading of plans from the strategic (sediment cell level) to the regional (sediment sub-cell level) to the local management units, defined largely on the characteristics of the shoreline.

Satellite-based systems offer great potential to provide data on morphology at the cell and possibly sub-cell level that can be inter-related with local data obtained from either traditional methods or using airborne techniques. Section 4 describes in detail such a project to obtain morphodynamic data using remote sensing techniques.

4. The COAST project

4.1 Background to COAST

COAST (COAstal earth observation of Sediment Transport) is a project that was funded as part of the BNSC (British National Space Centre) Applications Development Programme (ADP). The aim of the project was to develop a concept for the commercial use of sediment transport data from remotely-sensed sources, meeting with the wider aim of the European Space Agency (ESA) to promote commercial exploitation of data obtained from satellites.

The concept of extracting data on suspended sediment from Earth observation images had been proved in earlier satellite missions such as SPOT, Landsat MSS and Coastal Zone Colour Scanner (CZCS). Future sensors with improved spectral resolutions such as SeaWiFS offered the potential to provide quantitative information on sediment transport, suitable for commercial exploitation. As such, the prime aim of COAST was to bring the concept of measuring suspended sediment using data from space out of the research environment and closer to the market. The project ran for two years and was completed in 1996. The main components of the COAST project were:

1. To test and, where necessary, develop algorithms for the retrieval of surface Suspended Particulate Matter (SPM) concentration from SeaWiFs and CASI data.
2. To develop a method for calculating the concentration of sediment through the water column.
3. To demonstrate the utility of COAST data products in a GIS environment.

Item 3 was a less significant component of the work and so we shall concentrate on items 1 and 2.

The concept of the COAST project is shown in Figure 5. This shows how the atmospheric correction, SPM concentration retrieval and SPM concentration profile components of the work integrate together, and the ancillary data requirements. The method for calculating the SPM concentration through the vertical is very important from the perspective of coastal engineers as remote sensing techniques can only give a measure of SPM concentration in the first few metres of the water column. The output of the sediment profile algorithm can be combined with a current model to give sediment transport vectors or with a model such as PISCES to predict morphology.

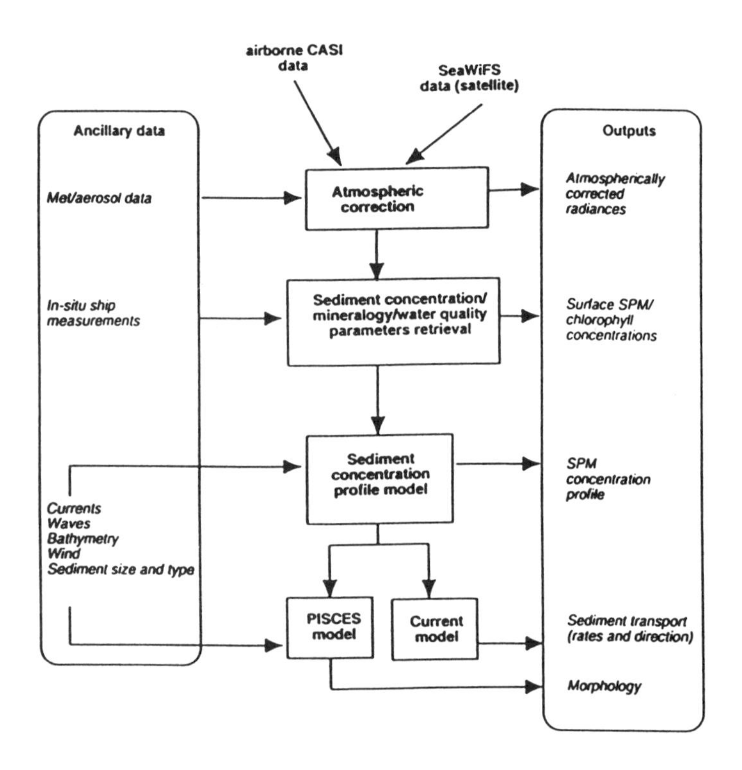

Figure 5 COAST concept

4.2 SeaWiFS and CASI

Processing algorithms developed during COAST were designed to operate with data from both satellite and aircraft, specifically:

(a) SeaWiFS: the Sea viewing Wide Field-of-view Sensor, part of the SeaStar platform. It was planned for launch in 1994, but was finally launched in August 1997 and the first data is expected around December 1997.

(b) CASI: the Compact Airborne Spectral Imager. In the UK there are two scanners. One is operated by the Environment Agency (EA), and the other is operated by the Natural Environment Research Council (NERC)

However the algorithms should be adaptable enough to work with spectral scanners that have the following minimum specification:

- one visible waveband,
- short near-infrared waveband (745-760nm)
- long near-infrared waveband (838-885nm) and
- ideally one waveband 705-715nm.

Therefore, although primarily designed for CASI and SeaWiFS data, COAST software should in theory be able to use data from alternate satellite sources such as MERIS (MEdium Resolution Imaging Spectrometer), MODIS (MODerate resolution Imaging Spectrometer) and alternative airborne sources such as DIAS (Digital Airborne Imaging Spectrometer) and VIFIS (Variable Interference Filter Imagine Spectrometer).

Adopting a two-sensor approach is necessary to ensure that the SPM data obtained can begin to match the spatial and temporal requirements of coastal engineers. This can be seen by examination of Table 5 which compares 'ideal' collection of SPM data with the collection that can be provided by CASI and SeaWiFS. Earth observation data can provide a wide-area overview, whereas aircraft data can provide data closer to the coast and in estuaries, where greater temporal and spatial resolution is required. It should be noted that the delay in the launch of SeaWiFS meant that suitable data was not available for use during the COAST project.

Parameter	'Ideal' measurement	Airborne scanner	SeaWiFs capability
Coverage	Wide area (~100km)	~ 10km	Wide area (~100km)
Repeat frequency	few hours	few hours	1 day
Number of measurements annually	~ 100s (seasonal and tidal variations)	~ 10 (several survey campaigns)	50-100 (based on period of orbit and cloud cover statistics)
Spatial resolution	~10-20m (to detect local detail and wide features)	1-10m	1km
Sediment concentration measurements	Accurate measurements of entire water column sediment concentration	Measurements of surface suspended sediment concentration (estimate of <10% error)	Measurements of surface suspended sediment conc. (estimate of <10% error)
Sediment transport measurements	Accurate measurements of total sediment volume transported in the water column	Measurements of surface suspended sediment transport (top few metres)	Measurements of surface suspended sediment transport (top 1-2 metres)
		Estimates of total sediment volume transport	Estimates of total sediment volume transport

Table 5 Sensor coverage

4.3 Atmospheric correction and product extraction

Atmospheric effects account for 80% to 90% of the visible signal detected by airborne and satellite sensors. In consequence accurate atmospheric correction is of great importance in obtaining quantitative data. Atmospheric scattering adds radiation to and removes radiation from the path sampled by the remote sensor. Rayleigh scattering is the result of radiation interacting with atmospheric molecules: it is inversely proportional to the fourth power of wavelength, hence is stronger in the blue bands. Mie scattering is due to atmospheric aerosols such as water vapour and dust particles: it is inversely proportional to wavelength.The Rayleigh component of scattering is relatively straightforward to calculate as the molecular composition of the atmosphere can be considered to be stable. The aerosol contribution is difficult to calculate because atmospheric aerosols are highly variable both spatially and temporally.

For Case 1 waters (stratified shelf seas and deep ocean), atmospheric correction theory is well established and is based on the assumption that the water leaving radiance in the near infrared (NIR) is zero due to high water absorption at these wavelengths. Thus all the NIR signal measured by a remote sensor can be assumed to originate from the atmosphere and the aerosol (Mie) component of scattering can be established by subtracting the calculated Rayleigh component. These results are then extrapolated across the other spectral bands. This approach to atmospheric correction is known as the 'darkest pixel' method.

For Case 2 waters (coastal waters that contain significant amounts of SPM) the 'darkest pixel' method cannot be used as there is a significant amount of water-leaving radiance in the NIR . As part of the COAST project researchers at Plymouth Marine Laboratory accordingly developed a new method for atmospheric correction in Case 2 waters known as the 'bright pixel' method. This approach uses a coupled hydrological and atmospheric optical model and requires a waveband in the range 705–715nm. During image processing, each pixel is processed initially to mask off land and cloud followed by calculation of the Rayleigh scattering component. A flag is then set based on the water reflectance at (710±5)nm which discriminates between Case 1 and Case 2 waters at SPM values as low as $1 mgl^{-1}$ over a range of atmospheric aerosols. The value of $1 mgl^{-1}$ is commonly accepted as marking the separation between Case 1 and Case 2 waters. Once images have been atmospherically corrected, values of suspended sediment can be calculated. Presently a simple band ratio algorithm is employed, similar to that used for CZCS data, although more advanced methods are under development.

Typical results obtained using the COAST processing algorithms are shown in Plate 9, (opposite page 175). These images were collected by the Environment Agency over Southampton Water on the south coast of England in August 1995. The River Hamble enters a third of the way Southampton Water. Plate 9(a) shows the total uncorrected radiance at 600nm. Plate 9(b) shows the atmospherically corrected image with the land masked off at 600nm. Plate 9(c) shows the water-leaving radiance at 865nm after atmospheric correction; the Case 1 (black central portion) and Case 2 (purple) areas can be clearly differentiated. Finally Plate 9(d) shows the derived SPM concentrations.

4.4 Vertical profile modelling

The distribution of suspended sediment through the water column depends on a number of variables, of which the most important is the grain diameter. Muds (fine clays and silts) , with grain sizes in the range 0–62μm are the most easily suspended by current or wave action, and hence are the most likely to be visible at the water surface. Grains of diameter 62–2000μm comprise sands . The coarsest sizes in the range 500–2000μm would rarely be suspended in the coastal environment, except possibly in the surf zone. Grains in the range 200 to 500μm would very rarely be suspended at a height greater than 1m, and so would not be visible at the water surface. Grains between 62 and 200μm will be relatively easily suspended, and could reach the water surface, although the concentration will be much lower than at the bed.

The behaviour of sands in suspension is relatively well understood, particularly for steady currents, *e.g.* rivers. The behaviour of muds, including their vertical variation in concentration, is not particularly well understood. However, the basis for all widely accepted theories for sediment concentration profiles is diffusion theory. This assumes that the downward settling of the grains is balanced by an upward diffusion of grains through turbulent processes. While it is beyond the scope of this chapter to venture into theoretical hydrodynamics, it is worth outlining the principles of diffusion theory and this predicts that the concentration profile forms a single curve, or narrow family of curves if it is plotted as

$$\ln [C(z)/C_s] \quad \text{versus} \quad (U^*/w_s) \ln [Z/h_s]$$

where

$C(z)$	is the concentration at height z above the bed,
h_s	is the height close to the water surface,
C_s	is the concentration at h_s,
w_s	is the settling velocity of the grains and
U^*	is the friction velocity of the flow.

The formulas derived for calculating the concentration profiles were incorporated into a software algorithm. The algorithm is capable of giving results of increased confidence, depending on the level of ancillary data available as specified in Table 6. The performance of the vertical profile model was tested against field data from Christchurch Bay on the south coast of England. These results are shown in Figure 6 for four sample survey sites. The results are in general impressive, particularly given the poor spatial sampling of the in-situ measurements through the vertical. However there are some notable dissimilarities, for example station 3. This in part is likely to be due to the low concentration of sediment (mud) found in Christchurch Bay.

Method	Sediment type	Data known
1	Unknown	Surface concentration and height, water depth
2	Mud	As 1, plus sediment type
3	Mud	As 2, plus site specific mud coefficient
4	Sand	As 2, plus site specific sand constant
5	Sand	As 2, plus d50 of bed material, d50 of suspended sediment, current speed, wave height and period

Table 6 Model options for profile predictions

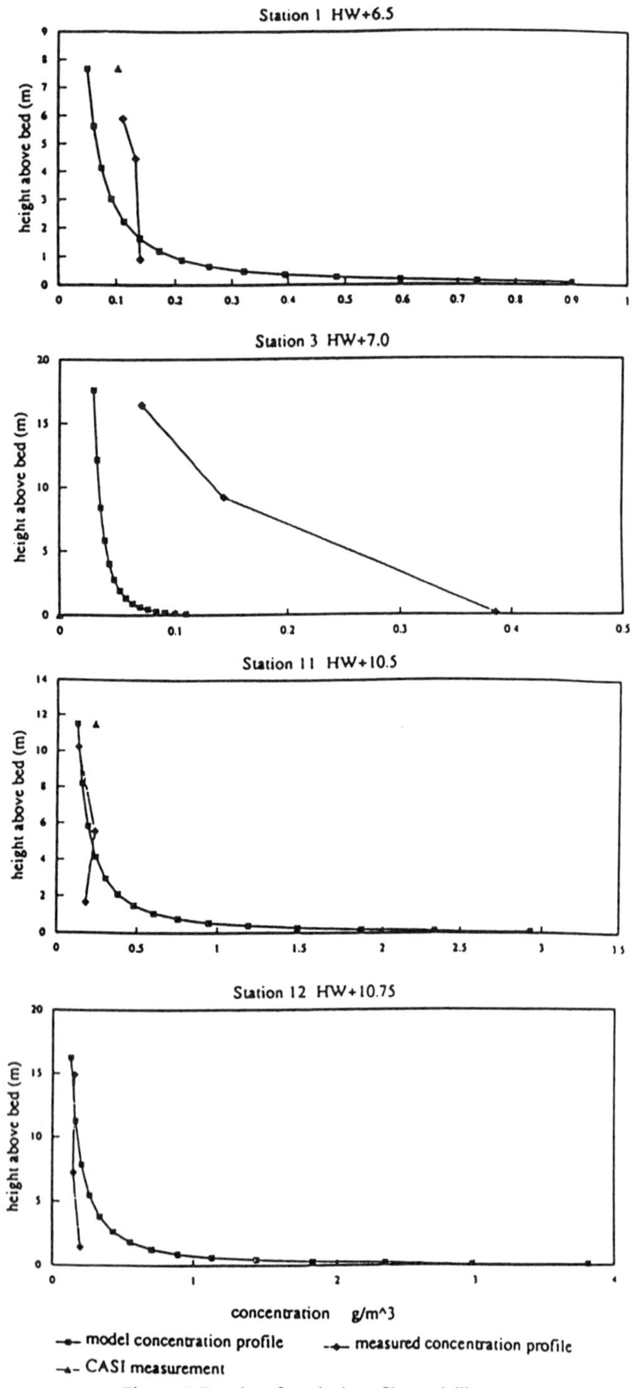

Figure 6 Results of vertical profile modelling

The important point to remember about this approach is that it can only be used under conditions where the sediment suspension is visible at the water surface. This excludes a range of sediments, primarily those with grain sizes larger than 200μm (medium sands upwards) since most of the sediment transport occurs close to the bed. Also, estuarine environments can contain bottom layers of fluid mud up to a metre thick that are sufficiently mobile to move at velocities comparable with the current speed, but do not experience any re-suspension throughout the water column. This transport mechanism is also not visible at the water surface. Even in conditions of visible sediment suspension, however, false conclusions about the volume of sediment transport can be drawn. Strong density stratification, e.g. in salt wedge estuaries, results in sediments in the upper layers being carried down rivers, isolated from those in the bottom layers which remain stationary.

5. Summary

From this chapter it is important to draw together some of the key points about coastal engineering, remote sensing and computer modelling. The physical processes that take place in the coastal zone, i.e. waves, tides, currents and their interaction with the seabed and shoreline and coastal structures are not completely understood; consequently, assessing how they will be modified by, for example, the construction of a new marina is difficult. To date the most common way to make such forecasts is by the use of mathematical models. These models, however, are only as good as (a) the level of understanding of the processes that are being modelled and (b) the accuracy of the data available to drive these models.

Coastal engineering is very data hungry but data collection by in-situ techniques on a large scale is expensive and time consuming. Accordingly good data to drive models (and correspondingly to provide feedback for the development of more accurate models) is not easy to obtain. Remote sensing, and Earth observation techniques in particular offer the benefit of providing 'data on tap'. That is, the data collection has already been performed and all that is required is image selection followed by the selection of an appropriate processing algorithm.

This is the promise that many coastal engineers felt that Earth observation would be able to make, but in reality this was not the case. EO data has demonstrated benefits in areas where spatial resolution of upwards of a kilometre and temporal resolution upwards of a day are acceptable, for example in meteorological forecasting and oceanography. When these scales are compressed to those required in coastal regions - spatial resolution upwards of a metre and temporal resolution upwards of an hour - Earth observation data from space has understandably failed to make an impact. Furthermore, satellites are regarded as unreliable and the common perceptions are that they are rarely launched on time and can fail midway through their mission life time.

Strategic management initiatives such as SMPs and CZM schemes have identified the need for wide area synoptic data of the type that can be readily provided by satellite. These can be supplemented by data collected in local areas by in-situ or aircraft techniques, where higher temporal and spatial data are required. In addition, research over recent years, stimulated by initiatives such as the European Commission's CEO (Centre for Earth

Observation), has identified new possibilities of providing data that are valuable to coastal engineers. Furthermore, future satellite missions promise both greater spatial resolution and improved accuracies for the extraction of data and accordingly the role played by Earth observation data in coastal engineering will increase.

Acknowledgements

I would like to acknowledge the following individuals and organisations whose background material I called upon in preparing these lecture notes. HR Wallingford (Reports EX 1345, 1346, 1768, 1986, 2533, TR4, 1996b COAST Sediment Profile Algorithm - Functional specification and 1997 Nearshore bathymetric surveying using remote sensing techniques), Plymouth Marine Laboratory, Smith System Engineering (1996 COAST, Final Report), The UK Environment Agency (Southern Region) and the following members of staff at HR Wallingford: Richard Soulsby, Howard Southgate and Alan Brampton.

Reference

Mason D, Davenport I and Flather R, 1996, Improved coastal zone management and flood forecasting using SAR-derived inter-tidal digital elevation models. Proceedings of the Second ERS Applications Workshop, London, UK, 6-8 December, 1995 (ESA SP-383, February 1996)

Remote sensing for coastal zone management – new techniques and applications in the Netherlands

Raymond Feron

Rijkswaterstaat, The Netherlands

1. Introduction

The North Sea coast of the Netherlands is made up of dunes, dikes, and other water barriers. Together they protect the Netherlands against the North Sea. Dune areas represent about three quarters of this line of defence, varying in width from less than one hundred metres to several kilometres (Plate 1, following page 174). Under the influence of the forces of nature this barrier is constantly moving, advancing seaward in some places and receding landward in others. The patterns of coastal accretion and coastal erosion are well known thanks to an extensive monitoring programme over the past few decades. The current policy with regard to coastal protection (dynamic preservation) utilises the natural processes along the coast wherever possible. This allows our coast to maintain its characteristic appearance with special natural values. Where necessary, sand nourishment programmes (annual costs DFL 70 million) are used as a preventive measure to stop structural coastal recession.

The introduction of innovative techniques, based on remote sensing, in the Rijkswaterstaat organisation is one of the mission statements of the Survey Department. In the Netherlands, coastal zone management is an important governmental activity which justifies strategic investment of resources in new techniques that are applicable in the coastal zone. Although several other remote sensing based techniques are being developed (e.g. multi-spectral scanners, digital photogrammetry, water quality monitoring and change detection of urban regions) this chapter will focus on three techniques. These three techniques are applicable in the coastal zone and result in accurate topographic information.

The first and most promising is airborne laser altimetry. Laser altimeter systems can deliver highly accurate digital elevation models (DEMs) with only a fraction of the costs of conventional techniques e.g. photogrammetry. In relation to aerial photogrammetry, either analytical or digital, laser systems have advantages based on more flexible working time and faster processing times. In the Netherlands laser technology has recently been introduced for operational coastal zone monitoring. Laser derived digital elevation models (DEMs) will supply coastal zone managers with accurate and reliable height information for the beach and coastal dune area. This new technique provides necessary data sets for coastal zone management, aiming both at the reduction of costs and a quality improvement in terms of accuracy, precision, and resolution of derived products. Laser scanning will provide an elevation model of the sandy coast. From the DEMs, profiles of the beach and fore dunes will be extracted on a yearly basis, replacing the costly and time consuming determination of profiles through photogrammetric means. At the same time the laser scanning data, covering an area in total, provides an opportunity to assess more accurately the volumes of sand having been eroded from beach and fore dunes.

The second is the so-called Bathymetry Assessment System (BAS) which is a good example of the integration of remote sensing with physical knowledge (models). An innovative method has been developed using radar satellite images, hydro-dynamical models and a limited set of in-situ depth measurements to interpolate between a limited number of depth profiles. Under favourable meteorological and hydrodynamic conditions, airborne or spaceborne SAR imagery shows features of the bottom topography of shallow seas to a maximum depth of approximately 30m. The imaging mechanism is explained, resulting in a set of equations that allow SAR image variations to be interpreted as variations in the bathymetry. Based on the physical mechanisms a suite of models has been developed.

The third technique that will be discussed is interferometric SAR or InSAR. This interferometric SAR technique can be utilised to derive height information, either absolute or relative, of the Earth's surface. Both techniques will be introduced with emphasis on the observational concept and underlying physical processes. Compared to optical sensors the use of radar for Earth observation has only just begun. The most important reason to use radar instead of optical sensors is its capability to penetrate through clouds and its day and night imaging capabilities. Radar interferometry might add another reason to use radar. Recently it has been shown, that with spaceborne SAR, deformations of the surface of the Earth can be measured with centimetric accuracy. This information becomes available using interferometric processing techniques on multiple satellite images.

The application of the techniques to coastal zone problems will be demonstrated. Where applicable, the results are discussed in relation to conventional techniques and quality assessment. Several case studies illustrate pros and cons. Since most new techniques deliver an increasing amount of data, efficient data processing will be discussed. Conventional techniques have a strong emphasis on the actual data collection. With the introduction of new techniques a functional shift occurs from performing measurements to the analysis and presentation of extremely large data sets. Finally, some conclusions and discussion will result in an outlook for future use of remote sensing in coastal zone management.

2. Laser scanning for height measurements

Airborne laser terrain mapping systems are less limited by environmental conditions than aerial photography. This advantage of airborne laser scanning technology combined with faster processing times makes it an attractive alternative to the traditional techniques for large-scale geo-spatial data capture. The recent development of commercial laser-based topographic terrain mapping systems is driven by the availability of compact solid state lasers, high speed scanners, precise airborne inertial navigation systems (INS), and low-cost multi-channel GPS receivers. Laser-based systems offer advantages over existing survey instruments in areas such as forestry or coastal zone monitoring. During the post-processing it may be possible to classify each data point as ground, building or vegetation. This classification is far from straightforward and is therefore subject to extensive research. Once the laser data has been successfully classified, a digital terrain model can be generated using straightforward interpolation. Other derived products like vegetation height, average building heights, and surface roughness can be derived from the laser data.

2.1 Principles of laser scanning

While technologically advanced in design and function, airborne laser mapping systems are based on simple concepts. It starts with a high accuracy laser range finder that scans beneath the aircraft. Depending on the scan angle, the laser scanner produces a swath over which the distance from the aircraft to the ground is measured for every laser pulse. Simultaneously with each laser pulse the scanner angles are measured. To correct for the aircraft's movements an inertial navigation system accurately records the roll, pitch, yaw and heading of the aircraft. Two or more GPS receivers are used to locate the aircraft's position within approximately 10cm. One receiver is installed in the aircraft, while the other GPS receiver should be located at a known ground location. The ground-based receiver is used to identify errors (troposphere, ionosphere, multi-path etc.) which can be used to correct the aircraft's position during post-processing. Figure 1 illustrates the basic concept and necessary elements of an airborne laser mapping system.

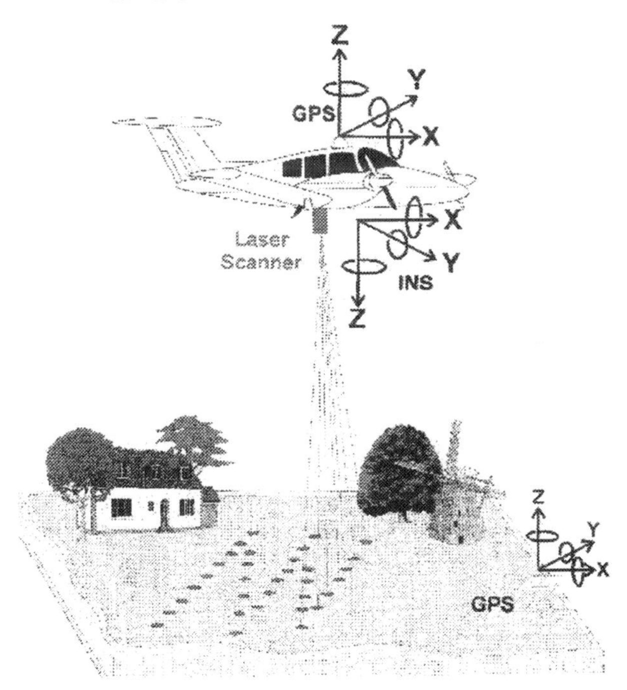

Figure 1. Principle of laser-scanning. A laser mapping system consists of a solid state laser, a high speed scanner, a precise airborne inertial navigation system (INS) and two or more multi-channel GPS receivers.

The two most important (characterising) parameters of a laser scanning data are the point density and the size of the foot print. These two parameters will be derived. The point density is determined by the average surface Ω that will be illuminated by one laser pulse

$$\Omega = D_{across} D_{along} \tag{1}$$

with D_{across} the across track distance and D_{along} the along track distance between subsequent laser points on the Earth's surface. The across track distance depends on the flight speed v and the mirror frequency f_m

$$D_{across} = \frac{v}{f_m} \tag{2}$$

This relation is valid for laser scanners that have a full rotation. The across track distance for scanners with a forward and backward scan is not constant. It will depend on the position relative to the turn-point of the mirror. The total scan width W_T is depending on the flight altitude H and the maximum scan-angle $\phi_{\max}$

$$W_T = 2H \tan \phi_{\max} \tag{3}$$

Defining W_X as the position within the scan line, relative to the turn-point, the across track distance is given by

$$D'_{across} = 2\frac{W_X}{W_T}\frac{v}{f_m} = \frac{W_X}{H \tan \phi_{\max}}\frac{v}{f_m} \tag{4}$$

The along track distance can be written as

$$D_{along} = W_T \frac{f_m}{f_p} = 2H \tan \phi_{\max} \frac{f_m}{f_p} \tag{5}$$

with f_p the laser pulse frequency. With these derived relations the point density Ω appears to be a function of flight altitude, flight speed, laser pulse frequency, and maximum scan-angle.

Another important parameter for laser-scanning is the foot-print of the laser pulse on the ground. This foot print is a function of the divergence of the laser bundle Θ and the flight altitude, H. The bundle divergence is determined by the optics in the laser system. The radius r of the footprint can be written in terms of Θ, H and the scan angle ϕ as

$$r = \frac{\Theta H}{2\cos\phi} \tag{6}$$

The receiving device basically consists of an optical frame which directs the received signal through a filter towards a photo diode. The filter allows only light with the transmitted wavelength to pass. The photo diode transforms the incoming light in an electric signal which is amplified and directed to a signal processing device. If the incoming signal is above some threshold the signal is stored and the arrival time is determined relative to its transmitting time. In order to assure an accurate distance measurement the sampling frequency of the receiving device has to be at least 1.5GHz. The combination of the filter and the threshold signal processing determines the signal to noise ratio. Depending on the application, the signal processor can be designed to register the first or last part of the reflected pulse, resulting in the vegetation or topographic height or ground level respectively.

2.2 Application of laser scanning data

The coast of the Netherlands has been monitored for over a hundred years. Since the mid-sixties, this has been done systematically and every year. The results are used to guarantee the safety of inland areas and for establishing the nature of changes in the morphology of the coastal zone. These height levelling measurements are carried out along sections perpendicular to the coast. The sections are positioned at intervals of 200 to 250m. At these sections the height/depth values are obtained up to a distance of about 800m sea-ward and approximately 200m land-ward from the first line of dunes. New technological developments, such as laser altimetry and the analysis of radar satellite images, will improve the spatial data coverage. It will be possible to measure the height/depth of large areas of the coastal zone within a short period of time. For the inner dune area laser altimetry data will allow surface changes to be monitored from year to year. These surface changes can be due to erosion and accretion of sand. However, the accuracy of the laser-derived DEM is affected by possible vegetation cover. The effect of this vegetation can in most cases be removed by filtering procedures. The laser DEM and thematic data will be placed in a GIS environment with specific tools to allow the user to combine depth and height measurements of the coastal zone, extract cross sections, assess the quality of elevation models derived and to generate bird's eye views of areas in which one is interested in.

In 1995 a number of experiments were performed to assess the precision and reliability of the laser technique for DEM acquisition as well as to gain insight into its operational aspects. Large project areas were chosen in several parts in the Netherlands along the coast and estuaries, the most important being the group of Friesian islands in the north. The total area involved several hundreds of square kilometres. The quality demands were specified as following:

- point density 1 per 15m^2
- a maximum systematic error of 5cm
- a precision of 15cm (standard deviation) for non-covered or sparsely vegetated areas
- a precision of 20cm (standard deviation) for areas covered with dense vegetation

In every project area, small patches were chosen where control measurements were carried out with a precision of a few centimetres. These validation measurements were used for checking on point density as well as assessing the geometric quality by simple statistical means. Furthermore, laser-derived coastal profiles were compared with the photogrammetric-derived coastal profiles. In Figure 2 the results of such a comparison are presented, made between 132 photogrammetrically surveyed coastal profiles of the island of Ameland, and the laser profiles. The latter are generated by interpolation from a regular grid of laser points to the planimetric coordinates of the profile points. For each profile differences are then calculated between the interpolated laser heights and the corresponding photogrammetric heights. From these differences, a systematic offset and an RMS error can be calculated for each of the 132 profiles. In Figure 2, the x-axis gives the position of a profile along the coast in kilometres from the most western point of Ameland. On the y-axis, the systematic offset and the RMS error are given in metres above Mean Sea Level. The RMS error is generally about 15cm in magnitude. The systematic offset starts low in the west (between kilometre 1 and 3), then suddenly increases to 15–20cm.

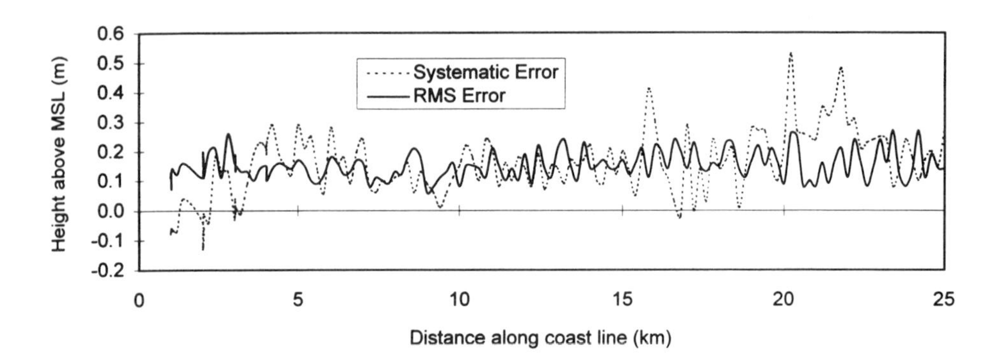

Figure 2. The results of a comparison made between 132 photogrammetrically surveyed coastal profiles of the island of Ameland and the laser profiles.

The results for 1995 were slightly disappointing, as in general the quality demands were not met. Systematic offsets larger than 10cm occurred, together with RMS errors of more than 20cm in magnitude. For assessing sand volumes at the beach, these are unacceptably high error values. Also, some parts of the delivered DEMs showed the presence of remaining large outliers that had not been filtered out. Apparently the filter algorithm was not able to handle the complexity of the terrain.

In trying to explain the presence of the errors encountered, several error sources could be traced – some of which were of such a nature that they allowed corrections to be made by reprocessing the raw laser data. This was true for errors related to the registrations by the Inertial Navigation System (INS) and over-saturation of the receiving system. Another systematic error source concerned the transformation between the WGS(1984) coordinate system and the Netherlands National Reference System, a transformation that was not defined with sufficient accuracy.

The dense vegetation cover in some areas and terrain type caused the largest loss of accuracy and point density. In dense vegetation areas the laser was simply unable to reach the ground. As a consequence, systematic offsets of several tens of centimetres were found to be present between laser data and control points. As these offsets varied for different control areas, even in the case for areas which had the same type of vegetation cover, no means were at hand to apply a valid correction for them. The terrain relief influences the degree in which the laser data can describe the terrain surface. With a point density of 1 per 15m^2, small-scale variations may be missed. Furthermore, sudden changes in the terrain surface, like steep dike slopes, were partly filtered out as the filtering program assumed these to be terrain surface anomalies. So here the filtering had been too severe, which once again shows that no standard filtering algorithm can be used for all parts and that additional terrain information must be added.

The laser data for 1996 has been validated with both GPS control measurements and with photogrammetric data. Table 1 gives the statistical summary of the comparison. The laser data performs excellently for flat areas with minimal vegetation cover. Vegetation has a significant effect on the offset. The results for dune areas, characterised by steep height variations and unpredictable vegetation variations, are again disappointing. These results have been analysed and are summarised in the following table. Height differences between photogrammetrically derived profiles and laser profiles more often have a systematic character rather than being random. Undersampling of dune areas may result in systematic errors. Dune-tops are missed, dune valleys are filled in, and dune slopes result in systematic height differences. The first two problems indicate that filtering has been applied too severely. The third can be explained either by large areas of vegetation or by a coordinate shift in horizontal direction i.e. planimetric accuracy.

	average offset (cm)	average st dev (cm)
GPS control data (hard topography)		
Raw laser data	4	6
Filtered laser data	2	9
Photogrammetric control data		
Filtered laser data beach	13	17
Filtered laser data dunes	15	48

For the future, filter procedures for the removal of noise, vegetation, and topographic objects still have to be improved. Purely statistical filters have to be combined with *a priori* information about the terrain. Improvements in the laser altimeter systems will probably improve the planimetric accuracy from approximately 1m to better than 30cm. This planimetric accuracy is of vital importance for dune areas. The use of the received signal amplitude (intensity) either in the processing or in the filter procedures may lead to further improvements in laser-derived digital elevation models.

3. Bathymetric Surveys using SAR Images

In the Netherlands, as of 1964, the topography of the coast is measured in profiles perpendicular to the coast line. The measurements consist of depth measurements of the near coastal zone and height measurements of the beach and fore dunes. In total 1700 such coastal profiles are measured yearly, through acoustic measurements and standard analytical photogrammetric means. Based on these combined observations it is decided where and how much sand is to be nourished. This dynamic maintenance of the coastline has been the guiding principle in the Netherlands for defending the lowland against the sea in the past five years. This relatively new strategy indicates that some freedom is allowed for the dynamic processes to take place. Knowledge of the actual depth of the sea bottom and shipping lanes and the height of the coastal defence works is fundamental to take corrective actions, if necessary. In practice this means that either sand nourishment or channel dredging is carried out. On a long term basis morphodynamical information is used to make predictions of future morphological changes. This type of information is used to formulate a long-term strategy towards coastal defence matters.

The traditional technique to measure sea bottom morphology, using *in situ* shipborne echo sounders, is expensive and time consuming. Based on nearly 10 years of fundamental and application research an innovative method has been developed using radar satellite images, hydro-dynamical models and a limited set of *in situ* depth measurements to assess a bottom depth chart. Under favourable meteorological and hydrodynamic conditions (moderate winds of 3 to 5ms^{-1} and tidal currents of about 0.5ms^{-1}), airborne or spaceborne

Synthetic Aperture Radar (SAR) imagery shows features of the bottom topography of shallow seas to a maximum depth of approximately 30m. Several demonstration studies have shown that an intelligent combination of SAR images and a limited amount of conventional echo-soundings can result in a bottom depth chart.

3.1 Principles of the Bathymetry Assessment System

Alpers and Hennings (1984) proposed and derived a theory, based on hydrodynamics, that explains in first order the imaging mechanism of bottom topography on radar sensors. We simply start with the relation they derived (assuming all the simplifications to be valid in first order). Alpers and Hennings (1984) have shown that the following relation can be derived between the amplitude or cross-section modulation and current gradient (given the fact that the advection time is large compared to the relaxation time)

$$\frac{\partial\sigma(x)}{\sigma_0} = -\frac{4+\gamma}{\mu}\frac{\partial U_x(x)}{\partial x} \tag{7}$$

where $\sigma(x)$ is the normalised radar backscattering cross section (NRCS), σ_0 is the constant background NRSC corresponding with areas where no variations in current velocity occur ($\partial U(x) = 0$), μ is the relaxation rate, U is the surface velocity, $U_x(x)$ is the surface velocity component in the radar look direction (x), and γ is a parameter varying between ½ for gravity waves and 1½ for capillary waves. The relaxation time τ_r, which is μ^{-1}, is the response time of the wave system to current variations. It is determined by the combined effect of wind excitation, energy transfer to other waves due to conservative resonant wave-wave interaction and energy loss due to dissipative processes like wave breaking. The relaxation time is difficult to measure; however, from theory we expect τ_r to be of the order of 10-100 wave periods. For typical wavelengths and wave periods observed with a SAR system this results in an estimate for τ_r to lie in the range of 1-100s. After some more derivations and restricting ourselves to a simple one-dimensional model for describing the bottom topography current interaction the following relation can be derived,

$$\frac{\partial\sigma}{\sigma_0} = \frac{4+\gamma}{\mu}|U_0|d_0\cos^2\phi\frac{d'}{d^2} \tag{8}$$

where U_0 is the constant equilibrium surface velocity, d_0 is the water depth outside the bank area, ϕ denotes the angle between the flight direction and the direction of the sandbank and d' is the slope of the bottom in the current direction. We can now conclude that an imaging radar system should image bottom topography best when the current flow is in the cross-track direction and when the topographic features are aligned parallel to the flight direction. Bottom topography will not be imaged by radar when the flow is parallel to the flight direction or when topographic features are aligned in the flow direction.

One of the main predictions of the above described imaging mechanism is that the image's intensity variations are proportional to the slope of the bottom topography divided by the square of the depth: d'/d^2. Figure 3 shows a schematic plot of the predicted relation between bottom profile and SAR image intensity variation, where several simplifications have been made. The steep slopes face the flow direction, and the steep slope regions are associated with strongly divergent surface currents in which the spectral energy density of the Bragg wave sharply decreases below its average value. Consequently, these surface areas show a much reduced radar back-scatter. Since the width of the steep slope regions is small, they appear as thin distinctive dark streaks in the image. However, the regions downstream of the sand wave crests, which are associated with weakly convergent currents, exhibit a slightly increased surface roughness. They will appear on the image with a slightly brighter grey tone level. In Figure 4(a) these grey tone variations due to large sandbanks are visible in an ERS-1 SAR image.

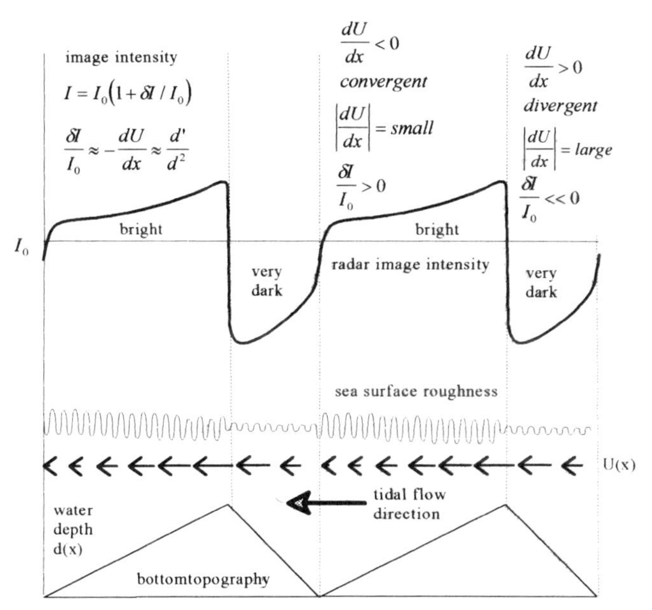

Figure 3. Schematic plot of the relation between an asymmetric sand wave profile and variations in tidal current velocity, surface roughness, and radar intensity. In this particular case the steep slopes of the sand wave face the flow direction. They are associated with strongly reduced images intensity. When the tidal flow direction is reversed the entire image intensity modulation pattern will be reversed resulting in a negative image (after Alpers and Hennings (1984))

The imaging mechanism of mapping sea bottom topography by imaging radar consists of three stages:

1. **Interaction between (tidal) flow and bottom topography** results in modulations in the (surface) flow velocity. This relation can be described by several models with an increasing level of complexity (continuity equation, shallow water equations, and the Navier-Stokes equations).

2. **Modulations in surface flow velocity** cause variations in the surface wave spectrum. This is modelled with the help of the action balance equation, using a source term to simulate the restoring forces of wind input and wave breaking.

3. Variations in the surface wave spectrum cause **modulations in the level of radar backscatter**. To compute the backscatter variations a simple Bragg model can be used.

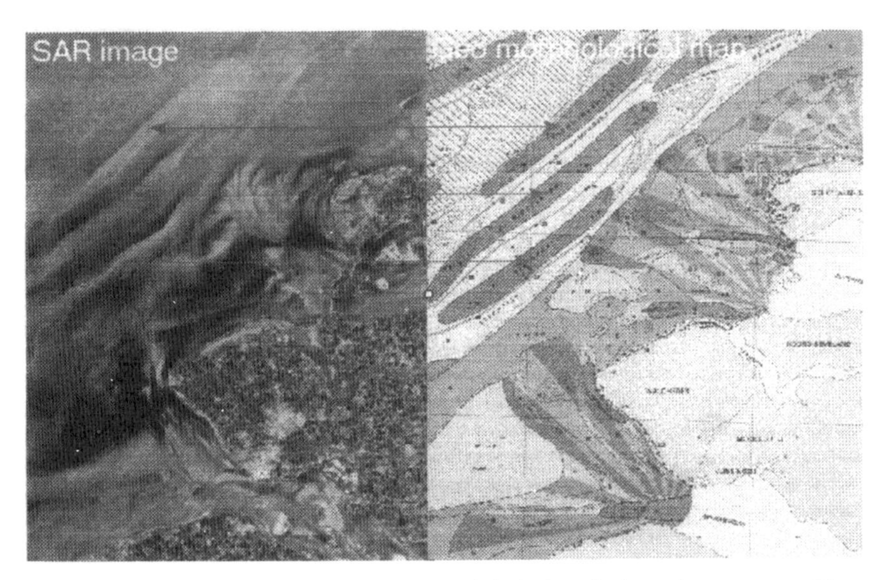

Figure 4. (a) SAR image with (b) corresponding geomorphological map in the south-western part of the Netherlands. Large sandbanks aligned parallel to the Netherlands coastline evidently appear in the SAR image as thin distinctive light/dark streaks in the image.

Based on the above three physical mechanisms, a suite of models has been developed. This suite of numerical models generates the radar backscatter, given the bathymetry and the wind. Because the parameter of interest is the bottom depth it is necessary to invert this depth-radar backscatter relation. Therefore, a data assimilation scheme has been developed, minimising the difference between the calculated and the measured radar backscatter by adjusting the bottom topography (Figure 5).

3.2 Application of the bathymetry assessment system

Over 9 demonstration case studies of this bathymetry assessment system have been conducted in the past five years. These demonstration studies show a significant progress in performance of this system for constructing depth maps. In 1995 a study was performed to test the method for accuracy, reliability and cost-effectiveness in a pseudo-operational situation. Test sites with various specific morphological features such as shipping lanes, inter-tidal flats and shallow water sand waves were chosen in the Waddenzee area and in the Netherlands foreshore area near Rotterdam harbour. The most important results of this recent test were

- The bathymetry assessment system can deliver depth maps with an average depth precision of 20–30cm between depth survey sections separated by 500m. With this precision it seems suitable for monitoring surveys.
- Multi-temporal analysis of SAR imagery improves the accuracy of depth assessments, up to a level comparable to that of traditional methods.
- The use of radar remote sensing techniques may reduce traditional monitoring survey efforts by a factor of approximately 50%.

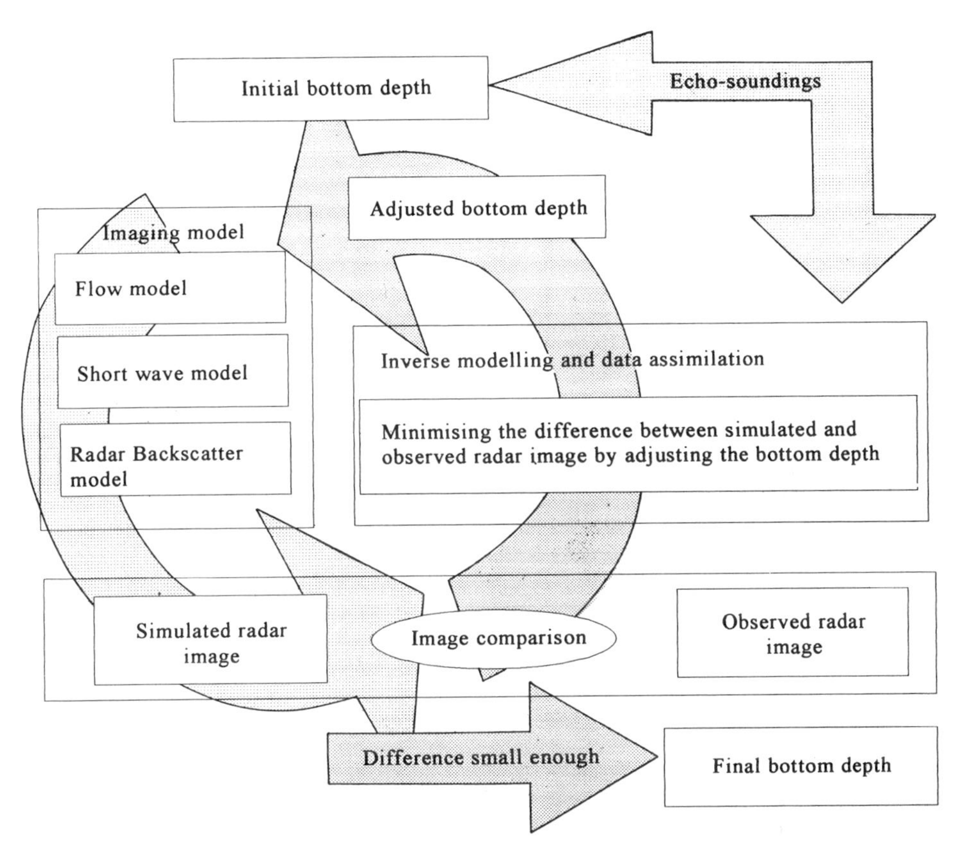

Figure 5. Outline of remote sensing based bathymetric assessment system

Limitations in the use of the bathymetry assessment system for operational applications are

- The operation still includes a lot of research-like tuning.
- Only radar remote sensing experts can recognise structures in SAR images that may be misinterpreted for bottom topography.
- Satellite systems only deliver a limited quantity of SAR images. Because a successful application of the system depends on hydro-meteorological conditions it can happen that over a period of months no suitable images can be collected.
- The resolution of satellite SAR systems is approximately 30m. Bottom structures smaller than 60m will therefore not be resolved but aliased to larger scales.

In the next few years the method will be implemented in the Rijkswaterstaat organisation and developed into an integrated conventional and remote sensing based observation method. The depth survey system based on SAR images will be developed following a three step process:

- Demonstration (1990-1995)
- Development of a pre-operational system and prototype (1996-1997)

- System operational at a limited number of Regional Departments (1998-1999).

Figure 6 shows a more detailed outline of the necessary steps in developing an operational system. Besides the above-mentioned three steps it also shows (ongoing) developments in the theory of the imaging mechanisms. One such development involves the upgrade from a quasi one-dimensional approach to a full two-dimensional model that will make the system applicable in more complex situations.

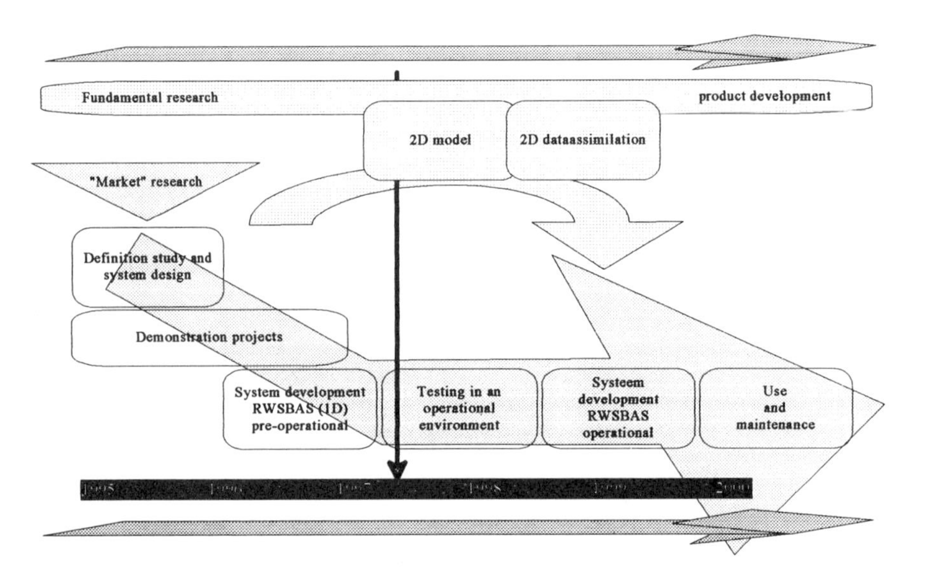

Figure 6. Towards an operational remote sensing based bathymetry assessment system. The system development consists of demonstration, prototype and operation. Parallel to this development trajectory should be research and development activities supporting the system and enabling state of the art techniques.

4. SAR Interferometry

Topographic maps are typically produced from stereo pair optical photographs in which relief causes the same terrain to appear in slightly different positions for different look angles. This shift in location is interpreted in terms of height of the terrain. Exactly the same concept as used in the optical stereo technique can be applied to radar images, resulting in a topographic map. In this section a relatively new technique, the interferometry approach, is described. In the interferometric case, the observable terrain shift is of the order of the radar wavelength (dm) rather than the resolution cell size (10m). Thus a major problem in both optical and radar stereo techniques, that of recognising the resolution elements in both stereo images, is circumvented.

4.1 Principles of SAR interferometry

In an airborne or spaceborne synthetic aperture radar (SAR), microwaves illuminate the terrain below the platform, and echoes scattered from the surface are collected. Subsequent signal processing performed on the echoes produces an image of the scene, where each pixel contains amplitude and phase information. The amplitude information, when displayed, produces the well known SAR images which look very similar to photographs. It is also possible to use the phase information to make a SAR interferogram. In SAR interferometry, two images are made of the same scene by two separated antennas. The antennas may be on the same platform, or the same antenna may be flown twice over a scene. If the antennas are separated in the range direction, perpendicular to the line of flight, altitude information about the surface can be deduced. This configuration is called across track interferometry (XTI). If the antennas are separated in the azimuth direction, parallel to the line of flight, motions of the surface such as ocean currents may be measured. This configuration is called along track interferometry (ATI).

Both methods (XTI and ATI) require the matching of two images, typically accomplished by resampling or interpolating one image to make it overlay the other one accurately. Then the phases corresponding to each pixel are recalculated and differenced, resulting in an interferogram. Phase unwrapping (the procedure in which phase ambiguities are resolved) and geometric rectification result in the desired altitude information or surface motion vectors. Interferometry can also be used in a so-called differential mode. Differential SAR interferometry can be used to detect very small elevation changes (a fraction of the radar wavelength), even over large areas (10–50km). In this method, two interferograms are made from three or more SAR images of a scene acquired at different times. After accurate co-registration (pixel match and recalculation) the interferograms phases are again differenced, resulting in a third interferogram. In this double differenced interferogram all the topographic information (i.e. altitude) has been eliminated. The new phase image contains non zero phases only in areas where the surface has been disturbed (changed) between the consequent times of the observations. These phases therefore contain information about the amount of surface motion that occurred.

The principle of along track interferometric SAR (ATI-SAR) on board of an aircraft is shown in Figure 7. The system has two antennas. The connecting line between the two antennas is the baseline B. Only the x-component of the baseline is important for ATI. Both SAR antennas alternately transmit and record radar radiation with wavelength λ. Suppose that part of the radiation is back scattered from an object moving with velocity U towards or away from the radar. During the very small time delay Δt in which the two almost identical images are recorded by the two antennas the object will have moved a small distance towards or away from the radar. This movement will cause a phase difference $\Delta\varphi$ between corresponding pixels of the images (11). This phase difference contains information about the velocity U of the object. The following relation can be derived between the velocity and the phase difference

$$\Delta t = \frac{B}{V} \tag{9}$$

$$\Delta r = U\Delta t \tag{10}$$

$$\Delta\varphi = \frac{4\pi}{\lambda}U\Delta t = \frac{4\pi}{\lambda}\frac{U}{V}B \tag{11}$$

with V the velocity of the aircraft. In practice, the movement of the aircraft (yaw and pitch) will result in baseline components in y- and z-direction which will give additional phase differences. Accurate attitude measurements can be used to apply corrections for this effect. Calibration of current velocities is necessary to derive absolute velocities from the relative phase differences.

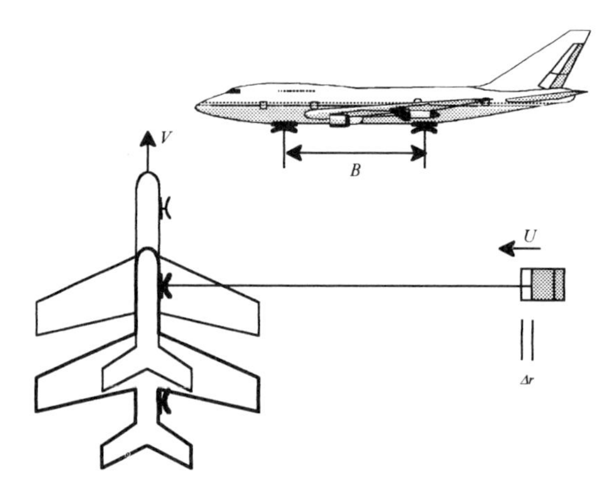

Figure 7. Concept and geometry for airborne along track interferometric measurements

For single pass across track interferometry (XTI), with two antennas separated by a baseline B, the radar instrument yields the line of sight distance, or slant range *r*, from the antenna to the target (Figure 8). This distance can be determined within the slant range resolution of the radar, which depends on the transmitted signal bandwidth. A typical slant range resolution for SAR systems is 4–8m. The phase difference $\Delta\varphi$ is a measure for the direction of the incoming radar pulse. It is given by

$$\Delta\varphi = \frac{2\pi}{\lambda}\left(B_y \sin\theta - B_z \cos\theta\right) \tag{12}$$

with θ the incidence angle and B the baseline separated into a horizontal component (B_y) and a vertical component (B_z), and *h*, the local height above the reference height is given by

$$h = H_0 - r\cos\theta \tag{13}$$

with H_0 some reference height and r the range. From these equations we can see that by measuring the range, incidence angle, the exact flight track and the timing of the measurements, the three-dimensional location of the scatter is determined. It is therefore possible to generate a height map or digital height model of the scene illuminated by the system. However, the phase difference is measured with an ambiguity of 2π. This means that the resulting height map will also contain height ambiguities. Due to the geometric configuration of the method it is not possible to distinguish between aircraft roll measurement errors and a slope of the illuminated terrain. Accurate attitude measurements are therefore of vital importance to airborne across-track interferometry.

Another possibility for across track interferometric measurements is shown in Figure 9. The geometry is quite similar to that of Figure 8, but in this case only one SAR antenna is available on the platform. The second SAR image is therefore obtained in a second repeated pass some time later. The horizontal and vertical separations between the passes define the baseline. The equations for the phase difference are

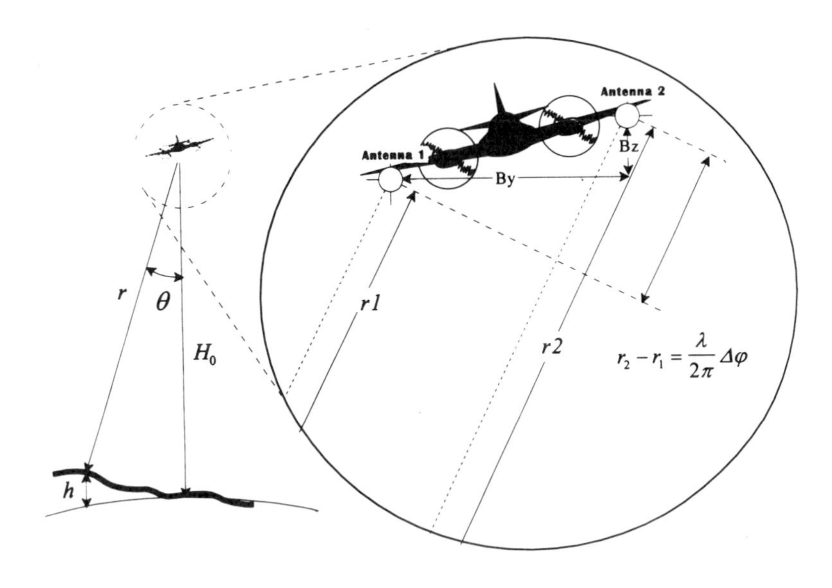

Figure 8. Concept and geometry for airborne single pass across track interferometric measurements

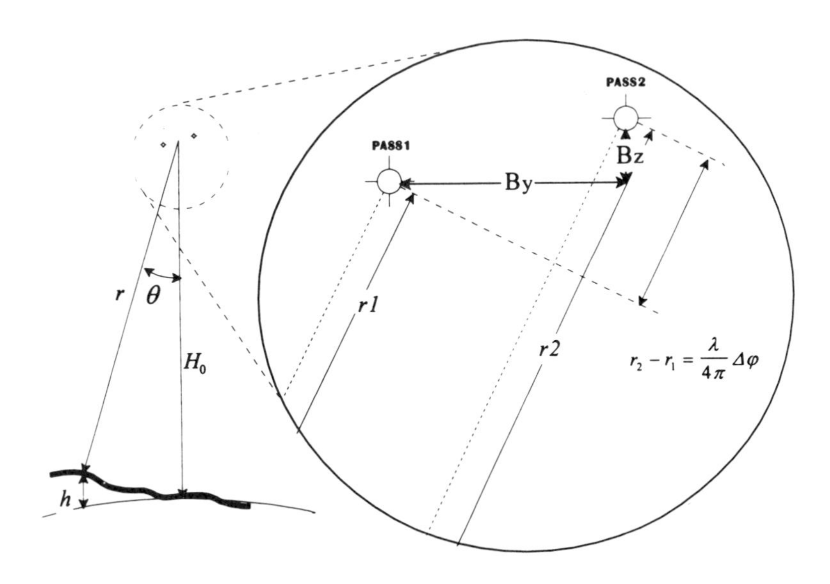

Figure 9. Concept and geometry for spaceborne or airborne multiple pass across track interferometric measurements

$$\Delta\varphi = \frac{4\pi}{\lambda}\left(B_y \sin\theta - B_z \cos\theta\right) \tag{14}$$

$$h = H_0 - r\cos\theta \tag{15}$$

These equations are almost identical to the single pass case, but now the images are taken separately which results in a phase difference that is twice as large. The concept shown in Figure 9 has been applied successfully to satellite systems like SEASAT and ERS-1 and ERS-2. In fact this technique combines across track and along track interferometer measurements. The (long) time interval Δt between the recordings of the SAR data gives a sensitivity for (very slowly) moving objects. Using this concept very small elevation changes or horizontal shifts between the recording of the images can be measured as a phase change of a corresponding cell in-between pass 1 and pass 2. Depending on the size of the baseline, the change in interferometric phase between different image pairs can be ascribed to topography or surface displacements.

4.2 InSAR processing steps

A first step in the InSAR processing is the precise orbit modelling and calculation of baseline separation. The Single Look Complex (SLC) images (containing both phase and amplitude) have to be verified, for example by comparing the pair of amplitude images. One of the images will be defined as the master image on which the other image (slave) will be co-registered. Correlation techniques will be used for a precise co-registration of the slave image to the master to an accuracy of sub-pixel level (one fifth of a pixel or better). After the slave image has been registered to the master, both images have to be filtered to eliminate possible de-correlation effects. The fact that both images have been taken from slightly different positions results in small changes in the received amplitude. This phenomenon which is called baseline or range de-correlation gives restrictions to the size of the baseline. If the baseline is taken too large, the reflections from a given object are not correlated between the two SAR images. Besides baseline de-correlation other factors can result in de-correlation between the interferometric image pair. Temporal de-correlation (when the surface has actually changed between the two images) and azimuth non-overlap effects also contribute. Filter techniques can be used to reduce this type of de-correlation effect. The two filtered images can now be combined to generate an interferogram. The removal of phase due to the Earth's curvature is called flat terrain filtering which requires information about the satellite orbit of both images. Following this procedure the InSAR processing results in several specialised image products including a coherence map and a precision geo-coded SAR image. Phase unwrapping is the procedure that resolves the phase ambiguities, i.e. attempts to estimate absolute heights from the relative phase values. Several (automated) algorithms are available for phase unwrapping. Residual phase trends in the interferogram can occur due to errors in baseline estimation. These phase corrections together with unwrapping errors may have to be corrected interactively. Figure 10 shows a flow-chart for InSAR processing.

4.3 Future developments and application of SAR interferometry

Although SAR interferometry is far from operational, several successful applications have been demonstrated by different research groups. The generation of digital elevation models from satellite based SAR image pairs results in height information with 20–30m horizontal resolution and approximately 5m height accuracy. Airborne systems can be used to generate elevation information with 2–5m resolution and approximately 1m height accuracy.

These specifications make InSAR derived digital height models of interest for remote areas which have not been surveyed before. The use of InSAR for deformation measurements (with accuracy of a fraction of the radar wavelength) has shown its potential in volcanic applications and in land-subsidence due to gas extraction in the Netherlands. This technique is also considered for precise monitoring of coastal defences like dams and dikes. Along-track InSAR can be used for detailed surface current mapping. Surface current maps can be useful for model calibration, coastal morphology, and harbour control. Given the successful demonstration projects InSAR will develop towards a highly precise observation technique for the monitoring and maintenance of large constructions in the Netherlands.

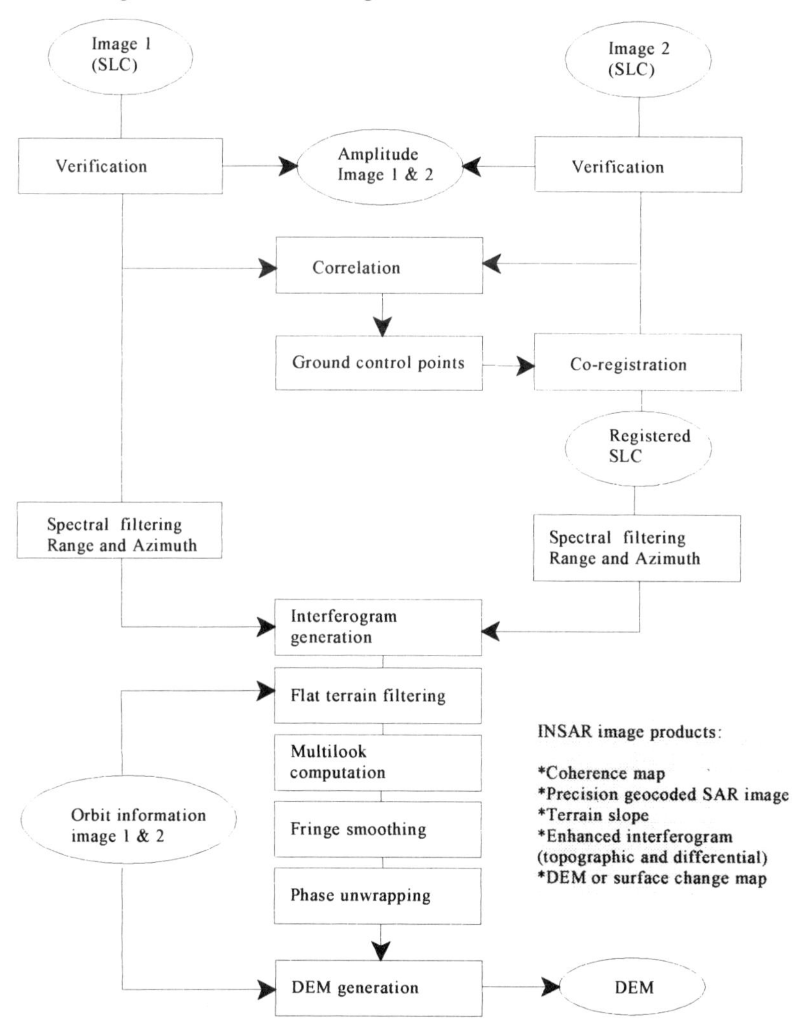

Figure 10. InSAR processing steps for XTI airborne or spaceborne configuration. A Digital Elevation Model (DEM) is an end-product but several specialised products can be derived in the process of generation a DEM.

5. Conclusions and Discussion: Outlook for the future

The current policy with regard to coastal protection in the Netherlands depends heavily on accurate topographic information. The so-called dynamic preservation which takes preventive measures in advance of structural erosion of the coastline needs regular and reliable information about coastal processes to ensure safety of drinking water supply, recreation, residential and industrial functions in the Netherlands. The development of efficient, flexible and fast new techniques within coastal monitoring and research activities is therefore a prime issue for the Ministry of Transport and Public works and Water Management, i.e. the Survey Department of Rijkswaterstaat.

Successful applications of innovative techniques based on remote sensing have been demonstrated. Laser altimeter systems are successful because they deliver the same product as conventional techniques (i.e. accurate height information) in less time and for lower costs. This condition might be a crucial factor to determine the success of remote sensing in being accepted in operational monitoring activities. Another factor is the ability to combine conventional techniques with remote sensing. This integration of different observation techniques often requires expert knowledge of the region and of physical mechanisms involved. The bathymetry assessment system is an example of such an integration type of new technique. It involves the integration of both conventional and new techniques together with numerical models. Interferometric SAR is an example of a new technique that will become operational in the near future because it circumvents certain problems in traditional stereo techniques. Monitoring surface change, as accurately as photogrammetry or levelling, will become feasible at only a fraction of the cost.

With the increase of remote-sensing-based operational applications in monitoring programmes, remote sensing is growing towards a competitive and possibly dominant technique for collecting geo-information for coastal zone management.

Reference

Alpers W and Hennings I, 1984, A theory for the imaging mechanism of underwater bottom topography by real and synthetic aperture radar, Journal of Geophysical Research, **89**, 10529-10546

Bibliography

Laser Altimetry

Gardner C S, 1992, Ranging performance of satellite laser altimeters, IEEE Trans. On Geoscience and remote sensing, vol. **30**, 5, pp 1061-1072

Lemmens M J P M and Fortuin E H W, 1996, Fouten-analyse van vliegtuig laseraltimetrie, Meetkundige Dienst, Rijkswaterstaat, MDGAR

Lillesand T M and Kiefer R W, 1994, Remote sensing and image interpretation, Wiley

Lindenberger J, 1993, Laser-profilmessungen zur topographishen Gelandeaufname, Deutsche Geodatische Kommission, Reihe C, 400

Measures R M, 1984, Laser remote sensing, fundamentals and applications, John Wiley, NY

Bathymetry

Calkoen C J, Wensink G J, Vogelzang J and Heinen P F, 1995, Efficiency improvement of bathymetric surveys with ERS-1, Report BCRS-95-01, Netherlands Remote Sensing Board, Delft

Van der Kooij M W A, Vogelzang J and Calkoen C J, 1995, A simple analytical model for the modulation of sand waves in radar imagery, Journal of Geophysical Research, **100**, 7069-7082

Vogelzang J, 1997, Mapping submarine sand waves with multi band imaging radar: 1 Model development and sensitivity analysis, Journal of Geophysical Research, **102**, 1163-1181

Vogelzang J, Wensink G J, Calkoen C J and van der Kooij M W A, 1997, Mapping submarine sand waves with multi band imaging radar: 2 Experimental results and model comparison, Journal of Geophysical Research, **102**, 1183-1192

SAR Interferometry

Gabriel A K, Goldstein R M and Zebker H A, 1989, Mapping small elevation changes over large areas, differential radar interferometry, Journal of Geophysical Research, **94**, 9183-9191

Goldstein R M, Zebker H A and Werner C L, 1988, Satellite radar interferometry: Two-dimensional phase unwrapping. Radio Science 23(4), 713-720

Hartl P, Reich M, Thiel K-H and Xia Y, 1993, SAR interferometry applying ERS-1: some preliminary test results. In Proceedings First ERS-1 Symposium -- Space at the Service of our Environment, Cannes, France, 4-6 November 1992, ESA SP-359, pp 219-222

Massonnet D, Feigl K, Rossi M and Adragna F, 1994, Radar interferometric mapping of deformation in the year after the Landers earthquake. Nature, **369**, 227-230

Massonnet D and Feigl K L, 1995, Discrimination of geophysical phenomena in satellite radar interferograms. Geophysical Research Letters, **22**(12), 1537-1540

Prati C, Rocca F and Monti Guarnieri A, 1993, SAR interferometry experiments with ERS-1. In Proceedings First ERS-1 Symposium -Space at the Service of our Environment, Cannes, France, 4-6 November 1992, ESA SP-359, pp 211-217

Schreier G (Ed), 1993, SAR geocoding: data and systems, Wichman

van Halsema D and Hanssen R, 1996 (January), Radar interferometry: A new tool for accurate height modelling. Geomatics Info Magazine **10**(1), pp 27-31

Zebker H A and Goldstein R M, 1986, Topographic mapping from interferometric synthetic aperture radar observations. Journal of Geophysical Research 91(B5), 4993-4999

Zebker H A, Werner C L, Rosen P A and Hensley S, 1994, Accuracy of topographic maps derived from ERS-1 interferometric data. IEEE, Transactions on Geoscience and Remote Sensing, **32**(4), pp 823-836

Pollution monitoring and clean up of coastal zones in the post-Soviet era

Serge V Victorov

National Institute of Remote Sensing, St.Petersburg.

1. Introduction

In the former USSR, the Hydrometeorological Service (in some periods it was called the State Committee and Agency) was responsible for monitoring the environment, including marine and coastal environments. Regional Administrations of the Service performed measurements, collected data and published them on a regular basis (monthly, four times a year or once a year) for selected users. Those materials contained raw data relevant to water quality, concentration of oil, heavy metals and other pollutants in sea water. Measurements were carried out on a network of stations providing systematic study of various parameters. Since the early 1990s, as a result of the collapse of the national economy, regular measurements at many stations especially in remote regions had to be stopped. Regular publications are not available because of commercialization of these activities, although nowadays there are no formal restrictions to the dissemination of environmental information.

In this chapter we shall present a wide scope of problems related to the monitoring of marine and coastal environments in the seas around the former USSR, rather than focus on certain pollutants in a selected sea. Information on 14 seas belonging to three oceans will be given showing how different their geographical and oceanographic characteristics are. It is interesting to note that, though the seas of the Arctic Ocean and the southern seas inside the country are so different and located tens of thousands of kilometres apart, they could still be put together for analysis and unexpected links found.

The general geographical and oceanographic description of the seas around the former USSR is based on Belinskiy and Istoshin (1956); there appears to be no other more recent book dealing with the general geographical aspects of those seas. This situation should be improved, but when speaking of the seas around the former USSR, one should not forget about some aspects of releasing the information about the seas in this country. Their depths, currents and other oceanographic parameters have always been considered as sensitive to the military security of the country. Therefore, few data were actually available to the public. Moreover, it was a hard task to study in detail the tremendous areas of the seas around the country. Severe climatic conditions and the low density of population in the coastal zone also prevented intensive mapping and research. Some seas will be covered in some detail, while other distant seas will be dealt with briefly.

While reviewing and discussing the whole scope of oceanographic and environmental problems of the open sea, shelf and coastal zone of the seas around the former USSR, we shall inevitably be touching some problems arising from the use of renewable and non-renewable natural resources along with their economic and political dimensions.

2. Seas of the Polar Ocean

2.1 General features

The Ocean between the Euro-Asian and North-American continents is also often called the Northern or Ice or Arctic Ocean. The length of coastline of the former USSR/Russia from the border with Norway to the Bering Straits is about 18,000km. There are 6 seas here (from West to East), namely the Beloye (White) Sea, the Barents Sea, the Karskoye (Kara) Sea, the Laptevikh Sea, the Vostochno-Sibirskoye (East-Siberian) Sea and the Chukotskoye (Chukcha) Sea. All these seas are the so-called young seas. In ancient times there was mainland here, which is why these seas are so shallow (200m, in some places 500m). This is the largest shallow offshore region in the world's oceans. The coastal zone was mapped only in 1733–43 and some regions were only mapped as late as in 1792.

These seas are interesting from many points of view. The so called Great Northern Route from Europe to Japan along the coasts of Siberia is 10,000km shorter than the traditional route from the Mediterranean Sea via the Suez Canal and the Indian Ocean. It has long been a dream of seamen to go by the Great Northern Route during one summer season. But it was only in 1932 that the Soviet icebreaker Sibiryakov achieved this for the first time.

The **Beloye (White) Sea** is the smallest among these seas; its area is about 90,000km^2 and its mean depth is only 7m. It can be considered as an estuary of the Dvina river. It is remarkable for diamond deposits located in the sea and on its coasts.

The **Barents Sea**, named after its explorer Barents, has an area of approximately 1,400,000km^2; its depth is 100–350m (very seldom does it exceed 400m). One of its bays is often called the Pechora Sea. This is a shallow sea, an estuary of the Pechora river which has an annual discharge of about 200km^3. The Novaya Zemlya (New Land) Island is situated here, with a length of about 900km and a width of about 90km. This island, or rather the archipelago of the same name, was the main USSR nuclear weapons test site. The annual mean temperature of the air is –5°C here. The sea is influenced by the Gulf Stream which brings 150km^3 of warm (4–12°C) water each day. For this reason the south-eastern part of the sea is never covered with ice. The water in the Barents sea is a composition of two water masses: Atlantic water with a temperature above zero and of high salinity and subzero polar water of low salinity. The tidal range (the difference between low and high tide) is up to 4m.

The **Karskoye (Kara) Sea** has an area of 883,000km^2 with a maximum depth of 620m. The great Siberian rivers Enisey and Ob bring their waters to this sea. The River Ob with its tributaries can be used for cargo shipping for 10,000km up the river. River inflow influences the ice conditions in the sea – even in winter there is drifting ice in this sea.

The **Laptevikh Sea**, named after its explorers the brothers Dmitry and Khariton Laptev, has an area of 650,000km^2 with a volume of 338,000km^3. Its depth is 0–50m in the southern part and up to 2,000–3,000m in the northern part; the mean depth is 519m. There are many islands in the sea. The big rivers Khatanga, Lena, Yana, Anabara, Olenek and

Omoloy bring their waters (and suspended matter) to the sea. There are also ancient river valleys on the bottom of this sea. The sea is becoming more shallow with (historical) time. Another interesting feature is coastal erosion which leads to the degradation of small islands composed of a mixture of ancient ice and soil.

Actually, the whole mainland shore and the Novosibirsk archipelago are composed of a frozen mixture of sand, clay and pieces of ice which is easily destroyed by the sea waves. The Vasilievsky island (4 miles long and 1/4 mile wide) was discovered in 1815 and was mapped in 1823. After its discovery nobody saw the island for 100 years, although it was still shown on the maps. In 1936 a sand bank at a depth of 3m was mapped on this spot. The legendary Sannikov Land (part of the Novosibirsk archipelago) was discovered and mapped in 1810. It was seen in 1886, but it could not be found in 1937. It is in this mixture of ice and soil that the ancient mammoths have been found.

The annual mean temperature of the air is −15°C here. The height of the tides in the western part of the sea is 2–3m, in the eastern part 40–60cm. Strong winds from the south lead to a decrease in water level in the coastal zone, while winds from the north and west lead to an increase in the water level up to 2.5–3.0m.

This sea is the most brackish among the seas of the Northern Ice Ocean, with salinity of 0–5ppm in the southern part of the sea. This is because every year the rivers bring $500km^3$ of fresh water to the sea. The river Lena fresh water plume, 150–200km wide, can be traced far into the sea. The mean river water temperature in June is 7.8°C, in July 14.5°C and in August is 13.1°C. There is a complicated system of currents in this sea. The Great Siberian Polynya (an area with a sea ice conditions anomaly) is located to the north of the Novosibirsky islands in the northern part of the Vostochno–Sibirskoye (East-Siberian) Sea.

The **Vostochno-Sibirskoye (East-Siberian) Sea** is the most severe polar sea with an air temperature in winter ranging from −25°C to −30°C and in summer the smaller range of 2–4°C. Its area is $900,000km^2$ and it is a shallow sea with depths of less than 200m. Salinity in the western part of the sea is low (0–28ppm) due to the influence of the rivers Indigirka, Alazeya and Kolyma.

The **Chukotskoye (Chukcha) Sea** has an area of $582,000km^2$ with a mean depth of 88m and a maximum depth of 2,000m. About 90% of the sea area is only 30–60m deep. In the vicinity of melting ice fields the salinity is about 24ppm. The water is very transparent in this sea (6–30m by Secchi disk).

2.2 The Barents Sea

2.2.1 Biological conditions

We shall consider biological conditions in this sea. "In the Barents Sea, the summer flow of water varying in temperature and salinity, the shallowness and the intensive summer sunlight add up to virtually optimal conditions for life during the six summer months. The edge of the polar ice-cap, in particular, is an important production area.

During the winter, nutrients are concentrated in the water below the ice as a result of the freezing-out of salt, which leads to an efficient vertical mixing of the water. When the ice melts, a nutrient-rich surface layer forms. This, combined with the fact that the water column is stable, means that phytoplankton grow very rapidly and in large quantities." (Northern Europe's Seas. Northern Europe's Environment 1989)

From Coastal Zone Colour Scanner (CZCS) images one can see that surface concentrations of chlorophyll-a are highest in the coastal water masses and in the Atlantic waters. High concentrations are also found along the coasts of Novaya Zemlya, along the Svalbard coast and along the ice-edge. Concentrations in the Barents Sea are low.

The seasonal and spatial distribution of phytoplankton is determined by physical factors, the most important being light and water mixing processes. About 50% of the species found in this region are diatoms. 50 to 60% of primary production in the Barents Sea is new production, taking place in spring. The spring bloom lasts for a couple of weeks. Production estimates vary from 90–120gCm^{-2} per year in the Atlantic water masses to 20–50gCm^{-2} per year in the Arctic water masses.

After the spring bloom, the plankton population is reduced considerably and most of the biomass can be found in the lower parts of the euphotic zone where it may not be detectable. From the viewpoint of detectability of plankton from satellites one should take into consideration climatic conditions and biological issues. How much of the production remains in the pelagic system depends on species composition and on the weather. In cold years, herbivorous zooplankton develops later than in warmer years. In such years grazing is less intensive and large parts of the phytoplankton biomass sink through the euphotic zone (and thus become invisible from satellites) and become incorporated in the deep pelagic and benthic food chains. Zooplankton comes next in the food chain, and they are the staple food of herring and capelin. A high density of phytoplankton, zooplankton and plankton-eating fish is the basis of the rich fauna of the area. There are few species, but a large number of each.

In the Barents sea there are large stocks of fish, especially capelin, herring, cod and Arctic cod, as well as vast numbers of sea birds – some of the largest populations in the world – and such mammals as the grey seal, common (or harbour) seal and harp seal, walrus, humpback whale, fin whale and polar bear. Sea birds play a key role as the link between the marine and the land ecosystems, by taking up large amounts of biomass from the sea and transporting it to land. Regarding sea birds in the coastal zone ecosystem, much could be done on this problem using remotely-sensed data.

The Barents sea itself, its shelf and its coastal zone, all with their rich fauna and food chains well adapted to the environment, are ecologically vulnerable. External influences can have serious effects. Discharges from heavy industry in the Murmansk area have caused water and air pollution by heavy metals and stable organic substances, especially at Kirkenes. It can be seen from satellite images that the forest around Murmansk is badly damaged by pollutants. The ever increasing scale of shipping – both civil and military – along the coast involves risks of oil discharges and other pollution. Nuclear-powered vessels, both ships and submarines, are another potential environmental hazard.

2.2.2 Contamination of seawater with radioactive isotopes

The problem of nuclear waste products, which originated from the Russian nuclear-powered merchant fleet (including icebreakers) and submarines, has been highlighted a few years ago and became a problem of international concern (including UNEP and also Greenpeace activists). Usually the Russian authorities have been blamed for their negligent actions in relation to waste products safety transportation and dumping in the deep sea. This is indeed a serious problem with both technological and moral aspects.

Against this background a new piece of information recently came from Canada. Fred Pearce published a short article entitled "Sellafield leaves its mark on the frozen north" in the British popular magazine *New Scientist* on 10 May 1997 (page 14). He stated that "A joint Canadian-Russian study of radioactive isotopes in the Arctic Ocean, completed in May this year, investigated contamination of the Kara Sea, where Russia dumped large amounts of radioactive waste in the 1980s. It found that radioactivity in seawater there comes mainly from Sellafield and fallout from atmospheric nuclear weapons tests prior to the 1963 partial test-ban treaty". "Radioactivity from Britain's Sellafield reprocessing plant in Cumbria has spread through the Arctic Ocean into the waters of northern Canada. The contamination, which has never before been picked up so far from Britain, is having a bigger impact on the Arctic than the Chernobyl accident, according to new Canadian data".

The new data show that a plume of iodine-129 from Cumbria has penetrated beyond Siberia to the north-western shores of Canada at a depth of about 200m. Radioactivity in the plume was "an order of magnitude greater than the background level from nuclear weapons fallout", said project leader Mike Bewers of the Bedford Institute of Oceanography in Halifax, Nova Scotia. According to Per Strand of the Norwegian Radiation Protection Authority, Sellafield's discharges of caesium-137 and iodine-129 into the Irish sea take a tortuous path through the Arctic: a plume of radioactive seawater works its way north, riding the Norwegian current around the north of Scandinavia into the Barents Sea. From there, part of it crosses the pole, and then works its way down the east coast of Greenland, before turning its southern tip and heading back north into Baffin Bay, while part of it goes east along the northern shore of Siberia and eventually reaches the Beaufort Sea on the north-western shores of Canada.

It is worthwhile to note that "the patterns of discharges from Sellafield, showing peaks in 1975, 1977 and 1980, were replicated four years later in the Barents Sea as the radioactive plume worked its way north". Strand estimated that the reprocessing plant in Cumbria "has released 40×10^{12} Bq of caesium-137, of which about 15×10^{12} Bq has reached the Arctic so far. This is between two and three times more than the contamination from Chernobyl, which is reaching the Arctic via the Baltic and North Seas. Contributions from Europe's second largest reprocessing plant at Cap de la Hague in France are much less."

On 21 June 1997 Tam Dalyell published a brief article in the *New Scientist* entitled "Something nasty in the waters (more comments from the Westminster)". It is generally believed that "many of the former Soviet Union's obsolete nuclear submarines laid up in Northwest Russia have deteriorated to a desperately dangerous state and represent a formidable environmental hazard" (Dalyell 1997). There is no reliable information on whether this is actually the case, and it is very difficult to quantify the above statement, to clarify what is 'desperately dangerous', to what extent it is dangerous, and what is a 'formidable' hazard in terms of risk analysis.

Doug Henderson, the UK Foreign Office minister with special responsibilities for relations with Russia, said that "Britain, with other European countries and the US, has been looking at ways of resolving the problems that these vessels pose". The objectives of joint efforts with the Russians are "to identify safe and practical ways to manage and dispose of the spent fuel and radioactive waste from them" (Dalyell 1997). The work is in progress now. A project funded by the European Union has compiled an inventory of radioactive waste in the Kola Peninsula and Archangelsk region. Several thousand spent fuel elements were reported to be stored aboard three service ships moored near Murmansk. Another recently completed project led by Britain's AEA Technology and France's SGN, has identified a technique for safely removing spent fuel from the storage tanks on one such ship, the former icebreaker 'Lepse'. The third project has developed transport and storage flasks to contain the spent fuel elements. The British company NNC is leading a study funded by the European Union on the radiological consequences of dumping redundant submarine reactors in the Kara Sea. It was shown that "although the reactors are dumped in water no deeper than 300 metres, doses from the radioactivity releases are likely to fall well within internationally accepted limits" (Dalyell 1997).

2.2.3 Monitoring of marine and coastal environments

The coastal zone in the Barents sea is sparsely populated. There are a few small Norwegian towns, a number of fishing villages and the Russian Murmansk area with a massive naval base and heavy industry. There is a routine programme of environmental monitoring of this sea with standard measurements of parameters relevant to hydrography, chemistry and biology.

There is also an advanced national programme of monitoring the Barents Sea marine environment with eight proposed test sites about 100km^2 each. The test site near Murmansk should account for the Pechenga-Murmansk chemical plants air pollution and west-european sources. The test site near the White Sea should account for contamination transported from that sea. The test site near Cheshskaya Bay and the river Indiga represents a relatively undisturbed marine and coastal environment and is regarded as a background protected area. The test site near the river Pechora outlet accounts for possible oil pollution transported from the mainland at the Komi oil fields. The Prirazlomnoye and Shtokman test sites will be used to monitor the forthcoming development of these areas with intensive oil and gas production. The test site located at the south-western coast of Novaya Zemlya deals mainly with monitoring the radioactivity of water and sediments. The test site in the central part of the Barents Sea is selected for collecting background information. Comprehensive studies of geological, tectonic and morphometric characteristics of the sea bed, coastline changes, sedimentation processes and underwater discharges are being planned. The advanced programme was launched several years ago and is used as a framework for research activities and fieldwork carried out by several national agencies.

2.3 Oil and gas in the Arctic

We start with a strategic assessment of the role of the Polar Ocean as a source of energy for Europe made by Niini (1996).

"Recent estimates show that there are more than 60 bln t of oil and gas in the shelf zone of the Barents Sea and Kara Sea. Together with Timano-Pechora province this region is capable of producing more than 300 mln t of oil and gas during the first decades of 21st century, while the resources of the North Sea will be limited at the end of this century."

In general the feasibility of the oil and gas industry in the Arctic is based on low costs of transportation of oil and gas products.

There are oil fields on the shore and in the sea. Development of land fields started before the activities in the sea, but both types of activity are closely connected with maritime operations. It is obvious that oil will be transported from sea platforms either by pipelines or by ships. As for transportation of oil products from the land-based platforms, there are two options: either pipelines laid on the land or sea terminals and ships. The latter option is being studied within the framework of the project 'Northern Gates'. The economical feasibility of this type of transportation needs to be proved. The transportation of products from oil and gas fields (located on the mainland in Siberia) thousands of kilometres from the sea by a special fleet is also being considered. One can use the rivers Ob and Enisey to carry oil and gas products to the sea and further along the coasts of the Russian Arctic.

Various methods of transportation of oil and gas products were experimentally studied in 1995, and data on the economical aspect of the use of different types of vessels have been obtained. A new option of going through the ice fields was suggested and tested: 'kormoi forward'. According to Niino (1996) the advantage of sea transport is that the fleet of tankers capable of working in ice conditions in the Arctic could be built within 2 years, while pipelines cannot be built during this period.

At the end of the 1980s there was an expansion of interest in oil and gas production on the continental shelf of the Barents sea.

Oil and gas fields with an estimated capacity of 306.3Mt of oil are located in an area of 7525km^2 in the northern part of the Nenets Autonomous Region of the Russian Federation. To provide the export of oil, according to the international project 'Northern Gates', a sea terminal connected by pipelines with oil wells is planned to be built. To transport oil a special fleet of tankers of icebreaker class is intended to be used. A set of domestic, foreign and joint venture companies are participating in the project. The project duration is 50 years with a total income of \$104.5x10^9. (For more details of this project see Rost 1995.)

Among the other major oil and gas fields one should mention Prirazlomnoye (south-eastern part of the Barents sea about 60km from the coast) with its geological resources of 250–300Mt of oil (50–60Mt of which can be extracted). It was discovered in 1989. The investment in this project is about \$1.3x10^9, the profit is expected to be \$(5–7) x 10^9. BHP Petroleum (Australia) is planned to be involved. Prirazlomnoye is planned to become a test area for the new investment plans and new technologies of drilling and transportation of oil and gas products.

Two local Russian industrial enterprises with a background in the construction of nuclear submarines will be contracted. There are more than 40 other prospective areas on the shelf of the Barents, Kara and Pechora seas which are currently on the agenda of Rosshelf and Gasprom Russian national companies. Total resources here are more than 1.25x10^{9}t of oil and 13.7Mm^3 of gas (Fattakhov 1995).

The pipelines will be laid on the sea bed and on the tundra land. The latter is a very vulnerable area. If one drives a truck here, the footprints will be seen for a decade. At low temperatures oil slicks on the surface of land and sea are hardly degraded by natural bacteria; this process lasts for years here. Though oil production is in its infancy here, in the long run there is a danger of acute oil discharges in conjunction with exploratory and commercial operations.

2.4 Oil slicks and ice

We shall discuss the problem of potential oil pollution of the Arctic seas in more detail following Northern Europe's Seas, Northern Europe's Environment (1989).

All types of oil discharge threaten sea-bird populations, especially the species living mainly on the surface of the sea and diving for their food. Recent disasters have also shown that many birds have died when a mixture of oil and water reached coastal areas. The protection of seabirds has not yet become a matter of national concern in Russia as it seems to have become in Britain. Oil destroys the natural waterproofing of a bird's feathers; it makes the birds less buoyant so that they may sink and drown. It also reduces a bird's insulation against cold seas. Even if a bird survives the cold and does not drown after oiling, it may die later from the poisonous effect of swallowing oil when preening. Oil damage to fur-bearing animals can be lethal, since it destroys the animal's thermal insulation. Oil is also a threat to fish-breeding areas. Eggs and fry, above all, are sensitive to oil products.

It has not been established how the edge of the polar ice-cap, which is biologically by far the most productive area in the Barents sea, would function if an oil discharge took place in this area. Oil discharges in cold waters are not necessarily more harmful to the environment than those in warmer waters. The decisive factor is whether the oil enters the open sea, areas with ice, or directly affects areas with large animal populations. If oil in the Barents Sea came into contact with ice before any major degradation or dissolution had taken place, the oil would probably be deposited on the upper or lower side of the ice and thereby break down more slowly. If so, flora and fauna beneath the ice would probably be killed immediately or noticeably affected in some other way, and this in turn would have repercussions on the production necessary to sustain life on the edge of the ice-cap.

As phytoplankton biomass varies considerably throughout the year, the effect of oil pollution will be dependent on its timing, with the most critical period being the period of spring bloom. As various species have their own tolerance limits, oil pollution can result in changes in phytoplankton community population structure and can lead to changes in secondary production. Bacterial degradation of oil can result in depletion of nutrients in the surface layer. In its turn this will lead to a decrease of new phytoplankton growth. The Barents Sea is regarded as particularly sensitive to pollutants with the same dispersion patterns as oil. It has been shown that phytoplankton at high latitudes contains more lipids than phytoplankton at lower latitudes. So arctic phytoplankton may accumulate toxic substances to a larger degree that phytoplankton from temperate regions. One should note that there are no decisive and comprehensive experimental data on this issue. In some experiments with oil spill under the ice no effects on species composition, cell density, chlorophyll-a concentration and productivity of ice algae could be found after only 2 days of recovery. So one can assume that the biomass of phytoplankton with its high growth rate will recover quickly after a local oil spill disaster.

There is limited experience in the world of drilling and laying down pipelines in this type of arctic region. There is a high probability of damage to the pipelines in permafrost conditions during the summer warming. At this point we quote a few words on the general situation with the oil and gas pipelines in Russia as a potential source of environmental hazards. "During the last year of the existence of the USSR the oil and gas industry required 800–850 thousand tons of pipes annually to maintain and restore the pipelines. But no more than 400–450 thousand tons have been bought. Thus each year the deficit increased by 400–450 thousand tons. Nowadays the countries of the former USSR (mainly Russia with its 600 thousand km of gas and oil pipelines which is more than 85% of pipelines in the former USSR) need 4–4.5 million tons of pipes. About 20% of pipes have been in use for more than 33 years. More than 50% have been in use for 20 years and more." (Neft i Kapital 1995). As a result of this situation more oil leakage can be expected in Russia. Oil leakage in the Komi region in October 1994 (90 thousand tons of oil) was given much attention in the mass media but this was just a single example among many other unknown environmental disasters.

The oil and gas industry in the former USSR has already become a factor of global economic and environmental importance. Just to illustrate the environmental significance of oil and gas industrial activities we consider the following example. The global atmospheric methane budget is being monitored by the NOAA Climate Monitoring and Diagnostics Laboratory (Boulder, Colorado, USA) at 40 locations. Dlugokencky *et al.* (1994) examined the dramatic low growth rates in terms of the global methane budget in 1992 (from +15 to −2 units), and concluded that this decrease could be explained by a 2% decrease in the total source. Further they suggested that the change in the northern hemisphere may reflect increased efficiency (decreased leakage) in the delivery of natural gas in the former USSR (Novelli *et al.* 1995).

This suggestion shows the global role of the gas industry of the former USSR (though it could not be backed by the assessment of the quality of gas pipelines in the country, which will be given below).

2.5 Space-based ice information system

Ship routing operations in this region are currently provided by a Russian space-based ice information system. The system was presented by Nikitin (1991) and in brief form was described by Victorov *et al.* (1993). This system consists of three main parts:

1. Satellites as the source of raw data.
2. A ground-based analytical centre.
3. Communication links (either ground-based or via satellites), including end-user equipment in the Arctic.

A brief description of the technology used in the system follows.

Raw information from various types of satellites is used as the input data (radar, visible and IR imagery from satellites of OKEAN, KOSMOS and NOAA series). Data processing and analysis are performed using interactive image processing technology consisting of the following steps:

- General analysis of the scene, quality assessment, selection of interesting zones.
- Optimisation of visual presentation, modification of histograms, palettes etc.
- Segmentation of tones, smoothing, threshold transformations, construction of isolines.
- Identification of special structures, spatial differentiation and lineament analysis.
- Integration into the chart format, transformation into a chart projection, superimposition of a co-ordinate system and a coastline.
- Thematic analysis and editing based on all the available operational data and the information from the data base.
- Annotation of the output images with explanations of their important eatures.

The output products of this system are sets of colour display frames and comments. They are transmitted over conventional satellite TV channels (EKRAN, Moscow) or electronic communication channels (GORIZONT, INMARSAT) and thus may be made available in nearly all parts of the World Ocean including the icebreakers operating in the Arctic region.

2.6 Basic research in the Arctic

Current research activities in the Polar Ocean are aimed at further analysis of the role of this region in the context of global climate change. The Russian Federal Centre for Arctic and Antarctic Research, AANII (St. Petersburg), is the national co-ordinator of various research programmes. We mention some of those following Frolov (1996):

As part of the international programme LOICZ (Land-Ocean Interactions in the Coastal Zone) a Russian project with international partnership LOIRA (Land-Ocean Interaction in Russian Arctic) was proposed in 1996.

The International Arctic Science Committee (IASC) runs the project BASIS (Barents Sea Impact Study) dealing with the study of various impacts on the Barents Sea environment.

The USA National Science Foundation and the Russian Foundation for Basic Research are running the project 'Russian-American initiative in the study of Land-Shelf system in the Arctic'. The objective is to determine the interplay between variations of the Arctic environment (river discharges, evolution of ocean halocline, sea level change, permafrost etc.) and components of global climate change. There are 3 main research areas:

- effects and feedback of climate change on hydrology, oceanography and biogeochemistry of the Euro-Asian system Land-Shelf in the Arctic region;
- responses and feedback of continental glaciers, sea-ice and permafrost in connection with climate change in modern and previous climate periods;
- adaptation and responses of man to climate change during the last 20,000 years.

Numerical models are being developed dealing with transport and degradation of oil products in the rivers and estuaries in Arctic conditions, intended to assess the impacts of oil spills on marine and coastal environments (Frolov 1996).

3 Seas of the Pacific Ocean

3.1 General features

There are 3 seas in the so-called Russian Far East (from South to North), namely the Japan Sea, the Okhotsk Sea and the Bering Sea. The total area of these seas is 4,800,000km^2 and the total volume is 7,000,000km^3. They stretch 5,000km from south-west to north-east.

The **Japan Sea** has an area of 1,000,000km^2 with a depth of 4035m. The total length of its coastline is 7,600km (3,700km within Russia). The coasts are smooth. In winter only its north-western part is covered with ice. The water salinity is high. The water is transparent (30m by Secchi disk in the Tsusima current) as there are no big rivers bringing suspended matter. The tidal range is greater than 3m.

The **Okhotsk Sea** is known for its high tides (up to 13m in Penzhinskaya Bay) with tidal currents of 15–17kmh^{-1} in places. Its coastline (including islands) is 10,450km long of which 450km belongs to Japan, its depth is 3647m. The coastline is smooth.

The Amur is the only big river. The rivers carry 585km^3 of fresh water each year, of which the Amur brings 370km^3. Only in winter is the sea is covered with ice.

The **Bering Sea** has an area of 2,304km^2 and its maximum depth is 4773m. The coastline is very indented. There are several large rivers including the Anadyr, the Yukon and the Kuskokvim. The sea is connected with the Polar Ocean by the Bering Strait which is only 50m deep and 90km wide; water exchange is small. The northern part of the sea is covered with ice for 9 months (November to May) but the southern part is free of ice during the whole year. Tides are 2–3m (up to 7m in places).

Vladivostok and Nakhodka are the main ports in this area providing shipping of timber and minerals and other cargoes from Russia to other countries.

Recently an accident with the tanker NAKHODKA took place. On 2 January 1997 during a severe storm the nose part of the tanker 40m long (about 1/3 of it's total length) was separated in the open sea between China and the Kamchatka peninsula. Out of 19,000t of fuel oil on board the tanker about 3,800t formed a 1km by 30km oil spill which drifted towards the coast of Japan. In a few days about 500km of coast were heavily polluted. According to the Russian mass media, it was only on the 9th day that the Russian local authorities formally offered their help to Japanese authorities. The accident caused fuel oil shortage in the Kamchatka as it needs regular oil supplies from outside. There are oil and gas fields in this region though oil production is still in its initial stage.

3.2 Sakhalin Island coastal zone research using airborne data

In 1996 at the World Exhibition *Oceanology International '96* held in Brighton, UK, the Russian National Institute of Remote Sensing Methods for Geology (VNIIKAM) for the first time ever offered its services and co-operation in off-shore and coastal zone research based on access to unique archives of multi-year high-resolution remotely sensed imagery of the seas of the Russian Far East, European and Arctic seas. VNIIKAM has 53 years of experience in airborne and space-based remote sensing of coastal zone and off-shore areas. During the period 1965–1990 VNIIKAM carried out the following research on the island of Sakhalin:

- The total area was surveyed with airborne side-looking imaging radar (wavelength 3cm).
- Thermal infra-red images were acquired in several areas of the island.
- A tailor-made airborne photographic technique was implemented to study the seabed in the coastal zone around the whole Sakhalin island and Maneron island.

These images were used to produce geological maps of the coastal zone and some environmental maps. A set of maps was produced detailing the coastal zone with up to 20m isobaths showing coastal terraces, old river deltas, water turbidity, sedimentation and coastline changes. Maps on the scale 1:50,000 and 1:25,000 were produced. Several coastal zone areas were photographed 5–6 times during the 20–year period. These maps and high resolution satellite imagery of this region can be used to study coastline change due to erosion and accumulation processes for the period since 1965 onwards.

We now discuss in a broader context the role of airborne data in monitoring the coastal zone. Since the 1960s much research aimed at the development of airborne methods of photography in the coastal zone has been carried out in this region by Sharkov and others at VNIIKAM. The most intensive airborne surveys and ground-based *in situ* observations of shores and coastal waters (including under-water observations) were carried out in Sakhalin island in 1965–1986 and in the Kurile islands in 1973–1986. More than 300 films were taken of coastal and offshore zones up to a depth of 20m on the original scale of 1:1000–1:100,000.

These activities and similar activities carried out in the Caspian Sea formed the basis of a special technology of airborne photography of the sea bed in shallow seas. Techniques for geological, geomorphological and landscape mapping of the sea bed in the coastal zone were also developed (Kildyushevskii, 1997, private communication).

Aerial photography of the sea bed in coastal zones should be regarded as the basis for a large-scale detailed study of these areas before actually planning any industrial development. Applications of aerial photography include

- geological and geo-engineering surveys,
- exploration of mineral resources (oil, gas, coal, coastal placers, building materials, ground waters etc.),
- hydrotechnical construction including shore-protection works,
- hydrographic support of coastal navigation,
- mapping of underwater landscapes,
- setting up and maintenance of aquaculture farms and
- reconnaissance of submerged archaeological sites.

Aerial photographs of the sea bottom can only be obtained for comparatively shallow sea areas. The best results can be achieved in clear seas. Photos should be taken in the periods of high water transparency, see Figures 1–4. Figures 5, and 6 show how coal-bearing areas can be determined.

Some aerial photos showing interesting features in the sea and in the coastal zone can be found in Gur'eva *et al.* (1968). Features on the sea surface in the water and in the atmosphere are apparent and these include wind stripes and sunglint at the sea surface, stripes of surface-active substance or foam caused by Langmuir eddies, stripes of foam at the boundary between turbid and clear water, plankton in the upper water layer, river plumes, oil slicks, turbid water caused by the eruption of a subsurface mud volcano and foamed water as a result of a subsurface gas eruption.

Time series of aerial photos can be used to study dynamical processes at the sea bottom and changes in the coastline. For example, it was possible to track the life evolution and estimate the amount of solid material released during the eruption of the mud volcano on Kumani island. This island appeared in the shallow water area 7m deep as a result of the eruption of a subwater mud volcano on 4th December 1950. The first aerial photograph of the new-born island was taken on 30th April 1951 when it had a rounded shape with a crater in the centre. The island, however, was subject to intensive wave erosion. The volume of this island at the date of 30th April was estimated at 560,000m^3. Later photographs of the island were taken on 11th September 1951, when its dimensions were 200m by 110m, and on 3rd June 1952 when its dimensions had decreased to 70m by 40m. 14 months after the first photograph about 480,000m^3 of solid material had been eroded. By 3rd August 1952 the island did not exist any more. Depth measurements carried out in 1959 showed a cone in this place with a water depth of 4.5m. Changes in a number of other islands have also been observed using time series of air photography.

Figure 1 Aerial *photograph* of complex dislocated sedimentary unit (sandstone and aleurolites) complicated by two disjuncted disturbances. At the top of the image the land is shown.

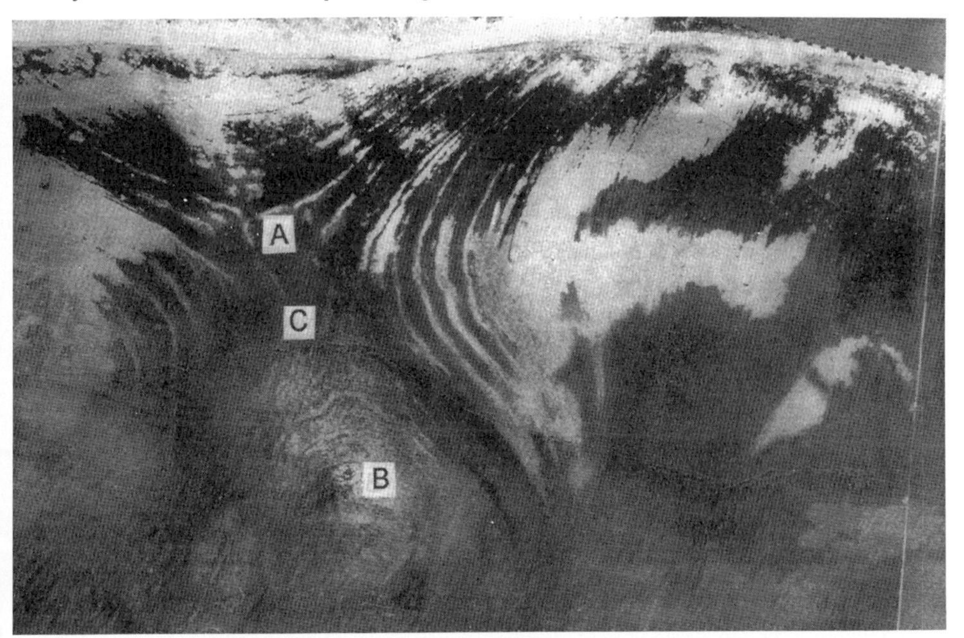

Figure 2 Aerial *photograph* of periclines of two anticline folds (A and B) separated by saddle (C). Such image provides determination of areas prone for structural oil and gas deposits. Light colours correspond to sandstone soils and dark colours are associated with bed rock strata covered with algae. Land is at the top of the image.

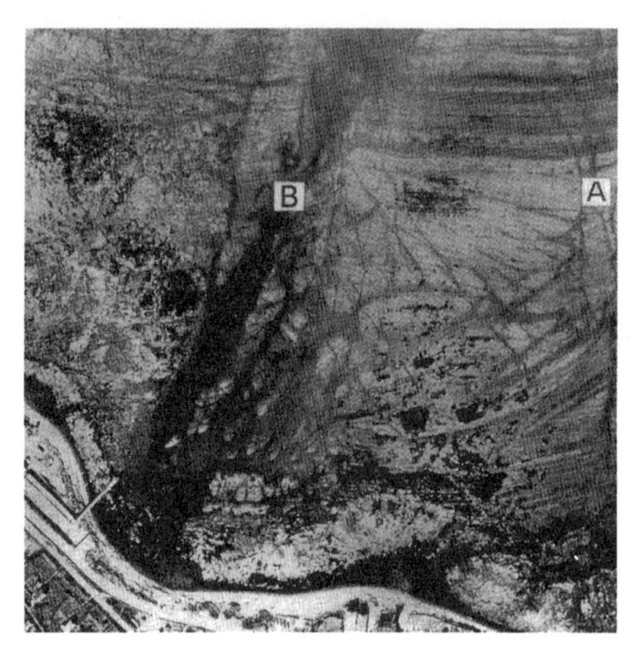

Figure 3 Aerial *photograph* of major disjunctive (B) and jointing (A) from the sedimentary bed rock thickness. Revealing such tectonic structures is very important for exploration of joint reservoir rocks and oil and gas deposits. Land is shown at the left bottom corner.

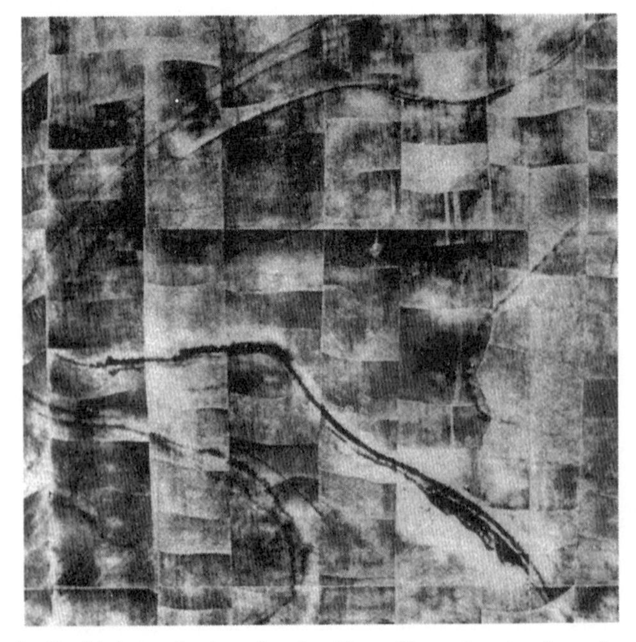

Figure 4 Mosaic of aerial photos showing submarine ridges of brown ironstone layers (iron ore deposits).

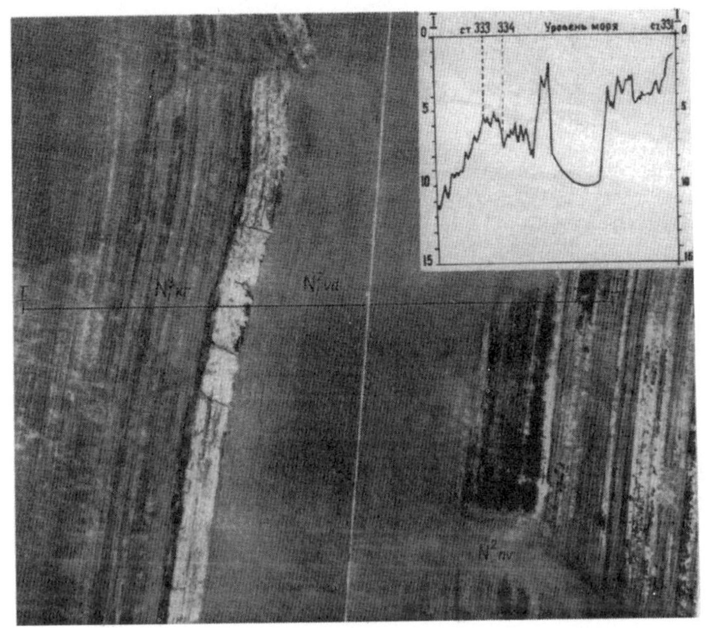

Figure 5 Aerial photograph of sedimentary formations with different substance composition indicating coal-bearing unit. Acoustic sounder profile across line 1-1 is shown at the top right corner.

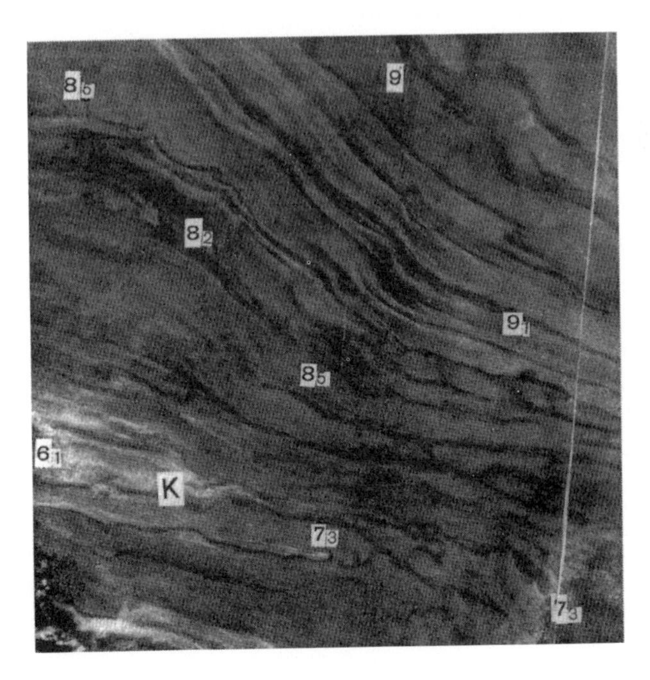

Figure 6 Aerial photograph of coal-bearing unit. (K - coal layers; numbers indicate depth in metres).

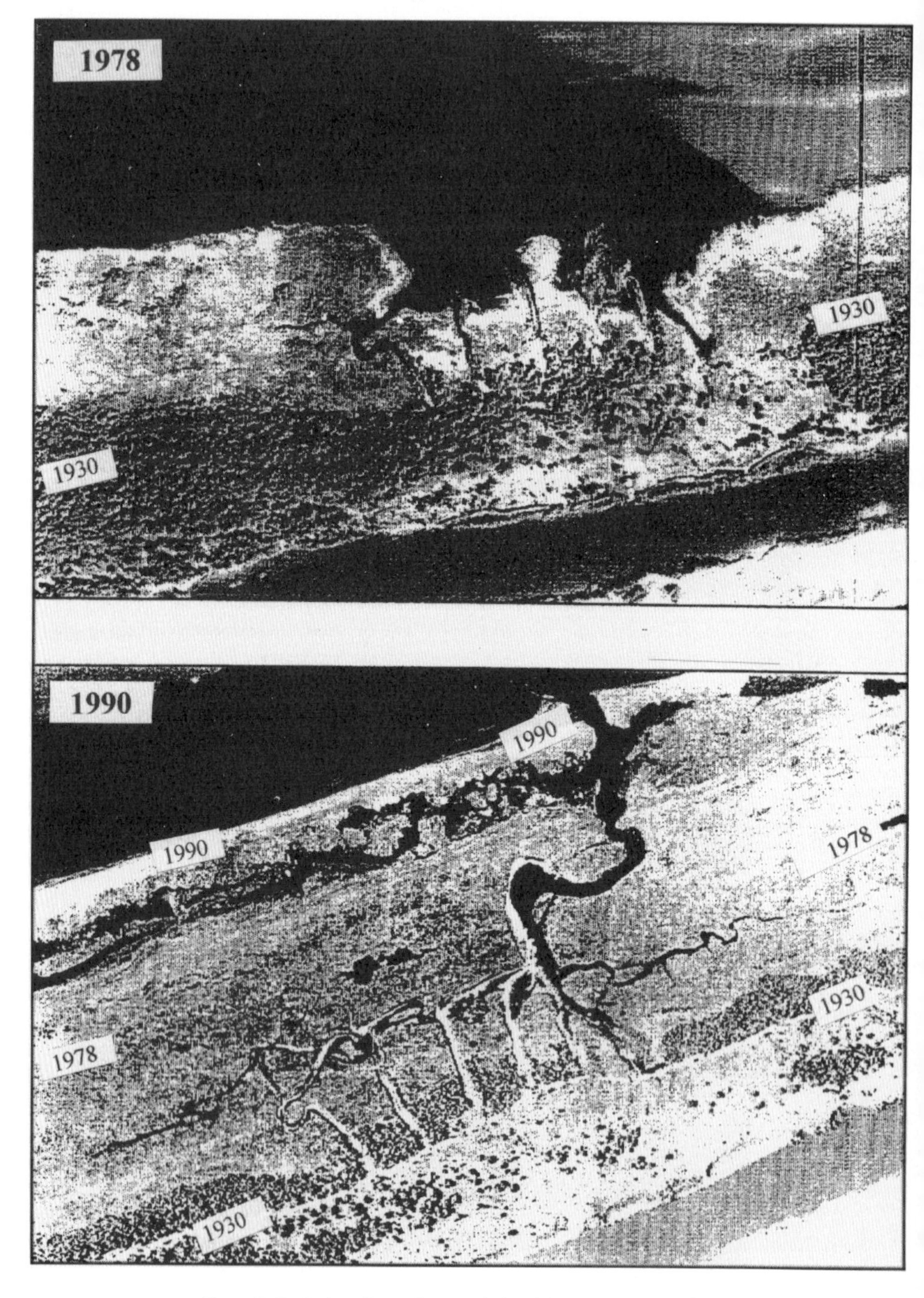

Figure 7 Evolution of coastline near industrial waste waters outflow.

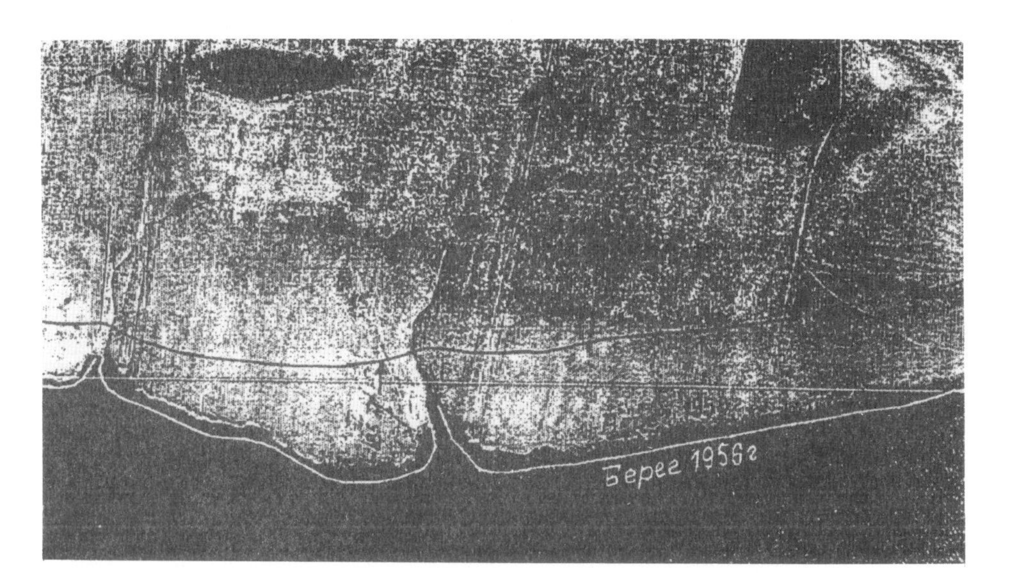

Figure 8 Coastal erosion as recorded in aerial photos.

Figures 7 and 8 are based on two pairs of aerial photographs from the VNIIKAM archives and they show accumulation and erosion processes in the coastal zone. Changes in the coastline in an area where industrial waste waters with high concentration of suspended matter were released are shown in Figure 7 for the period 1930–1990. Sedimentation caused visible change of this coastal zone. Coastlines of 1930, 1978 and 1990 are shown in aerial photos taken in 1978 (top) and 1990 (bottom).

Coastal erosion and changes in the coastal zone for the period 1956–1986 are shown in Figure 8. Coastlines of 1956, 1958 and 1986 are shown in aerial photos taken in 1958 (top) and 1986 (bottom). During this 30-year period the coastal zone lost a shore band some 185m wide.

Vegetation is an important part of underwater landscapes. The types of algae are indicators of sea bed soils. In tidal seas one can easily check the results of analysis of aerial photography during low sea periods.

Satellite imagery is incapable of providing detailed patterns of the sea bottom in the coastal zone. Hence it seems reasonable to complement satellite images with airborne imagery of selected areas.

4. Seas of the Atlantic Ocean

4.1 Baltic Sea

The Baltic Sea is a semi-enclosed sea with high human impact. Its area is 422,000km^2 (the Straits excluded), of which the area of the Gulf of Bothnia is 117,000km^2, the Gulf of Finland is 30,000km^2, and Gulf of Riga is 19,000km^2. The volume is only 22,000km^3, as it is a shallow sea 60–115m deep (with a maximum depth of 459m). It is a young sea (some 10,000 years in its present shape) formed after the last Ice Age.

Due to the large amount of fresh water brought by many rivers, the Baltic is a sea with very low salinity. Usually the following types of water are considered here: fresh water, brackish water with low salinity (< 10 ppm) and brackish water with high salinity (> 10 ppm). (Note that the term brackish already means water with low salinity).

The water is markedly stratified, with both thermoclines and haloclines. The sea is a sample-card of varying salinity. In the northern part of the Bothnian Bay salinity in the bottom water is 3–4ppm, in the Bothnian Sea 4–6ppm, in the Gulf of Finland 5–9ppm and in the Baltic Proper 6–8ppm. Salinity rises in the Straits from 8 to 20ppm, and reaches the almost oceanic value of 35ppm between the Kattegat and the Skagerrak.

The Danish Straits are the Baltic's only link with more open seas and more saline water. If a certain small influx of salt water through the Straits were not possible, the Baltic Sea would long ago have been transformed into a gigantic freshwater lake.

The water balance here is as follows: the surface-water flow of brackish water out of the Baltic is roughly 960km^3 per year, while the influx of salt water from the Kattegat and the Belts is about 475km^3 per year. As the Straits are shallow, only true storms, usually in the winter and autumn periods once every few years,, can give rise to major saltwater intrusions. These events become milestones in the history of the sea. The supply of new oxygenated saline water is not in proportion to the oxygen consumption. The risk of oxygen depletion, hydrogen-sulphide formation and benthic death is an ever-present one in the Baltic Sea, but the situation has been greatly exacerbated by over-fertilisation. Water exchange between surface and bottom varies over the sea from just days or weeks in the Straits to 4–6 years in the Bothnian Bay and Sea (Northern Europe's Seas. Northern Europe's Environment, 1989).

Some 70 million people live in the drainage area of the Baltic Sea with several industrialised metropolitan areas located in the coastal zone. The pollution load in the Baltic area is tremendous. The main routes are rivers and the atmospheric deposition. Tables giving annual inputs of pollutants by type, source and country are given by Victorov (1996). Statistical data for the last 5 decades gave integrated numbers for the former USSR; it is difficult now to extract separate data for Russia, Latvia, Estonia and Lithuania.

St. Petersburg with its 5 million citizens is the largest metropolitan area in this region. Moreover St. Petersburg in many senses links the Baltic region with the north-west of Russia and integrates economic activities in the Baltic Sea, the White Sea and the Barents Sea.

This region is the natural route for the transport of oil and gas from the Barents Sea fields (Timano-Pechora) and Western Siberia to Europe. There is a large oil-processing plant in Kirishi near St. Petersburg which is connected by pipelines with some oil provinces in Siberia. There is a plan to construct a new pipeline Kirishi-Primorsk-Porvo. Primorsk is a small port in the Gulf of Finland near the border with Finland where Porvo is situated. This idea means that an oil terminal should be constructed in the coastal zone and the pipeline should be laid through the recreational zone near St. Petersburg. Another project suggests that a new pipeline Kirishi-Petrozavodsk-Murmansk be constructed. The advantage of the second project is that it uses the only non-freezing Russian port in the Polar Ocean. Both projects have encountered strong opposition from environmentalists.

We shall deal in more detail with the state of marine and coastal environments in the St. Petersburg area later.

4.1.1 Remote sensing of the Baltic Sea

A review of satellite observations of the sea is presented here following Victorov (1996). As an object of study by means of remote sensing observations, the Baltic Sea is a very difficult task for the following reasons: (a) the spatial scale of phenomena and processes here is rather small and generally only high resolution satellite data are useful, (b) the shape of the sea is very remarkable – it is elongated from the south to the north, with the Gulf of Bothnia spreading far to the north, and the Gulf of Finland spreading far to the east – thus leading to a rather distinct variability of climatic conditions based on a mere geographical dimension, (c) the coastline is not a smooth one, with a lot of small bays and capes each playing their role in generating the overall pattern of oceanographic fields and (d) the total number of cloud-free days is not very high, which makes it difficult to obtain time series of satellite imagery of visible and infra-red bands providing the best spatial resolution.

Elements of water dynamics in the Baltic as recorded in satellite images of various bands can be analysed from two viewpoints: as pure oceanographic phenomena and in the ecosystem context – as the mechanisms by means of which the nutrients, biota, sediments and pollutants are being transported, dispersed and exposed to various driving forces such as the solar irradiance or the atmosphere. The marine ecosystem approach needs a better understanding of the biophysical and biochemical processes and much more sophisticated models (in terms of their spatial and temporal resolution) than the relevant marine sciences are able to provide now. One may say that the observational capabilities of remote sensing are in some cases ahead of the marine ecologist's capabilities to assimilate the observations. Thus, bearing in mind the possible future ecosystem approach, in this section we will mainly discuss the oceanographic meaning of the remotely-sensed data.

The following dynamical processes and phenomena can be monitored in the off-shore areas and coastal zone of the Baltic Sea.

Seasonal alongshore fronts can be observed during the spring heating and the autumn cooling of the water. The peculiar feature of the Baltic Sea water, its low salinity, creates preconditions for the formation of density fronts which are similar to the thermobar phenomenon in large lakes. Such density fronts appear as a result of the non-homogeneous heating and cooling of the nearshore and offshore waters in spring and autumn.

Upwelling events in the oceans are well known, and the physical nature of this phenomenon was understood many years ago. Strong alongshore wind plays the main role in generating the so-called Ekman transport effect. The Ekman transport in the upper water layer leads to the deviation of the drift current from the wind direction (to the right in the southern hemisphere). The coastal upwelling events are of importance as the bottom water masses coming to the surface bring nutrients, thus affecting the marine environment and biological productivity; upwelling zones are known to be rich in fish.

The Baltic oceanographers have been studying the upwelling events for many years, but it is not so easy to study them using routine techniques since the lifetime of upwellings in the Baltic is short and their dimensions are sometimes too small to be detected by rather scarce coastal hydrometeorological stations. For these reasons the study of coastal upwellings in the Baltic has benefited so much from regional satellite oceanography.

The first systematic study of the upwelling zones in the Baltic Sea based on satellite data was made by Bychkova and Victorov (1987). The following tasks were set :

- to determine the zones of upwelling events,
- to estimate the frequency of the appearance of upwellings in each zone,
- to find out the spatial dimensions of the zones and the directions of fronts,
- to estimate the intensity of upwellings (the temperature difference between the upwelling zone and the neighbouring waters, and the value of the horizontal gradient of the temperature within the zone),
- to study the role of the bottom topography and the coastline configuration in the generation of upwellings and
- to describe the development of an upwelling in time and to determine its main phases and the inertial period (the delay in time between the setting of a favourable wind and the arrival of cold waters at the surface).

The results obtained were presented in many original articles and have been summarised by Victorov (1996). There are two peculiar features of eddies in the Baltic Sea. The majority of the eddies recorded in satellite imagery of the Baltic Sea are of the cyclonic type. This corresponds to the general cyclonic circulation of water in the Baltic Sea, with the existing quasi-permanent currents being close to the coast. In the infra-red band, satellite imagery of the so-called 'mushroom-like' dynamical eddy structures is recorded actually consisting of two eddies – cyclonic and anti-cyclonic. Sometimes only the cyclonic part could be seen in the infra-red band satellite images, thus leading to a bias in calculations.

Another peculiar feature of the eddy structures in the Baltic Sea is the generation of long chains of eddies. They are recorded in satellite imagery of various parts of the sea.

It should be noted that with the development of satellite sensors for sea surface observations the still finer spatial structure of hydrophysical fields such as sea surface temperature, sea roughness, or patchiness in the visible band can be seen and analysed. One of the characteristic features of spatial order of these fields is the eddy patterns which represent the universal type of distribution of matter in nature, on scales from the Milky Way to microscopic viruses.

Hence, when moving step by step from an AVHRR image in APT mode (spatial resolution of 3 to 4km) to an AVHRR image in HRPT mode (1.1km), then to a Landsat TM image (120m in IR band), to MSU-E (30m) and to satellite photography (several metres) one can see in the Baltic a still more complicated pattern of inter-connected eddy structures of various types, shapes and dimensions. This pattern resembling astrakhan fur seems to be a specific indicator of a water mass internal structure. One can see more and more eddy structures of a smaller and smaller size at each step of satellite sensors development. So the pre-satellite days when the detection of each eddy was considered worthwhile to be announced in a separate article have passed forever. That is why in regional satellite oceanography it seems more reasonable to discuss the feasibility of a study not of separate eddy structures but of a local eddy field in time, that is of a kind of climatology of eddy fields.

Now we shall consider the problem of satellite monitoring of biological phenomena and pollution in the Baltic Sea. We start with seasonal blooms. As was shown above, algae (algal blooms) can be used as a tracer to detect and study dynamical phenomena in the Baltic Sea. The algae are of still greater importance as a biological indicator of the state of the marine environment and the state of the entire marine ecosystem of the Baltic Sea. The remotely-sensed concentration of phytoplankton gives information on the eutrophication of a certain part of the sea. In the 1970s the phytoplankton bloom intensified considerably, which is the evidence that the nutrition income has already caused the eutrophication of the Baltic. The role of algae as a biological indicator is based, *inter alia* on the fact that their photosynthetic activity is strongly dependent on the external factors. Unfavourable external conditions may depress the photosynthetic activity which can have serious ecological consequences. Both the concentration of phytoplankton and its type are the indicators of the state of marine ecosystem. Some years ago the diatomic algae were dominating in the Baltic (an example of their detection in Kiel Bight using remotely-sensed data was presented in 1984). In recent years the blue-green algae pushed out the diatomic ones and caused changes in the whole trophic chain . The diatomic algae are preferable for the health of the sea, but the present dominating blue-green algae *Aphanizomenon flos aquae* and *Nodularia spumigena* are more accommodative to external factors and are able to fix molecular nitrogen directly from the air and transform it into organic material. Thus they play a considerable role in the balance of nitrogen in the Baltic Sea. Nielsen and Hansen (1983) calculated that each seasonal bloom of the Baltic Sea produces 2,000t of nitrogen which amounts to 20% of its annual income with the Swedish rivers discharge.

Blue-green algae are of micrometre dimension, but they tend to form clusters shaped in long filamentary structures, which makes them detectable from aircraft and satellites. The analysis of multispectral visible band satellite images recorded in favourable atmospheric conditions, aside from the sunglint, enables one to determine optical inhomogeneities of the upper layer of the sea caused by considerable concentrations of suspended substances of mineral and organic origin including algae.

It is important to discuss some peculiar features of the study of the phytoplankton (algae) fields in the Baltic Sea using satellite images. The use of satellite data in the study of algae is based on the effect of visualisation of plants at a certain stage of their development. Strictly speaking, a rather uncertain bio-optical parameter is being registered in satellite imagery. This term reflects many uncertainties in the process of detection and the analysis of the data. The appearance of blue-green algae at the surface itself is an uncertain phenomenon. Gas cavities of multicellular blue-green algae allow them to accumulate in the upper layer during calm weather and leads to the formations of fields with dimensions and optical properties which enable their detection from satellites. The process of floating or sinking is governed by biological factors. It is impossible to detect algae if they are located too deep, even if their abundance is very high. On the other hand, when the concentration of algae is very high, their clusters may have positive buoyancy and, as was stated by Nielsen and Hansen (1983), "the upper surface of algae is located above the water and it is almost dry". But in this case there are no remote sensing techniques at all to determine dry algae concentration. Even in the case of wet algae, there is uncertainty in the vertical distribution of the algae which in principle makes it impossible to determine their concentration on a quantitative level. In general the problem of quantitative analysis of satellite imagery in the Baltic Sea is a difficult one due to specific optical properties of the sea water. Hence, in the Baltic Sea regional satellite oceanography addresses two separate problems: (1) the detection of algae fields and determination of their instantaneous visible dimensions as a basis for the study of their temporal and spatial variability on various scales, and (2) determination of plankton concentration and assessment of its biomass as a basis for the study of its variability. There is another uncertainty which can be a problem for both types of studies, namely the cloudiness which may impose a bias on any time series of satellite data. It is obvious that only routine standardised technology of satellite data acquisition, processing and analysis can make it possible to study interannual variability of blooms (though with the restrictions imposed by the above-mentioned uncertainties).

With this discouraging information in mind, let us turn to the realities of the analysis of anomalous patterns in satellite images as applied to the study of bio-optical characteristics of the Baltic Sea. The phenomenon of seasonal blooms in the Baltic Sea has been observed accidentally in satellite imagery since 1973 (for a review see Victorov 1996).

"Analysis of series of satellite images and sea-truth *in situ* shipborne data collected during the periods of summer bloom in the Baltic Sea for several years enables us to study the year-to-year variability of this phenomenon – appearance frequency of regions of bloom, their duration and intensity."

This long-term task was formulated and set in 1987 (Victorov *et al.* 1988). This was partly done by Rud and Kahru (1994) who obtained access to multi-year sets of digital satellite data and sophisticated technical facilities at Stockholm University. They used 135 archived pre-selected sufficiently cloud-free AVHRR NOAA scenes recorded in HRPT mode (1km resolution) for the summer (30th June – 24th August) period of years 1982 to 1993. For each year a chart of the late summer bloom of nitrogen fixing filamentous cyanobacteria (dominated by the species *Nodularia spumigena* and *Aphnizomenon flos aquae*) was compiled; the total area covered by the accumulations was also determined.

The following standardised technique was applied to satellite imagery to detect algae patterns. AVHRR channels 1,2,4 and 5 data were involved. Channel 1 data were analysed and three clusters were formed: pure water, cyanobacterial accumulations (with empirically determined albedo range between 2.3% and 4%) and clouds. To account for the internal structure (texture) of filamentary fields of algae, a 3 by 3 pixel digital filter was applied. At the next step, pixels with albedo in channel 2 exceeding the albedo in channel 1 by 0.2% were rejected. Additional threshold filtering was applied using channels 4 and 5 (thermal data) to account for clouds (Rud and Kahru 1994). It is clear that the algorithm for algae bloom patterns retrieval was tuned: it had to provide detection of bloom features with results comparable to those obtained by an expert manually. Texture analysis could have been used for this purpose. It is difficult to say whether an average of about 10 satellite images per year is a representative amount of data. CZCS archived data have been analysed in a similar way within the framework of the OCEAN (Ocean Colour European Archive Network) Project. The OCEAN Project brought additional information on seasonal blooms.

Results obtained by Rud and Kahru in 1994 showed that there were no shifts in the period of bloom during 1982–1994. The interannual variations of surface cyanobacterial accumulations were considerable with two peaks in 1982–1984 and in 1990 onwards. AVHRR data surprisingly showed no blooms at all during the summer periods of 1987 and 1988; only tiny areas of blooms were recorded in places in 1985 and 1986. The reasons for these variations are not yet clear.

The transport of sediments and pollutants is an important issue for coastal zone monitoring and management. River discharges spreading in the open sea containing suspended and dissolved substances can be traced far from the source. The river Nemunas-Kurshi Bay turbid waters are in some cases visible for a hundred kilometres to the north along the Latvian coast as a 70km wide light band. Turbid waters from a source located near St. Petersburg can be traced in the Gulf of Finland at a distance of 60–100km.

In some regions of the Baltic Sea even harmless (at the first sight) suspended sediments may carry contamination. In the Neva Bay of the Gulf of Finland, for example, there is a correlation between the concentration of suspended sediments and water pollution with heavy metals, chlororganic substances and bacteria (Sukhacheva and Victorov 1994). So while tracing the transport of suspended matter one is able actually to obtain some information on the transport of pollutants.

The local marine and coastal environment may be also affected by thermal pollution from man-made powerful sources of heat. Atomic power stations' cooling systems produce water with temperature 6 to 13 °C higher than that of the surrounding sea water and require about 150m^3 of water per second (Hupfer 1982). There are several atomic power stations in the Baltic region. Some of them are located on the shore. There are two nuclear power stations on the coasts of the Gulf of Finland. Hari (1984) presented the results of the study of the marine environment in the vicinity of the Loviisa power plant located on the northern coast of the Gulf of Finland. He showed that there was a considerable thermal anomaly in the water within a radius of only a few kilometres from the plant. These anomalies can be detected in summer time using satellite sea surface temperature (SST) information. In winter time the anomalies in the ice cover caused by thermal outlets can be traced in visible band satellite imagery.

There are many satellite images of the Gulf of Finland and of the Gulf of Riga covered with ice. In favourable conditions in these images it is possible to detect the power plants' thermal outlets into the sea, including the nuclear powerplant located in Sosnovij Bor (60km from St. Petersburg, at the coast of Koporskaya Bay) by local anomalies in the ice cover. (More reliable results were obtained during airborne thermal surveys when an IR radiometer showed an increase of up to 10 °C in the distribution of the SST near the outlet of a power plant in Koporskya Bay.)

In terms of radiation safety near nuclear power stations and monitoring of the environment in the Baltic region, detailed studies of the sea water dynamics in the vicinity of potentially dangerous objects seems to be important.

4.1.2 Operationalization of oceanography on a regional and local scale

Among the most important problems of monitoring the coastal zone is the problem of operational collection and delivery of data. In recent years satellite oceanography obviously showed a tendency to split into two branches, namely global satellite oceanography and regional satellite oceanography. Regional satellite oceanography is an interdisciplinary (oceanography/remote sensing) science which is based on satellite imagery analysis and deals with processes and phenomena mainly in the seas, coastal zones and separate parts of the ocean, while global satellite oceanography deals mainly with processes and phenomena in the world's oceans as a whole. It is regional satellite oceanography that is relevant to the subject of this book. Regional satellite oceanography (see Victorov 1996) is focused on the following problems:

- operational marine forecasting,
- monitoring of the marine environment,
- study of water masses dynamics within the water body,
- study of spread of river discharge,
- monitoring of water quality and transfer of pollutants,
- study of biological productivity in local areas,
- complex monitoring of coastal zone,
- study of the interaction between the coastal zone and off-shore waters,
- the study of ice cover morphology, etc.

We shall consider the problems of regional satellite oceanography and its operationalization.

It is believed that regional satellite oceanography is capable of providing operational/non-operational information in the following three areas (the concept of these areas is taken from ESA 1995):

Mapping and Monitoring

- Sea surface wave patterns
- Surface slicks
- Sea-ice extent, type, concentration and thickness
- Sea-ice movement
- Sea surface temperature

- Sea water colour anomalies (water quality)
- Shallow water bathymetry

Process Understanding

- Ocean features (eddies, upwellings, jets, frontal boundaries)
- Sea-ice related processes

Modelling for Forecasting

- Regional circulation
- Estuarine processes
- Coastal sediment transport and erosion
- Climate-change-related predictions.

A detailed review of the achievements of regional satellite oceanography, including the case study of the Baltic Sea, is given by Victorov (1996).

In recent years the issue of requirements for satellite data attracted the attention of many science and technology communities, international organisations and their committees and panels. The correct balance between the desired requirements, the real needs and the realistic capabilities of technology, engineering and information was the goal of numerous discussions. There recently took place (27–30th May 1997) at UNESCO Headquarters in Paris a discussion, where this problem was tackled during the parallel meeting of the Commission for Basic Systems/Working Group on Satellites of the World Meteorological Organisation and the Integrated Global Observing System/Space Panel (SCOT Conseil and Smith System Engineering Limited 1994, Polar Orbiting Satellites and Applications to Marine Meteorology and Oceanography 1996, Victorov 1996).

Operational marine applications require rapid processing and interpretation of data and dissemination of information to regional users. It is recognised that demonstration of the potential to extract information from satellite data is not enough. Relevant infrastructure, technical facilities and services should be, and are currently being, established in some regional seas. Operational sea-state forecasts are being made available to the marine users community. Shipping companies (optimal routing) and oil and gas companies (logistics of seismic surveys, mooring strategies for the connection of oil tankers to riser pipes and large-scale engineering works for offshore operations etc.) are notable here.

Nowadays many research groups and teams are working in the area of Baltic regional satellite oceanography. In many cases they do not exchange their ideas, approaches and results (or they only do it with great delay). The new Baltic Marine Science Conferences should provide a platform for these contacts in future. Nevertheless among many other regional seas the Baltic Sea should be considered as an area with advanced operational oceanographic applications of satellite data.

Satellite information is being incorporated into operational sea-ice monitoring and forecasting services and routinely used by ice centres in Sweden, Finland and Russia. ERS SAR-based oil-slick services which have been operational in Norway since 1994 (Pedersen *et al.* 1996) are being developed and integrated with existing services in other countries. Space-based algae bloom monitoring and forecasting services are being developed in the Baltic countries.

In the above-mentioned activities the following sensors/satellites are being used in a quasi-operational mode: NOAA AVHRR and ERS-1, ERS-2 SAR. However, gaps exist, and will exist in the future, due to unfavourable meteorological conditions and technological limitations. The solutions are: (i) to use models to interpolate between two consecutive satellite observations and (ii) to use operational technologies with satellite data assimilated into models. Thus, strictly speaking, we only have quasi-operational regional satellite oceanography, at the best. According to Johannessen *et al.* (1996), "There has, over the last decade, been a rapid development of various data assimilation methods which can be used with ocean and ecosystem models. At present, none of these methods are used operationally". This was a general statement. As for the Baltic Sea the HIROMB (High Resolution Operational Model of the Baltic Sea) (Kleine 1996) which, according to Funquist (1996), "has been in a so-called pre-operational mode since autumn 1995", is probably the most promising one. At least, sea-surface temperature simulation results obtained with this eddy-resolving model will be compared with relevant satellite patterns (Funquist, 1996).

What are the prospects for the operationalization of regional satellite oceanography from the satellite/sensors side? Civilian satellites and sensors with contributions to operational regional oceanography include:

- ERS-2 (1995–2000) with altimeter, scatterometer, SAR (Synthetic Aperture Radar) and ATSR (Along-Track Scanning Radiometer)
- ENVISAT (1999–2004) with altimeter, SAR, ATSR and the new instrument MERIS (Medium Resolution Imaging Spectrometer)
- METOP (2002–2017) with advanced scatterometer
- NOAA series with modified scanning radiometers AVHRR-3 (1994–1999) and VIRSR (1997–2000)
- RADARSAT (1995–present) which is currently providing data on sea state and ice.

The SeaStar satellite with the SeaWiFS colour scanner was launched in August 1997. A similar instrument EOS COLOR will be launched on the EOS satellite in the late 1990s. ADEOS was launched on 18th August 1996 with another ocean colour instrument OCTS. OKEAN-O missions with a real aperture radar RAR and an MSU-SK scanner as well as RESURS-O missions with MSU-SK, MSU-E scanners will be continued in the late 1990s.

Though the above list is not complete (for comprehensive information see Polar orbiting satellites and applications to marine meteorology and oceanography 1996), it is obvious that there will be no drastic improvements in terms of operational satellite oceanographic systems in the next decade. Various pilot operational and quasi-operational single satellites and series of satellites will still complement each other, and it will probably be difficult to harmonise data from various sources. Data handling (acquisition, storage, processing, fusion, interpretation, assimilation and dissemination of output information products) will definitely become a crucial point in Baltic regional satellite oceanography in the nearest future.

From the technological point of view, one centre can provide satellite data acquisition, processing and dissemination of output information products for the whole Baltic Sea region. It is possible for this centre to run the sophisticated ocean models (now and in the immediate future) and ecological models (in the distant future) in an operational mode. It probably could be the best option in terms of costs and benefits. Analysis of the state of oceanography in the Baltic Sea shows that:

- Current achievements of regional satellite oceanography in the Baltic Sea are encouraging and Baltic operational satellite oceanography exists, though mainly in a quasi-operational mode. The level of operationalization of oceanography in the Baltic Sea varies considerably for different oceanographic parameters.
- Further progress in satellite sensors and space-based ocean observing systems is needed to meet the requirements of regional operational satellite oceanography.
- There is high scientific potential for co-operation in the Baltic operational satellite oceanography which is so far unrealised. Hence the Baltic Marine Science community has ben invited to start discussions aimed at (a) designing a joint centre of operational oceanography in the Baltic Sea,(b) designing a joint regional satellite monitoring of the (open) Baltic Sea and (c) finding mechanisms for co-operation in the development of methods of satellite monitoring of the Baltic coastal zone (with an emphasis on coastal erosion, sedimentation, shallow water bathymetry and water quality)

4.1.3 Numerical modelling of processes in the coastal zone

Here we discuss numerical modelling of processes in the coastal zone, based on experience gained in the Baltic Sea. In recent years much attention has been paid to numerical modelling of currents and sediment transport in coastal waters on a regional and local scale. With the development of computer technology the grids used in these modelling exercises are becoming fine enough and comparable to the spatial resolution of satellite sensors. Still the efforts to combine the benefits of fine-grid models with those of high resolution satellite imagery seem to be insufficient. In many cases, the results of numerical modelling cannot be proved or checked by *in situ* measurements. The modelling community argue that there are no adequate measurements with proper spatial and temporal resolution to validate their models. There appears to be a gap between numerical hydrodynamic (dispersion) models and the remotely-sensed data, and the modelling community seems to hesitate (or not be able) to assimilate the remotely-sensed data in their models.

Victorov *et al.* (1991) invited the national experts in hydrodynamic modelling "to use satellite data to establish or improve interior and boundary conditions and some parameters of numerical hydrodynamic models". Blumberg *et al.* (1993) wrote: "there is a need to make better use of available observations. Models require data to establish interior and boundary conditions, to update boundary fields, to validate the model physics and to verify the simulations...One needs to blend the results from both circulation and water quality models with the available data to provide for the best estimates of how water and materials are transported throughout a coastal system. The data assimilation, that is the process of this blending, is undoubtedly the most powerful tool presently available for extracting information and insight from the sparse coastal ocean data sets and the imperfect model results".

Allewijn (1996) wrote: "With the present computers it is possible to incorporate an increasing number of physical processes and in this way the models give a more realistic simulation of environmental processes and their expected consequences. The problem in developing realistic numerical models is the lack of synoptic and detailed spatial information for model calibration, validation and updating. In the past two decades Earth observation satellites have proven to be capable of supplying this information on a regular basis. Particularly during the past few years the number of satellites and advanced sensors has increased. After an experimental phase, remote sensing data is now a valuable information source in addition to in situ observations. Combining both types of information with numerical models needs extensive development effort in order to optimise the use of the large amount of data for policy and environmental questions. This effort consists of developing data-assimilation and data-integration techniques, together with data handling of the increasing amount of data.

Geographical Information Systems (GISs) have proved to be a powerful tool for analysing and presenting spatial information. Building a GIS shell around different types of observations and numerical models (assimilated with remote sensing and in situ data) is a logical next step. Such an information system should consist of several complexity levels in order to translate model output and observations to parameters which can be used for policy decision making. The concept of sustainable development has stimulated the development of GIS-based expert systems. This type of policy decision support method will become an important tool for environmental management."

At this stage we should mention one of the sophisticated models and relevant software packages named CARDINAL (Coastal Area Dynamics Investigation Algorithm) and the experience gained in modelling currents and transport of pollutants using this package. CARDINAL is a user-friendly program for the simulation of non-steady flow, dispersion of pollutants and transport of sediments using curvilinear co-ordinates with options for two-dimensional or three-dimensional conditions (CARDINAL User's Manual 1993).

The appearance of high-powered personal computers made it possible to develop software for solving various problems within the field of geophysical hydrodynamics via a user-friendly interface. CARDINAL provides tools to solve quickly and effectively very complex problems. A wide spectrum of users may use the software easily without any knowledge of programming. These include coastal engineers and environmentalists who may be considering outfalls of treatment plants or the protection of coastal structures. Also, this program can provide an excellent tool for students to develop a highly visual understanding of important features within hydrodynamics. CARDINAL may be useful in theoretical researches providing quick and simple testing of analytical solutions. The equation of dispersion of pollutants is solved by the third and first order implicit up-stream conservative finite-differences. This method of solution does not diffuse pollutants much, but does prevent numerical oscillations. The input information includes geometry and bathymetry, locations of the open boundaries and boundary conditions, loading of pollutants, time- and space- variable wind field, bottom roughness, relations for empirical coefficients and initial conditions. The output information includes velocity and flux vector fields, isolines and a bird's eye views of free surface and concentrations, time histories of all variables on selected locations, spectrum characteristics of free surface oscillations and a vertical distribution of all variables on selected cross-sections and locations.

The following application examples show the wide variety of problems that were studied by means of the CARDINAL software:

- **Lakes and Reservoirs**. Plescheevo Lake to the north of Moscow: simulation of dynamics as part of a study aimed to stop lake euthrophication. Kiev Reservoir in the Ukraine and

Lake Imandra on the Kola peninsula in Russia: dispersion of radionucleides after accidents at nuclear power stations.

- **Rivers.** Volga and Oka near Nizhniy Novgorod and Viatka at the north of the Russia: simulations of accidental spill of pollutants and propagation in the rivers. Kolima in Siberia: propagation of storm surges into the river in the presence of a hydroelectric power station for the estimation of navigation conditions. Angara near Baykal Lake, Ingoda and Velikaya in central Russia: estimations of the influence of the development of large sand quarries in the river beds on hydraulic and turbidity regimes.

- **Bays.** The Neva Bay at the head of the Gulf of Finland: simulations of dynamics and dispersion of pollutants for the choice of optimal locations of treatment plants outfalls and study of the impact on the environment of the flood barrier, now under construction to protect St.Petersburg against floods.

- **Seas.** The Baltic Sea: simulations of storm surges for the implementation of the storm forecast system for St.Petersburg. The White Sea: tidal motion and transport of pollutants from river estuaries. The Sea of Japan: simulation of tsunami. The South-China Sea: simulation of typhoon propagation.

Currently the CARDINAL software package is being used together with a tailor-made database BASBAY as part of the activities aimed at the development of a regional Project 'Integrated Water Management in Northwest Russia'. The present phase is restricted to the St.Petersburg region and covers Ladoga Lake, the river Neva and the eastern part of the Gulf of Finland.

BASBAY is a user-friendly database for the storage, visualisation and statistical analysis of temporal and spatial variability of chemical, biological, health and hydrological properties of water in the Neva Bay and the Eastern part of the Gulf of Finland. BASBAY has been developed at MORZASCHITA, a special office set up for the St. Petersburg flood barrier construction. The sources of data are the Hydrometeorological Service and the Sanitary-Epidemiological Service. Chemical data (29 constituents in 9000 records) and hydrological data (18 constituents in 7400 records) have been collected since 1968, biological data (55 constituents in 2200 records) since 1977, and sanitary data (11 constituents in 3000 records) from 1987. Nowadays the database comprises more than half a million units of information (6.5 Mb of disk space). The BASBAY package is divided into 5 main windows: data input, area, depth, stations and analysis. The data input window is used to load measured data into one of the following arrays: hydrology, chemistry, health or biology. The area and depth windows allows one to assign boundaries and depths of a water body. The stations window is used to assign new monitoring stations, move, rename or delete them. More than 100 stations are present in the database. In the analysis window different types of data analyses are available. Among them are time histories of selected constituents at the selected stations, averaged time histories with a chosen period of averaging, spatial distributions of selected constituents using a chosen period of averaging, tables of various statistical (including Quantile analyses) and water quality characteristics.

The database is currently used by many local, regional and national institutions and by international organisations which are involved in the evaluation of the St Petersburg flood barrier's impact on the marine environment and the assessment of the local environmental situation.

An attempt is being made to build an integrated water quality index based on the datasets recorded in BASBAY. (An international conference 'Environmental Indices: Systems Analysis Approach' was held in St.Petersburg on 7–11 July 1997).

The CARDINAL software package and BASBAY database are likely to become the key elements of the project on 'Integrated Water Management in the St.Petersburg Region'. This Project can be considered as a follow-up of a series of previous research programmes aimed at studying of the state of marine and coastal environments in the Neva Bay and the eastern part of the Gulf of Finland carried out in the 1960s–1990s. The Project is based on the results of those programmes and should provide a step forward in the environmental management of an urbanised coastal zone on a regional/local scale.

4.1.4 Monitoring of marine and coastal environments in the Neva Bay

A brief review of this topic will be presented here; for more details see Victorov (1996). Besides being the response to an urgent environmental issue in a 'hot spot', the materials to be presented below are also related to two important coastal ocean problems:

- What is the build-up of man-mobilised materials (pollutants and natural) in the coastal zone? And how can these materials be used to trace natural processes ?
- What are the sedimentation rates in the coastal ocean (shelf, estuaries, continental rises, etc.)?

The Neva Bay is situated in the easternmost part of the Gulf of Finland. In the context of environmental monitoring of coastal waters the multi-year studies of the fields of suspended matter in the Neva Bay will be presented. Of special interest here is the attempt to separate the natural and man-made factors in the satellite images in connection with the construction of a major engineering structure, the St. Petersburg flood barrier.

The River Neva with a water discharge of 1800–3000 $m^3 s^{-1}$ enters the Gulf of Finland and forms its delta where St.Petersburg is situated. The Neva Bay is a shallow water body with low flow velocities. The St.Petersburg metropolitan area with its 5 million inhabitants, hundreds of industrial enterprises with poor filtering systems and inadequate municipal waste water treatment facilities, produces about 5 M m^3 per day of waste water which comes into the Bay. Nowadays about 30 % of the waste water is untreated. As a result of insufficient water treatment the environmental situation in the Bay, especially in terms of bacterial pollution and concentration of heavy metals, remained critical for a number of years. This information was made accessible to the public only in the 1980s as a result of general political changes in the former USSR. It happened that this period coincided with the construction of the flood barrier in the Neva Bay, which was intended to protect the city against floods coming as a long wave from the west. From 1703 till 1990 the city has survived 284 floods (defined as 160 cm and more above the standard mean sea level).

The flood barrier is a 25.4 km structure consisting of soil-filled dams with 6 water discharge sluices to allow for river outflow and 2 navigational passages for large and small ships. In the normal situation all the watergates are open and the barrier works as a bridge. At the alarm situation all the sluices and passages can be closed and thus protect the city against floods for 3 days which exceeds the longest historical flood period. Construction work started in the beginning of the 1980s and was scheduled to be completed by 1995. But when the barrier was 70 % completed all further construction work was halted by governmental decree (and in practical terms has not been resumed until now) under the pressure of public opinion.

The concerned citizens related the general poor environmental situation in the Bay with the construction of the barrier which, in their opinion, was a cork in the bottle preventing the outflow of polluted waters from the Bay and did not allow fresh water from the Gulf to wash through and flush the Bay. Analysis of satellite imagery of the Bay helped to provide current data on hydrology and marine and coastal environments in the Bay which have been used in lengthy discussions with the involvement of scientists and general public.

We consider the data sources and methods used in this study of the Neva Bay. Since 1980 the multispectral images from the METEOR'-30 satellite were analysed; later on, information of better quality from the satellite KOSMOS-1939 became available. Since 1988 data of medium ground resolution of 175m by 200 m from the scanner MSU-SK and in summer 1989 high resolution images from the MSU-E device (30m by 45 m) were obtained for the Neva Bay. A limited number of images from the USSR photographic satellites were also used in monitoring activities. Data from airborne observations of the Bay carried out in the State Oceanographic Institute in St.Petersburg were also included in the analysis. One SPOT image was also received (courtesy of Alain Cavanier, IFREMER, Brest, France).

Nowadays there are two major databases relevant to the Neva Bay which are to be integrated in a regional GIS 'The Neva Bay' (Usanov *et al.* 1994). The database of conventional hydrological, hydrochemical and hydrobiological data BASBAY at MORZHASCHITA was described above. Our database of remotely-sensed data consists of some 300 images from satellites, airborne survey charts and synchronous or quasi-synchronous in situ measurements relevant to remotely-sensed imagery since the end of the 1970s. The GIS 'The Neva Bay' is meant to become a consistent part of an integrated water management system in the St.Petersburg region. In the analysis of satellite imagery the following additional data were used:

- navigational maps of the Bay,
- data on mean velocities, general structure and schemes of currents,
- characteristics of suspended matter including size distribution of suspended particles,
- data on seston concentration and the regression characteristics, water transparency versus concentration of suspended matter,
- data on the sources of suspended matter,
- historical data on the spatial distribution of suspended matter in various hydrometeorological conditions,
- data on the meteorological situation before, after and at the moment of satellite overpass, including wind, atmospheric pressure and horizontal visibility,
- data on water level at different points in the Bay,
- in situ measurements of water transparency, salinity and temperature,
- airborne measurements of sea-surface temperature and optical properties of the water upper layer.

In the analysis of satellite images visual and instrumental techniques were used (Sukhacheva 1987). Later, with the progress in satellite imagery quality a method was developed which enabled us to interpret the images in a quantitative way and determine concentrations of suspended matter. To tune the method, in situ data on water transparency and concentration of suspended substance were used, which were collected during the periods 1959–1960 and 1982–1989. Actually this method could provide quantitative estimates of suspended matter concentration limits at each of the clusters recorded in a satellite image (Victorov et al.1991). Image processing techniques were used to map the spatial distribution of suspended matter in the Neva Bay with 6-cluster classification and presentation of results in false colours.

It was proved that for this water body more than ten levels of concentrations can be determind in high resolution satellite imagery. But for the sake of consistency we used a 6-level scale throughout the study. Waters with concentrations of less than 10 mg l^{-1} were considered clean; they are usually located in the central part of the Bay. Often extremely high concentrations exceeding 200 mg l^{-1} were measured from ships.

Many examples of satellite imagery of the Bay in a historical context were presented by Victorov (1996). Almost all the scenes show the Neva Bay with the city of St.Petersburg in the right part and the island of Kotlin in the centre. The flood barrier connects the island with the northern and the southern coasts of Neva Bay. Bright patches indicate areas with high concentration of suspended matter.

Various patterns of spreading from the sources and spatial distribution of suspended matter have been recorded in satellite images. Suspended matter can be used as a tracer to visualise flows and currents with their fine structure, as well as dynamical phenomena such as eddies and mushroom-like structures. To study the features of fields of suspended matter and the dynamics of water masses in the Neva Bay, satellite images were selected which showed various patterns in hydrometeorological conditions connected with the winds and the water level changes. With respect to characteristics of water level change, at least four types of hydrological situations can be studied:

- smooth change of level (slow decrease),
- level rise,
- sharp level drop,
- period of change of phase (a level drop after a steady rise).

Provided that the locations of the sources are frozen (which is not the case), each type of hydrological situation can be related to a specific distribution pattern of suspended matter and the current field as recorded in satellite imagery. Satellite images clearly showed that long before the construction of the flood barrier started, complicated patterns of the fields of suspended matter existed in the Neva Bay and were occasionally recorded in satellite imagery. Depending on the location of the source, one could find dominant areas of suspended matter fields either at the southern or at the northern coasts, or at both. There are four main sources and mechanisms that make the water in the Bay turbid:

- wind and wave mixing,
- dredging, bottom-deepening and ground-filling operations, dumping of spoil,
- untreated waste waters,
- dirty ice melting,

of which the first two are dominant in the Neva Bay.

The analysis of remotely-sensed images stored in our database showed that there is a very high variability of pattern characteristics of suspended substance fields on a synoptic scale according to the above-mentioned four types of hydrological situations. As for the seasonal variability, for a number of years in the 1980s (when routine bottom-deepening and ground-filling operations were being performed on a regular basis), there was a tendency for the total area of turbid water zones to increase from spring to the end of autumn. The reasons might be the increasing activity of dredging and other operations, including the dumping of spoil, in the summer period and the seasonal growth of wind and wave activity. It is difficult to show any tendency in the annual variability; no reliable remotely-sensed data exist which could show any barrier-dependent changing in general patterns of suspended matter fields before and after the construction of the flood barrier.

Analysis of a series of remotely-sensed data enabled the study of multi-scale spatial and temporal variability of hydrophysical fields in the eastern part of the Gulf of Finland caused by various natural and man-made factors, namely, annual, seasonal and synoptic variability connected with hydrometeorological conditions, or caused by anthropogenic factors, such as dredging and ground-filling operations to obtain new land for the city development, dumping of spoil (including similar operations during the period of the Barrier construction), etc. We found that ground-filling operations carried out since the early 1960s were the main source of suspended materials. The larger the volume of these operations, the more important was the wind and wave mixing factor. Besides, there was a kind of inertia; in autumn even after they finished work, the zones with high concentrations of suspended matter could still be seen. In spring when they had not yet started, no zones with high concentrations were observed practically in the whole Neva Bay and the eastern part of the Gulf of Finland.

Since 1990 (the date of the collapse of the former USSR followed by the collapse of the national economy) the intensity of ground-filling operations started decreasing and practically they came to an end in 1993. Water pollution with suspended matter decreased substantially as compared with the period of the 1980s. At the same period and for the same reason the impact from industrial wastes decreased. As a result the state of the marine environment seems to have improved, which was shown with remotely-sensed data and in situ measurements.

We found that the observed decrease in concentrations of heavy metals, bacterial and other types of water pollution (data of in situ measurements) was, to a certain degree connected with the decrease of concentration of suspended sediments. The available data show correlation between these two groups of parameters. (Victorov et al 1996).

4.1.5 The suspension of work on the flood barrier.

In the Autumn of 1987 concerned citizens appealed to Secretary General Gorbatchev. They felt that the pollution of the Bay was aggravated by the construction of the flood barrier and they protested against its completion. In response, several sessions on this subject were held in the Academy of Sciences of the USSR. In short succession three commissions of Soviet experts were nominated and set to work. They submitted their reports early in 1989. As the three national commissions came to contradicting conclusions on the Barrier and its role in the environmental situation in the Neva Bay and the region, in June 1990 on a governmental level a decision was made to invite an International Commission of experts. Of major importance was its conclusion that the ecology of the system Ladoga Lake – River Neva – Neva Bay with the eastern part of the Gulf of Finland has already, over a long period of time, been adversely affected by human activities, including principally the discharge of untreated or inadequately treated waste water, the destruction of wetlands for urban expansion and the dredging and dumping of spoil. During the 10 years that the flood barrier was under construction, its effects on the water quality of Neva Bay have been negligible compared to the above impacts.

Nowadays possible economic developments in the St.Petersburg region affecting its marine and coastal environments could be:

- new harbour developments outside the old city, but protected from flooding and well connected to road and railway transportation,
- reconstruction and relocation of industrial activities outside the old city,
- planning and execution of an intensified programme for upgrading of the sewerage system and municipal waste water treatment,
- construction of an orbital highway, including the Western route over the flood barrier and the island of Kotlin.

Taking those projects into account, the regional policy should include the following issues of primary importance :

- the St.Petersburg water system is one system and should be studied and managed as one in all of its physical, chemical and biological aspects,
- as the scientific evidence and the international expert opinion showed that the completion of the flood barrier would have no negative effect on the water quality of the Neva Bay, the Flood Protection Barrier should be completed to protect the city, the infrastructure and future investments in the developments,
- the environmental conditions of the waters surrounding St.Petersburg should be improved and water quality standards should be developed in the framework of the integrated Water Management Project.

The Integrated Water Management Project for St.Petersburg was launched in 1996.

4.2 The Black Sea

The area of the Black Sea is area is 423,000 km^2 and its volume is 537,000 km^3. Its mean depth is 1000 m, with a maximum depth of 2243 m. Its maximum length is 983 km and its maximum width is 531 km. Its coastline is rather smooth, except for the northern part. There are mountains on the eastern coast. There are only three small islands in this sea. There are some large rivers bringing about 500 km^3 of fresh water each year. They are the Danube (Dunai), Dnepr (Dnieper), Dnestr (Dniester) and Boog.

The Black Sea is connected with the Mediterranean Sea via the Bosphorus Strait. It is a remarkable fact that as late as in 1882 there were discussions on the direction of currents between the two seas. The crucial observation was made by the Russian Admiral S.O.Makarov, who used two barrels with different loads to trace their drift and thus showed that in the upper water layer the current is directed from the Black Sea to the Mediterranean Sea, while in the deep layer the direction of current is opposite.

The climate of this area is of Mediterranean type; the water temperature in the south-eastern part of the sea in August is 24 °C. The peculiar features of the sea are a high concentration of hydrogen sulphide and oxygen depletion in water layers deeper than 150–

200 m. Water transparency is 27 m in the open sea, while it is only 2–4 m near the rivers outlets. One can easily see river plumes in satellite imagery.

There are high waves in the sea, and coastal protection is needed in many places. In winter there are sea ice fields in the north-western part of the sea.

There are many ships of river-sea type which are used to reach many Danube river ports in Eastern Europe. Before the collapse of the former USSR Odessa was used as the most busy port in the Black Sea to provide transport links of the USSR with America, Asia and Europe. Now Odessa belongs to the Ukraine, so Russia faced the problem of using another port. Novorossiisk is planned to be used as the main Russian port in this sea. This port is even more important for Russia as it has oil terminals which could be used to transport oil from the newly independent states of Central Asia. For this purpose the oil and gas terminals are under reconstruction now. In this context one should mention the problem of the Bosphorus Strait. There is a discussion on the transport capacity of this strait and possible restrictions on transport of oil products. Some international commissions recommended the installation of modern navigational systems here and the enforcement regulations of shipping between the Asian and European parts of Istanbul. Thus the strait could be used more safely by all kinds of vessels including oil tankers, so there are no doubts in the important role of Novorossiisk as a major Russian oil port in the Black Sea.

The Crimean peninsula is a remarkable area. For decades it has been used mainly as a resort area for citizens of the former USSR. There are many resort places of international importance situated in the coastal zone along the peninsula. The whole economy of this area was oriented on tourism. With the collapse of the USSR the Crimea has lost its importance as a resort. Another remarkable feature of the Crimea is its naval bases, particularly Sevastopol. This town was a battlefield during the Crimean War in the 1840s (Britain, France and Turkey against Russia) and in World War II (Germany against the USSR).

Regarding the USSR Black Sea Fleet, as part of the former USSR Navy, there were arguments between Russia and Ukraine on how to divide the warships and coastal bases and the town of Sevastopol. After several years of disputes an agreement was signed by the Presidents of Russia and Ukraine in May 1997, according to which Sevastopol will be rented by Russia for a 20-year period.

The water quality in Odessa and other coastal towns is far from being high. Sewage waters are not properly treated; there are many direct outlets for untreated effluent. Some beaches are closed in the summer as the bacterial situation is dangerous for public health. There is a shortage of drinking water in many towns along the coast of the Black Sea.

There are some international programmes dealing with aspects of the marine and coastal environments in the Black Sea. The OCEAN Project run by the EC Joint Research Centre in Ispra (Italy) brought interesting applications of CZCS imagery to the problems of oceanography and ecology of the sea. An integrated approach to the problems of coastal zones in the north-western part of the Black Sea was demonstrated recently in the framework of a Dutch-Ukrainian Project (Allewijn 1996). Resulting from the EARSeL workshop on Remote Sensing and GIS for Coastal Zone Management (which was held in Delft in October 1994), a project was initiated by the Survey Department in the Netherlands and the Ukrainian Marine Hydrophysical Institute. This study, entitled an Integral System for Observation and Management of the N-W Black Sea, was partly financed by NATO, in the framework of a collaborative research grant. This study was completed in 1996. The final report contained a review which stated that the shelf zone of the N-W Black Sea was of high economic interest with respect to fishery and tourism. At the same time this coastal area was also subject to pollution caused by the inflow of the Danube and the Dnieper rivers. It was stated further that industrial activity and high urbanisation in the coastal zone caused the input of polluted waste water into the coastal sea. Oil spills in the shelf zone of the N-W Black Sea led to serious environmental degradation of this area. The major findings of this project are:

- regional-scale satellite observations (e.g. Landsat MSS) can be used to study the inflow of warm turbid water from the Danube, Dnieper and Dniester rivers, leading to pollution, eutrophication and algal blooms in the N-W Black Sea shelf zone,
- the spatial distribution of water pollution in near coast urban and industrial centres (e.g. Sevastopol) can be derived from high resolution (thermal and optical) satellite imagery,
- in addition to the visual inspection of oil spills by legal authorities, radar imagery (e.g. ERS SAR data) would be very helpful for deriving oil slick statistics,
- low resolution (thermal and optical) satellite observations (e.g. NOAA AVHRR) can be used to validate numerical modelling and experiments of water circulation and transport of polluted river water into the Black Sea,
- the numerical models available proved useful to simulate realistic flow patterns necessary for studying different pollution scenarios.

To understand the physical system and solve the complex environmental problems in this area, integral observation and analysis methods were required. The combined use of satellite remote sensing, field surveys and numerical modelling was needed to understand the physical processes in the coastal and offshore zone of the N-W Black Sea. The institutes involved in this project, (now completed), hope that at a later stage there will be (financial) possibilities to continue with this important topic on the application of remote sensing and GIS for environmental management of the Black Sea. Probably other EARSeL members are already involved in projects in this region and by joining forces further development and implementation of the ISOM Black Sea system can be achieved (Allewijn 1996).

4.3 The Azov Sea

The Azov Sea is a small sea which can be considered as the estuaries of the River Don and the River Kuban. The area of the sea is 37,600 km^2, its volume is 303 km^3 and its mean depth is only 8 m (maximum 14 m). Its length is 360 km and its width is 175 km. The sea can also be considered a Gulf in the Black Sea. The shores are flat. There are many small bays and salt marshes here. The two rivers bring about 40.7 km^{3}of fresh water into the sea each year. There are strong winds here which can cause up to 442 cm difference in water levels for inflow/outflow situations. There are ice fields in winter in shallow parts of the sea.

In the 1950s the Volga-Don Canal was built to provide cargo shipping direct access to many places in the inner part of European Russia. including river and sea ports in the Baltic Sea and the Beloye Sea water basins. The canal and the Tsimlyanskoye Reservoir with its 22.6 km^3 volume have affected the hydrological and environmental situation in this sea (see also Secton 5.1 on the Caspian Sea for comparison)

The state of marine and coastal environments is far from being good. They are badly damaged with heavy metals and toxic chemicals (waste products of industry) and pesticides. In some places beaches in recreational areas are closed.

Recently an international Project has been conducted aimed at setting up an Integrated Water Management System for the sea and the basin involving Delft Hydraulics, from the Netherlands, and local research teams.

5. Enclosed seas

5.1 The Caspian Sea

The Caspian Sea is not a sea at all, but the largest lake on the Earth, although it was described as a sea in all the textbooks in Geography which were being used by many generations of students in the Russian Empire and the former USSR. Some years ago after the collapse of the USSR a large question mark appeared near the word sea and international discussions and negotiations on the sea/lake issue were initiated. According to international laws (namely, agreements between the USSR and Iran of 1921 and 1940), the Caspian Sea is an inland water reservoir (lake) and should be jointly used by all the coastal states. But if one considers it as a sea, then one should take into account the international document on the Law of the Sea, which deals with such issues as the 12 miles limit, the Exclusive Economic Zone and Neutral Water. The present status of the Caspian Sea is far from being clear; for more details see Druzenko (1995) and Akimov and Zharkov (1994).

So the Caspian Sea or Lake has an area of 400,000 km^2, its width is 200–550 km. Its coastline is 6380 km long of which 992 km belongs to Iran. Its mean depth is 185 m and there are two deeps of 790 m and 980 m. In the middle part of the sea or lake (near the Caucasus mountains) there are areas about 200 m deep, while in the northern part the mean depth is only 5 m (with no places deeper than 20 m). There are many bays and islands in the sea. In the northern part of the sea there are large coastal areas only several centimetres deep. It is very difficult to say whether this is water or shore, because depending on the winds the coastline in this very shallow region is moving for several kilometres. This is also a problem for satellite and airborne observations as the actual border between wet land and shallow water is hardly recognisable. Severe storms can cause floods of low territories which are normally situated some tens of kilometres from the coast. In some places the water level can change up to 2 m within hours. (Thus oil drilling machines located on the land on the north-eastern coast should have barriers against possible sea floods).

There are eight rivers which bring each year 330 km^{3}of fresh water to this sea. The largest are the Volga (78 %) and the Ural. In the 1950s the deltas of these rivers gave 90 % of the world's catch of the most expensive fish, the sturgeon (osetr), sterlet (sterlyad) and stellate sturgeon (sevryuga).

The salinity of the water in the northern part of the sea is very low, it reaches 12–13 ppm in the upper 6 m layer in the middle part of the sea; in Kara-Bogaz-Gol Bay it is as high as 200 ppm. For many years this Bay was used as a source of table salt and some chemicals. The hydrological front in the northern Caspian Sea separates less saline waters influenced by the Volga river discharge from the water mass of the Middle Caspian Sea. In satellite images of visible bands, such as Figure 9, this front can be seen as the border between turbid shallow waters (light tone) and clear waters (dark tone). The river Volga delta is at the top left of the image. The hydrological front in the northern Caspian Sea is recorded in satellite imagery, with sea surface temperature and suspended matter (turbidity) acting as tracers. The location of the front is governed by river discharge, winds, currents etc. As a rule, the front is registered between the 10 m and 20 m isobaths.

Figure 9 METEOR image of the northern Caspian Sea from 19 May 1984. Scanner MSU-S, 600–700 nm

5.1.1 Sea level variations

We turn to the problem of the sea level and the related problem of the regulation of the discharge of northern rivers, taking part of their water and bringing it to the Caspian Sea.

This 40 year old story should not be forgotten and lessons should be learnt. The lessons could be useful for environmental management of coastal zones especially in the context of assessments of consequences of the possible global climate change.

Variations of sea level are of major importance for millions of people around the Caspian Sea as the level and coastline variations affect their lives, their homes, their agricultural lands, roads, pipelines, industry etc. The vital interests of Russia, Turkmenia, Kazakhstan, Azerbaijan and Iran are all affected by these changes. To some extent this story could be used as a model of the expected rise of the World Ocean level.

According to geological and palaeo-chronological data, the level of the Caspian Sea is a parameter with high temporal variability. Quasiperiodical changes of the sea level are called Caspian transgressions. Based on historical recorded data during the last 2000 years, the highest sea levels were recorded in the fifth century BC (–21.7 m on the absolute scale which means 21.7 m lower than the World Ocean level), in the first century AD (–22.3 m), in the fourteenth century (–22.0 m), at the end of eighteenth century (–22.7 m). Before the new transgression of 1929 the level was –25.5 m. This makes it possible that the level could rise for as much as 7 m, compared to its lowest value of –29.02 m recorded in 1977, to reach the –22.0 m value already recorded many times in the past. In this situation the whole New Caspian terrace (Dagestan) might be flooded which might be a disaster. The recent Caspian transgression which started in 1977 has already caused changes in the structure of the national economy of Dagestan which is closely linked to natural conditions. Plans for developments in the coastal zone are being re-considered. The structure of investments is being changed. In some places houses are being moved (Yusufov 1992). A similar situation caused by the threat of flooding can be seen in other coastal areas around the Caspian Sea.

In the beginning of the fifteenth century the level was –22.0 m, then it fell to –26.0 m. During the next century there were just minor fluctuations of level, then there was a steady decrease of level till the minimum value of –29.02 m in 1977. In this year the rise started and by 1991 the level reached the value of –27.43 m. According to Yusufov (1992) the rise of level since 1977 was not predicted by any scientists with only one exception (M S Eigenson in 1963 pointed out that the level should be considered together with data on global climate and the solar cycle and predicted the rise of level).

Recently some measurements of Caspian Sea level have been carried out within the framework of the TOPEX/POSEIDON mission. Satellite altimetric data were plotted together with in situ data showing seasonal variations of sea level.

The transgression which started in 1977 caused the abandonment of the plans aimed at compensation for the decrease of the Caspian Sea level by means of transport of water from the rivers which flow into the Polar Ocean. While since 1977 people generally have become deeply concerned about the rise of level, before 1977 society was very much concerned about the decrease of level of the Caspian Sea.

During the period 1930–1960 the level decrease of more than 2 m caused economic losses in the fishing industry, sea transport and other branches of the national economy especially in the coastal zone. It was understood that the reasons for this situation are both of natural origin (climate change, namely warming in the Caspian Sea basin) and human origin (too much water from the River Volga was taken for irrigation purposes in the lower Volga region, the construction of a cascade of hydroelectric power stations on the Volga led to regulation of water discharge in spring as the water had to be accumulated near those stations; moreover the water accumulating reservoirs were so large that there was considerable loss of water from their surfaces due to evaporation.

Many scientific institutions contributed to the solution of the problem with many ideas generated. It was believed that the value of –25 m recorded in 1930 was the most favourable for all the branches of the national economy. Analysis of sea level trends showed that in order to achieve this level one should bring some 1400 km^3 of water to the sea (that is 6 times the annual discharges of the Volga) which was impossible. Still it was considered possible to try to smooth the process of sea level fall by bringing an additional 37 $km^3$of water each year (there was also a proposal to add 74 km^3). The sources suggested were the rivers Pechora and Vichegda. Also to save Caspian water it was suggested that a regulated barrier should be constructed between the Caspian Proper and its Kara-Bogaz-Gol Bay. The level of water in this Bay is lower than that in the sea, so Caspian water flows into the shallow Bay where the water is being intensively evaporated. To prevent these water losses, estimated to be 10–12 km^3 per year, only a limited amount of Caspian water had to be allowed to reach the Bay. (The value of 5 km^3 was considered enough to provide quasi-natural hydrographical and hydrochemical conditions in the Bay).

Only one part of this plan was fulfilled, namely the construction of the barrier in the 1970s (though without the regulating watergates). The most ambitious part of the plan was the transport of water from the rivers Pechora (see Section 2.2 on the Barents Sea) and the Vichegda to the River Volga. For this purpose barriers had to be constructed on the rivers

Pechora, Kama and Vichegda and the canals Pechora-Vichegda and Vichegda-Kama had to be dug. 700 million m^3 of ground works had to be carried out and 1.3 million m^3 of reinforced concrete structures had to be constructed. Barriers on the rivers had to form reservoirs with a total area of 15,500 km^2 and a total volume of 235 km^3. According to the masterplan during a 7-year period some 60,000 people and 9,000 households had to be moved and more than 70 million $m^3$of timber had to be cut before flooding. More than 70 km^3 of water had to be stored in the Pechora-Kama-Vichegda super-reservoir of which the major part had to be transported to the Volga via its tributary the Kama. Though the masterplan was preliminary adopted in 1960 it has never become a reality and the barrier to restirct the flow of water into the Kara-Bogaz-Gol Bay was demolishd in the 1980s.

As for the Caspian transgressions it is still unknown what are the reasons for level rise and fall and whether these variations are predictable. The Caspian regional situation is as uncertain as the global climate change.

5.1.2 Oil and gas production and oil slick monitoring

Oil nowadays is the major driving force which stands behind all the developments in the coastal zone and offshore zone in the Caspian Sea. What one should understand is that any gas and oil project in this region will to some extent use the coastal zone in the course of industrial activities in the exploration and production phases.

In the former USSR the Neftyanye Kamni (Oil Stones) oil field was used to produce oil (see below). The sea depth there is 10–20 m. Now new oil fields are on the agenda. The Azeri, Chirag, Guneshli and Kalaz oil fields (more than 500 Mt of oil) are all part of one oil-bearing local province. The sea depth here is about 200 m which requires western high technologies to drill oil wells and lay down pipelines here.

In the 1970s the problem of monitoring the marine environment was put on the agenda in the former USSR. Monitoring of oil slicks on the sea surface was to be carried out using remote sensing techniques. Pilot projects, field work and laboratory experiments have been performed in the Laboratory for Satellite Oceanography and Airborne Methods of the State Oceanographic Institute in St. Petersburg (formerly Leningrad). Special methods for airborne control of the sea surface have been developed and a set of sensors has been designed and manufactured. These sensors have been installed onboard a specially equipped long-range low-flying aircraft IL-14.which was used to test the methodology of airborne monitoring of oil slicks and to determine procedures and manuals for routine operational monitoring of the seas in the former USSR. The Caspian Sea was selected as the test area, with the marine oil platform clusters located at Neftyanye Kamni being used as test sites.

A programme of experimental airborne environmental monitoring of the Caspian Sea was worked out. It consisted of 4 seasonal aircraft surveys of all the major oil producing off-shore regions each year with correlated *in situ* sampling of oil slicks. A standard navigational scheme of flights was to be used. Current meteorological data from a local network were collected as well. In 1975–1990 the programme was carried out in a semi-operational mode. Output products consisted of processed aerial photographic films and paper material (charts of oil slicks based on quick-look analysis of films and data of manual observations), magnetic tapes and paper tapes containing original records of sensors. Thus a unique comprehensive archive of oil slick pollution airborne data for the Caspian Sea was created, which should be regarded as a valuable source of background information on the state of the marine environment on regional and local scales.

In the course of the collapse of the former USSR, which started with perestroika, all the aircraft surveys as part of environmental monitoring of the Caspian Sea were stopped. The special unit which ran this work does not exist and much data was completely lost. Aerial films are still being stored (but in a room with no temperature and moisture control) in the form of 20-cm wide film rolls. The quality of the films has been deteriorating for many years. No efforts are planned to save this archive.

The archive contains original documents in which the actual situation on oil slick location (their patterns) has been recorded for a 15-years period along with relevant *in situ* data on oil concentrations and meteorological situations. It is impossible to repeat those observations. It is difficult to estimate the cost of acquiring this archive. If we take into account only the total aircraft flight time which was spent to collect data, the estimated cost would be several million US dollars. The actual value of this archive could also be considered from another point of view. The historical information and the assessment of water pollution with oil, based on multi-year regular aircraft surveys, could be of value as a source of background information to industrial companies which are reported to be starting some activities in the off-shore oil business in several regions of the Caspian Sea. It is practically impossible to use this archive in its present form and rather modest efforts should be applied in order to re-arrange the archive and update its form to a modern GIS format. Provided this is done, the database could be used as a basis for further monitoring activities.

Typical aerial photographs of oil slicks in the Caspian Sea show oil discharges from each hole at platforms forming thin oil film fields on the sea surface covering areas of tens of square kilometres. The fields are shaped according to the wind direction. In favourable conditions oil pipelines on the sea bottom can be seen in aerial photos (see Figure 10).

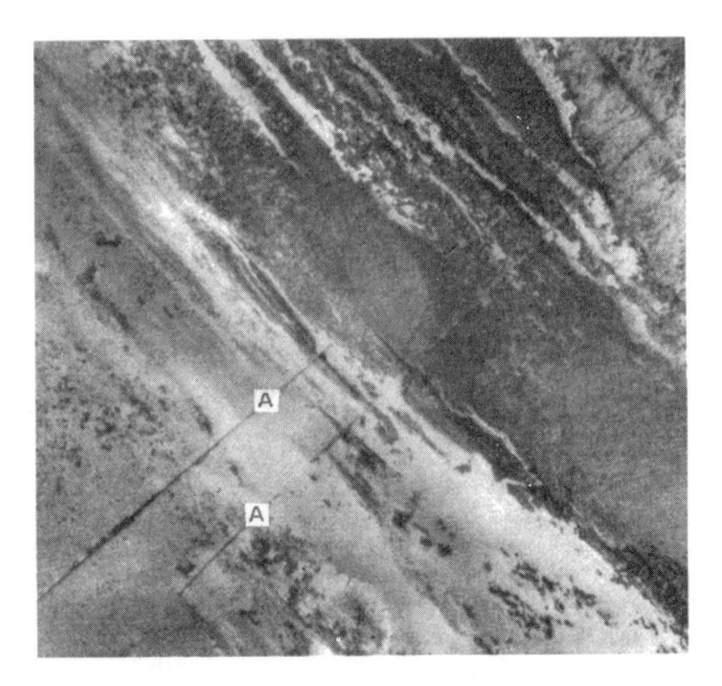

Figure 10 Two parallel oil pipelines (A) on the sea bottom.

5.1.5 The Kazakhstan sector of the Caspian Sea

We give some examples to illustrate the scale of industrial activities to be performed here in the near future.

In 1993 a consortium of international companies (British Gas, BP/Statoil, Italian Agip, USA Mobil, UK-Dutch Shell and French Total) signed an Agreement with Kazakhstan for geological, geophysical and environmental exploration of the Kazakhstan sector of the Caspian Sea.

In Near-Caspian Region the old Tengiz and Karachaganak oil provinces are located. The cost of the latter (which was discovered in 1979 and was developed mainly by Russian experts, and since 1990 by Kazahkstan experts) is estimated at 6 billion US dollars; it contains 2 billion barrels of oil and condensate. The British Gas company is planning to extract 500 million m^3 of gas and 243 Mt of oil during the next 46 years. This amount of gas is far more than enough for Britain for a period of over 10 years (Potter 1995). The Baiganinskoye oil province is situated 200 km north-east of the Caspian Sea. Its area is more than 12,000 km^2. The independent British company Enterprise Oil and the Spanish state company Repsol are involved in the exploration of this oil province (Potter 1995).

5.1.6 The Russian sector of the Caspian Sea

The oil field Inchke-Mor on Dagestan shelf has 20–60 million barrels of oil. The British company JKX Oil and Gas Ltd. is planning to start drilling together with a local Dagestan company and the Russian state company (Chernoye Zoloto/Black Gold 1995). This relatively small project will not add much to the environmental problems in the Caspian Sea when oil production increases as planned.

5.2 The Aral Sea

The Aral Sea is definitely not a sea, although it is called a sea on the maps; it is a lake. The area of this lake is 64,000 km^2 and its dimensions are 428 km by 282 km. Its water surface is 52 m higher that that of the World Ocean. There are many bays and islands and sand banks in the lake. Its depth is 10 to 25 m with a maximum depth of 68 m. Two rivers, the Amu-Darya and Syr-Darya bring 57.8 km^3 of water each year which contained 80 Mt of suspended substances and 33 Mt of mineral salts.

For comparison with concentration of particles in other rivers known for their turbid waters: the Ind brings 2500 g of solid particles per m^3; the corresponding figures for the Gang, the Amu-Darya and the Syr-Darya are 1986, 1593 and 850, respectively. The general direction of currents is clockwise, (while in the majority of seas in the Northern hemisphere it is anti-clockwise). The water temperature in summer is 24–27 °C; in winter usually there are ice fields near the coast.

The above-mentioned data on the sea level, sea dimensions and river discharges are from 1955 and have nothing in common with the present situation. Since 1955 over one-third of the Aral, Sea has disappeared and the remainder has effectively been divided into two parts, the Great Aral and the Small Aral (see Figure 11). Since the 1960s the marine and coastal environments in the Aral Sea have been changing drastically as a result of the following factors: (1) river discharge is decreasing, (2) the sea level is falling and (3) the coastal zone is changing as the former sea bottom becomes dry land area. Both surface and ground waters are affected, as well as soils, vegetation and habitats.

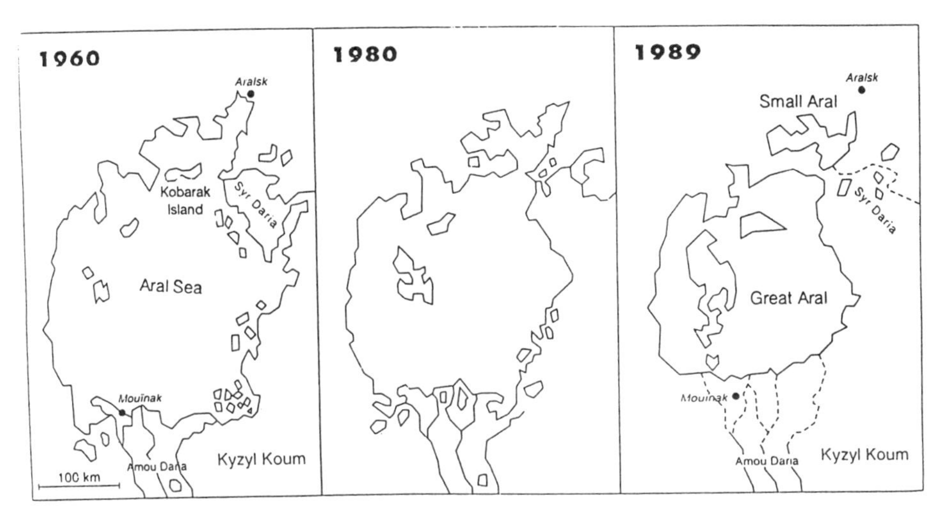

Figure 11 Evolution of the Aral Sea from 1960 to 1990 (Pieyna and Sehmi 1995)

We consider some recent data on the river Amu-Darya (Objedkov and Shmakov 1995). In the 1960s the flooded area in spring was 7,000 km^2, the area of constantly flooded lands was 2,650 km^2 and the total water mineralisation was less than 0.5 mgl^{-1}. In the 1990s there were no spring floods at all, the river water level decreased by 3.5-4 m, its chemical composition changed and total mineralisation reached a value of 1.3 mgl^{-1}. In 1951 the sediment concentration near Nukus was 471 mgl^{-1}, in 1968 it was 841 mgl^{-1} and in 1981 it was 1200 mgl^{-1}. During this period, due to wind erosion of over-fertilised agricultural fields and the former sea bottom, the concentration of salts in rainwater increased 5 to 10 times. In the southern part of the region the dry aerosol deposit is (3–9)tha^{-1} per year.

Since the 1950s the whole system of ground water in this region has been affected by intensive irrigation, over-watering of fields and leakage in the irrigation networks. Fertilisers and pesticides from irrigated cotton and rice fields reach the ground water and contaminate it. Ground water discharges reach the Aral sea and in their turn contaminate it.

Satellite images show not only turbid water and bottom features but also the changes in the coastline configuration at the southern and eastern parts, the increase of the area of the island Vozrozhdeniya (due to the sea level decrease). The former island of Lazarev is no more an island, the Ajibai Bay has disappeared and the Small Sea bay has actually become a separate water body (Sukhacheva and Victorov, 1990). In the analysis of the satellite imagery of the Aral Sea one should take into consideration the changing coastline, the new emerging features of the bottom topography as well as a specific aerosol content (salt dust from the former sea bottom).

There are several international projects dealing with monitoring the environmental situation in the Aral Sea region. There are some theories pointing out possible connections between the waters of Caspian and Aral Seas. The role of lowland area located to the east of the Caspian Sea as the accumulator of water in past geological periods is discussed. Some experts say that the present rise of level in the Caspian Sea might be related to degradation of the Aral Sea. There are scientists who say that the main reason for the Aral Sea degradation might be of climatic origin, meaning a decrease in precipitation and an increase in evaporation in the sea. Pieyna and Sehmi (1995) noted that if it were so, it would be very difficult to explain the rise of the adjoining Caspian Sea. They insist on the analysis of the situation in the catchment basin of the Aral Sea including the production of water in the snows and glaciers of the mountains of the upper catchment of the two rivers (about 80 % of the river flow is from snow and glaciers). In their opinion 'to save the Aral Sea' means first to make an assessment of the real situation with water production and use in the Aral Sea Basin.

As some 98% of the region's gross domestic product is water-dependent, the Aral Sea and the five independent states of Kazakhstan, Kyrgyzstan, Tajikistan, Turkmenistan and Uzbekistan (the former Soviet Central Asia) are inexorably interdependent as regards their water resources (Pieyna and Sehmi 1995). It seems worthwhile to present here the opinion of WMO experts on the state of water-resources monitoring in this area. Originally (before the collapse of the USSR) the regional network consisted of about 900 hydrological stations, 300 meteorological stations, 400 snow-depth measuring points monitored by helicopter surveys and 3 high-latitude research sites. This network was divided among the five independent states, two of which being essentially water suppliers and three being water consumers. During the short period since independence, the network has decreased dramatically. In the Amu-Darya headwaters in Tajikistan the number of meteorological stations has decreased from 25 to 10, and the number of hydrological stations has decreased from 7 to 2. Only sporadic data are received in Tashkent, the regional data centre. In the Syr-Darya headwaters the number of meteorological stations has decreased from 20 to 12, and the hydrological stations have remained at 15. Snow measurement by helicopter has ceased and about 30 snow-measuring sites have been closed. "The overall impression is one of rapid decay... What is urgently needed is to re-equip the data-collection network, especially in the mountainous areas... with a system of satellite-based data-collection platforms and data-loggers, and to upgrade the computer capacity for database development and analysis" (Pieyna and Sehmi 1995).

6. Conclusion

In some countries coastal zone monitoring and management has become a matter of national concern. There are several programmes of this kind in the USA and in the Netherlands (some of them have been briefly reviewed by Victorov 1996). There is a national Coastal Zone Society in the USA founded in 1975. Its annual meetings serve as nation-wide platforms for broad discussions of problems related to the coastal zone. Recent USA/ONR Coastal Mixed Ocean/Optics and the UK COAST Project could serve as examples of good experiments and research programmes carried out in the coastal zone.

In the absence of any national programme for the monitoring and management of the coastal zone of Russia, and in view of no declared intentions to initiate and develop such a programme on the national level in the foreseeable future, the following question arises:

The Russian Coastal zone – Whose Concern ?

The answer is that it could well become a matter of concern for international industrial and science communities.

It seems reasonable to put the problem of Russian coastal zone monitoring on the agenda of all the negotiations currently being carried out with international companies planning to be involved in the oil and gas industrial developments in the off-shore and coastal zone of Russia.

Field work being very expensive and territories being so large, remote sensing techniques should become an important tool in collecting data on marine and coastal environments. In this context it seems reasonable that industrial companies from developed countries involved in the development of the Russian shelf and coastal zones bring with them high technologies related to monitoring the environment. Those could include time series of remotely-sensed data (both airborne and satellite-derived) covering vast areas of coastal zones and seas where industrial activities are being planned. Those data sets should complement the databases of historical information that was collected in the Soviet era. Some of these were unique and very costly and represent enormous past investments. (Two examples of such archives were mentioned above: the Caspian Sea oil slick pollution airborne imagery for a 15-year period and the Sakhalin island coastal zone airborne photography archive covering a 25-year period).

It seems reasonable that data from all the sources be integrated in regional GISs tailor-made for coastal zone monitoring and management. Setting up and running these regional GISs also could become a subject of international concern and co-operation.

Data tackling, analysis of information and assessments of the state of environment in coastal zones should also become issues where international co-operation in various forms could bring fruitful results to the benefit of the local population and to the international industrial and science communities.

References

Akimov V and Zharkov S 1994, Kazakhstan: Diameter of a pipeline limits export. Chernoye Zoloto (Black Gold) (in Russian), October, pp. 42–45

Allewijn. A, 1996, EARSeL Newsletter, December.

Belinskiy N A and Istoshin Yu V, 1956, Seas washing the coasts of the Soviet Union (in Russian), Voenizdat, Moscow, 210 pp

Blumberg A F, Signell R P and Jenter H L, 1993, Modelling transport processes in the coastal ocean, Journal of Environmental Engineering, **1**, 31–52

Bychkova I A and Victorov S V, 1987, Elucidation and systematization of upwelling zones in the Baltic Sea based on satellite data (in Russian), Oceanology, **XXVII**, 218–223

Bychkova I A, Victorov S V, Demina M D, Lobaniv V Yu, Losinskij V N, Smirnov V G, Smolyanitskij V M, Sukhacheva L L and Brosin H-J, 1990, Experimental satellite monitoring of the Baltic Sea in the 1980s. ICES Paper C.M. 1990 (Hydrography Committee, Session P)

CARDINAL User's Manual, 1993, Nevskiy Courier Publishers, St.Petersburg, 36 pp

Chernoye Zoloto, Black Gold 1995, March, pp 46–47

Dalyell T, 1997, Something nasty in the waters, New Scientist, 21 June 1997.

Dlugokencky E J, Masarie K A, Lang P M, Tans P P, Steele L P and Nisbet E G, 1994, A dramatic decrease in the growth rate of atmospheric methane in the northern atmosphere during 1992, Geophysical Research Letters, **21**, 45–48

Druzenko E, 1995, Azeri, Chirag and deep part of Guneshli as a new object of world geopolitics. Neft i kapital (Oil and capital) (in Russian), **1**, 4–11

ESA, 1995, New Views of the Earth. Scientific Achievements of ERS-1, SP-1176/1, p.152

Fattakhov A, 1995, Chernoye Zoloto/Black Gold, September, p.15–16

Frolov I E, 1996, AANII Annual Report

Funquist L, 1996, Modelling of Physical Processes in the Baltic. In: Abstracts of the Baltic Marine Science Conference, 22–26 October, Ronne, Denmark, p 31

Gur'eva Z I, Petrov K M, Ramm N S and Sharkov V V, 1968, Geological and geomorphologic study of shallow seas and coasts based on aerial photography, Manual (in Russian), Nauka, Leningrad, 220 pp

Hari J, 1984, Correlation between the area of increased temperature around a thermal power plant and the ice conditions of the Gulf of Finland. In Proceedings of the 14th Conference of Baltic Oceanographers, Gdynya, Poland, September 1984, Paper P–22

Hupfer P, 1982, Baltic - small sea, big problems (in Russian), Leningrad, Gidrometeoizdat, 136 pp

Johannessen O M, Bjorgo E and Petterson L H 1996, Proposed strategy for the use of satellite remote sensing in EuroGOOS. In: Conference Pre-prints of the First International Conference on EuroGOOS, 7–11 October 1996, The Hague, pp93–114.

Kleine E, 1996, HIROMB – a High Resolution Operational Model of the Baltic. In: Abstracts of the Baltic Marine Science Conference, (22–26 October, Ronne, Denmark), pp. 30–31

Neft i Kapital (Oil and Capital) 1995, (in Russian), 1, p.26

Nielsen A and Hansen P, 1983, Plankton distribution as shown by satellite pictures, Paper presented at the Workshop on the Patchiness experiment in the Baltic Sea (Tallinn, March 1983)

Niini M, 1996, Northern Sea Route – alternative for export. Neftegaz (Oil and Gas) (in Russian), 2, pp. 27–34

Nikitin P A, 1991, Satellite-based monitoring of sea ice. In Proceedings of the 5th AVHRR Data Users Meeting, Tromso, 25–28 June 1991, Darmstadt-Eberstadt, METEOSAT, EUM P 09, pp 411–414

Northern Europe's Seas. Northern Europe's Environment, 1989, Report to the Nordic Council's International Conference on the Pollution of the Seas, 16–18 October 1989, pp 156–158

Novelli P C, Conway T J, Dlugokencky E J and Tans P P, 1995, Recent changes in carbon dioxide, carbon monoxide and methane and the implications for Global Climate Change. WMO Bulletin, 44, 1, pp 32–38

Objedkov Yu L and Shmakov A I, 1995, Study of hydrometeorological structure of river deltas using GIS (in Russian), The Study of the Earth from Space, No. 2, pp 50–56

Pedersen J P, Seljelv L G and Bauna T, 1996, Towards an operational oil spill detection service in the Mediterranean? The Norwegian experience: A pre-operational early warning detection service using ERS SAR data. In: Proceedings of the ERS Thematic Workshop on Oil Pollution Monitoring in the Mediterranean, Frascati, Italy, ESA.

Pieyna S A and Sehmi N S, 1995, Water-resources monitoring system for the Aral sea, WMO Bulletin, 44, 1, pp 64–66

Polar Orbiting Satellites and Applications to Marine Meteorology and Oceanography, 1996, Report of the CMM-IGOOS-IODE Sub-group on Ocean Satellites and Remote Sensing, Marine Meteorology and Related Oceanographic Activities, Report No.34, WMO/TD No. 763

Potter N, 1995, British companies look at the former USSR with a bit of fear. Chernoye Zoloto (Black Gold) (in Russian), March, pp 40–45

Rost A, 1995, Timano-Pechora contract: somebody should learn stepping back. Neft i Kapital (Oil and Capital) (in Russian), 1995, 1, pp. 82–85

Rud O and Kahru M, 1994, Long-term series of NOAA AVHRR imagery reveals large interannual variations of surface cyanobacterial accumulations in the Baltic Sea, in Proceedings of the EARSeL workshop 'Remote sensing and GIS for coastal zone management' (Delft, The Netherlands, 24–26 October 1994), pp 287–293

SCOT Conseil and Smith System Engineering Limited, 1994, Use of satellite data for environmental purposes in Europe, Study Contract No. ETES-0039-D, Commission of the European Communities, Directorate General for Science, Research and Development, Joint Research Centre, Ispra Establishment.

Sukhacheva L L, 1987. Using satellite data for determination of optical inhomogeneities in the upper layer of the seas, in Instruction manual for complex usage of satellite data in studies of the seas (in Russian), Leningrad, Gidrometeoizdat, pp 49–58

Sukhacheva L L and Victorov S V, 1990, Analysis of remotely sensed data for the Aral Sea and Caspian Sea (in Russian) (unpublished).

Sukhacheva L L and Victorov S V, 1994, Remote sensing and 'The Gulf of Finland Year 1996'. In Abstracts of papers presented at the 19th Conference of Baltic Oceanographers (Sopot, Poland, 29 August–1 September), p 93

Usanov B P, Mikhailenko R R , Victorov S V and Sukhacheva L L, 1994, Regional environmental GIS 'The Neva Bay': databases of conventional and remote-sensing data. In Abstracts of papers presented at the 19th Conference of Baltic Oceanographers (Sopot, Poland, 29 August–1 September), p 1063

Victorov S V, 1996, Regional satellite oceanography, (Taylor and Francis: London), 312pp

Victorov S V, Karpov A V and Nikitin P A, 1993, Ocean satellite programme in the Russian Federation. In Proceedings of the Commission for Marine meteorology Technical conference on Ocean Remote Sensing (26 April 1993, Lisbon, Portugal), Report No 28, WMO Technical Document No. 604, pp 9–22

Victorov S V, Losinskii V N and Sukhacheva L L, 1988, Studies of seasonal bloom of the Baltic Sea and elements of eddy dynamics of the upper layer based on satellite data of visible diapazon. In Proceedings of the 16th Conference of Baltic Oceanographers (Kiel, FRG, September 1988), 2, pp 1063–1078

Victorov S V, Sukhacheva L L, Lobanov V Yu, Lebedeva N I and Nekrasova A N, 1991, Distribution features of suspended substances in the Neva Bay under various hydrometeorological conditions (in Russian), Meteorology and Hydrology, 7, pp 80–85

Victorov S V, Usanov B P, Sukhacheva L L and Bychkova I A, 1996, On the role of remote sensing in the syudy of the Baltic Sea. Abstracts of papers presented at the International Seminar 'Gulf of Finland', St.Petersburg, 15–17 October, pp 35–37

Yusufov S K, 1992, Natural environment of Dagestan and its changes in connection with the rise of Caspian Sea level, Ph.D. thesis, St.Petersburg, 17 pp.

Participants

- Dr Giulia Abbate
  ENEA
  Casaccia, S.P. 77
  via Anguillarese, 301
  00100 Roma
  Italy

- Mr Dimitry Akimov
  NIERSC
  Korpusnaya str. 18
  St Petersburg 197110
  Russia

- Mr Peter Albert
  Free University Berlin
  Inst for Weltraumwiss
  Fabeckstr 69
  Berlin 10405
  Germany

- Ms Gwen Bayne
  Dept of APEME
  University of Dundee
  Dundee DD1 4HN, UK

- Mr Marco Benvenuti
  C.E.S.I.A.
  Cesia-Accademia Dei Georgofili
  Via Giovanni Caproni 8
  Firenze 50145
  Italy

- Miss Anna Black
  University of Edinburgh
  18 Edderston Road
  Peebles EH45 9DT, UK

- Miss Kasia Bradtke
  Institute of Oceanography
  University of Gdansk
  Al. Pilsudskiego 46
  Gdynia 81-378
  Poland

- Miss Amanda Brady
  Dept of APEME
  University of Dundee
  Dundee DD1 4HN, UK

- Mr V E Brando
  Dept de Science Ambientali
  Universita de Venezia
  Dorsoduro 2137
  Venezia 30123
  Italy

- Miss Dorota Burska
  Institute of Oceanography
  University of Gdansk
  Al. Pilsudskiego 46
  Gdynia 81-378
  Poland

- Mr A Cargill
  Dept of Geography
  University of Dundee
  Dundee DD1 4HN, UK

- Dr George Chronopoulos
  Dept of Applied Physics
  University of Athens
  Panepistimioupolis, Bld Phys- V
  15784 Athens
  Greece

- Prof A P Cracknell
  Dept of APEME
  University of Dundee
  Dundee DD1 4HN, UK

- Dr P S Crawford
  Dept of APEME
  University of Dundee
  Dundee DD1 4HN, UK

- Dr A Di Iorio
  PSTd'A
  Scientific Park of Abruzzo
  Via Santa Giusta no. 10
  L'Aquila
  Italy

- Dr P Dong
  Dept of Civil Engineering
  University of Dundee
  Dundee DD1 4HN, UK

- Dr R W Duck
  Dept of Geography
  University of Dundee
  Dundee DD1 4HN, UK

- Mr Francisco Durand
  Free University, Berlin
  Pestalozzi Str. 73
  Berlin 10627
  Germany

- Dr Raymond Feron
  Survey Dept
  Directoraat-General Rykswatersaat
  Delft
  The Netherlands

- Dr A Folkard
  Dept of Civil Engineering
  University of Strathclyde
  John Anderson Building
  107 Rottenrow
  Glasgow G4 0NG

- Mr M W Freeman
  Dept of APEME
  University of Dundee
  Dundee DD1 4HN, UK

- Mr Jose Antonio Gomez-Sanchez
  INTA
  C/Camino de Las Fuentes, 28
  Marchamalo 19180
  Guadalajara
  Spain

- Dr Rajashree Gouda
  Berhampur University
  C/0 Sri D P Gouda
  Lig-21 Gajapatrinagar
  Berhampur 760 010
  Orissa India

- Mr Yan Gu
  Dept of APEME
  University of Dundee
  Dundee DD1 4HN, UK

- Mrs Anita Jacob
  NERSC
  University of Bergen
  Edvard Griegs vei 3A
  Solheimsvik N-5037
  Norway

- Mr M Jimenez Michavila
  INTA
  Carretera De Ajalvir Km 4
  Torreson de Ardoz
  Madrid
  Spain

- Dr Najad A Kabbara
  National Centre for Remote Sensing
  PO Box 11-8281
  Beirut
  Lebanon

- Mr Michael Kleih
  Joint Research Centre
  ISPRA
  Via Fermi 1
  ISPRA 21020
  Italy

- Prof V Klemas
  Center for Remote Sensing
  University of Delaware
  The Grad. College of Mar. Studies
  Newark
  Delaware 19716-3501, USA

- Dr Janek Laanearu
  Estonian Marine Institute
  Paldiski Road 1
  Tallinn EE0031
  Estonia

- Dr Ian Lee-Bapty
  DERA Space Department
  A8 Building
  Farnborough
  Hampshire GU14 0LX, UK

- Miss Nadia Lo Presti
  University of Palermo
  Piazza Acquasanta, 12
  Palermo 90142
  Italy

- Mr Katsu Maeno
  Graduate School of Fisheries Sc
  Hokkaido University
  3-1-1, Minato-cho Hakodate
  Hokkaido 041
  Japan

- Dr T Malthus
  Geography Dept
  Edinburgh University
  Drummond Street
  Edinburgh EH8 9XP, UK

- Dr J McManus
  Dept of Geology
  University of St Andrews
  School of Geology & Geography
  Purdie Building
  St Andrews KY16 9ST,UK

- Mr David H Z Mhango
  Dept of Civil Engineering
  University of Strathclyde
  John Anderson Building
  Glasgow G4 0NG, UK

- Mr K Millard
  H R Wallingford
  Howbery Park
  Wallingford OX10 8BA, UK

- Miss Lyssania M Morales
  National University of Mexico
  Foot Ball 118 Int. 3
  Col. Churubusco Country Club
  Coyoacan
  Mexico, D.F.

- Miss Suwannee Naruekhatpichai
  32/289 Thai Airways Int. Vill. Soi 6
  Chaengwatta Road
  Pakkred
  Nonthaburi 11120
  Thailand

- Dr Fabrizio Nerozzi
  ISTEA - CNR
  Via P Gobetti 101
  Bologna 40129
  Italy

- Ms S K Newcombe
  Dept of APEME
  University of Dundee
  Dundee DD1 4HN, UK

- Mr Mohan L Nirala
  Dept APEME
  University of Dundee
  Dundee DD1 4HN, UK

- Prof Jin E Ong
  Centre for Marine & Coastal Studies
  Universiti Sains Malaysia
  Penang 11800
  Malaysia

- Dr S M Parkes
  Dept of Maths
  University of Dundee
  Dundee DD1 4HN, UK

- Dr Maria Concetta Pizzoferrato
  C.E.S.I.A.
  Via Podgora, 5
  67035 Pratola Peligna (AQ)
  Italy

- Mrs O Puentedura
  Dept Teledeteccion y Aeronomia
  Insti Nacional de Tecn Aero (INTA)
  CTRA Ajalvir km 4
  Torreson de Ardoz
  Madrid 28850, Spain

- Dr W G Rees
  Scott Polar Res Inst
  University of Cambridge
  Lensfield Road
  Cambridge CB2 1ER, UK

- Mrs Anu Reinart
  Inst. of Environ. Physics
  University of Tartu
  Tartu EE2400
  Estonia

- Mr E Ricchetti
  Dept of Geology & Geophysics
  University of Bari
  Via Orabona 4
  Bari 70125
  Italy

- Prof A C B Roberts
  Dept of Geography
  Simon Fraser University
  Burnaby
  British Columbia V5A 1S6
  Canada

- Ms E S Rowan
  Dept of APEME
  University of Dundee
  Dundee DD1 4HN, UK

- Ms Isabel Sargent
  Dept of Aeronautics Astron.
  University of Southampton
  Highfield
  Southampton SO17 1BJ, UK

- Miss Ansela Schwiebus
  Institute fur Hydro. u Meter.
  Technical University of Dresden
  Pienner Str 9
  Tharandt 01737
  Germany

- Mr D Sirjacobs
  Fac. of Agronomical Sci. of Gemblox
  25 rue de la Station
  7540 Kain
  Vezon (Tournai) 7538
  Belgique

- Mr M Slater
  Dept of APEME
  University of Dundee
  Dundee DD1 4HN, UK

- Dr D R Sloggett
  Anite Systems Ltd
  Genesis Business Park
  5 Albert Drive
  Woking GU21 5RW, UK

- Mr Vembu Subramanian
  Dept of Earth Science
  Memorial Univ. of Newfoundland
  St John's
  Newfoundland A1B 3X5
  Canada

- Mr M Tahir Kavak
  Dept of APEME
  University of Dundee
  Dundee DD1 4HN, UK

- Dr A O Tooke
  Dept of APEME
  University of Dundee
  Dundee DD1 4HN, UK

- Miss Irene M J Van Enckevort
  Dept of Physical Geography
  Utrecht University
  PO Box 80115
  TC Utrecht
  Netherlands

- Dr R A Vaughan
  Dept of APEME
  University of Dundee
  Dundee DD1 4HN, UK

- Dr Serge Victorov
  Head, Laboratory for Coastal Zone Res.
  VNIKAM
  6 Birzhevoy proezd
  St Petersburg 199034
  Russia

- Miss S F Wewetzer
  Dept of Geography
  University of Dundee
  Dundee DD1 4HN, UK

- Ms M Zoller-Shimoni
  Remote Sensing Lab.
  Tel-Aviv University
  Heura Hadasha 3/10
  Tel Aviv
  Israel

Index